Beyond the Double Bind

BEYOND

THE

DOUBLE

BIND

Women and
Leadership

KATHLEEN HALL JAMIESON

OXFORD
UNIVERSITY PRESS

OXFORD
UNIVERSITY PRESS

Oxford   New York

Athens   Auckland   Bangkok   Bogotá   Buenos Aires   Calcutta
Cape Town   Chennai   Dar es Salaam   Delhi   Florence   Hong Kong
Istanbul   Karachi   Kuala Lumpur   Madrid   Melbourne   Mexico City
Mumbai   Nairobi   Paris   São Paulo   Singapore   Taipei   Tokyo
Toronto   Warsaw

and associated companies in
Berlin   Ibadan

Copyright © 1995 by Kathleen Hall Jamieson

First published by Oxford University Press, Inc., 1995
198 Madison Avenue, New York, New York 10016

First issued as an Oxford University Press paperback, 1997

Oxford is a registered trademark of Oxford University Press

Library of Congress Cataloging-in-Publication Data
Jamieson, Kathleen Hall.
Beyond the double bind / Kathleen Hall Jamieson
p.   cm.
Includes bibliographical references and index.
ISBN 0-19-508940-5 (cloth)
ISBN 0-19-511572-4 (Pbk.)
1. Sex discrimination against women.
2. Women—Social conditions. 3. Double bind (Psychology)   I. Title.
HQ1237.J36   1995   305.42—dc20   94-42812

1 3 5 7 9 10 8 6 4 2

Printed in the United States of America
on acid-free paper

This book is dedicated to
Walter H. Annenberg

Acknowledgments

This book started on a holiday flight with my husband and our two sons to visit my sister in Hawaii. To vanquish my conviction that flying is an unnatural act, I tend to spend my travel time submerged in best-sellers, from schlock novels to scholarship.

This flight was no different. At thirty thousand feet, I opened a best-seller given to me by a friend—Susan Faludi's *Backlash*. To my surprise, I found myself disagreeing with page after page. By the end of the twelve-hour flight, I had retrieved a note pad from my flight bag and was outlining objections. Among the things that troubled me was the book's pervasive pessimism. If progress is inevitably followed by backlash, a few steps forward—a few steps back—why continue to pursue women's rights?

A post-vacation computer search confirmed some of my suspicions. In key places, about economic progress during the Reagan years and focal court cases on fetal protection, the book's evidence of backlash was either outright wrong or seriously misleading. But if not backlash, what?

Beyond the Double Bind is a call to conceive the history of women's rights in the United States as a collection of historical acts in which strategic choices by women and their advocates dispatched false choices. Here is a world not of progress undone by backlash but of binds broken, a world that does not parse neatly into victims and victors, a complex and fluid world in which women have gained the institutional means to ensure that progress made is progress sustained.

This is an optimistic work recounting what has been accomplished and how. It also chronicles what has and has not worked in overcoming the false choices women have been offered, including womb or brain, silence or shame, equality or difference, femininity or competence, and youth or invisibility.

In talking through all of this in coffee shops and walks in Fairmount Park, friends worried with me about issues I have not faced in writing about presidents and elections. A woman studying the presidency does not confront the question, when should I write as if I am part of this story? Out of habit and because I have experienced some of the binds that are the subject of the book but not others, I have written in the third person impersonal, as if I were somehow outside the story it tells, observing it all, unaffected by the outcome.

Yet my almost adrenal optimism and my visceral rejection of the backlash hypothesis have a basis in autobiography as well as biology. I had access to education denied my great grandmother. Although she was the valedictorian in her high school class, my mother never attended college. She applauded as her three daughters completed college and two of them went on to graduate school.

Where the first offer of a university job I ever had was withdrawn when the department chair learned (because I had the poor judgment to tell him) that I was scheduled to deliver our older son seven months later, the same university (different chair) saw no problem in hiring me two years later despite my declaration that we planned to have another child.

By then, of course, the world had changed. I was what some would call "an affirmative action hire." Protections not available when I was pregnant with Robert were in place by the time Patrick arrived. And with the advent of family and medical leave, mothering and fathering are more compatible with full time out of the home employment than they were then.

Who is hired is not the only change I have experienced. There were no female professors on the graduate faculty in the University of Wisconsin's Department of Communication Arts when I studied there. Now a woman chairs the department. And I serve as dean in an Ivy League university headed by a woman.

Where in 1967 I was told by a Woodrow Wilson Fellowship Committee that my engagement ring effectively disqualified me from support, in 1989, with the same ring on hand, I spent a rewarding summer in D.C. as a Woodrow Wilson fellow.

Although subtle and overt forms of sex discrimination persist, there are now forms of redress, some institutional, some rhetorical, available not just to the educated and well-to-do. I chart them in the book.

Any book is a collaborative effort and this is no exception. For service beyond the call of duty, I wish to thank the reference staff at the Biddle Law Library and the Bio-Medical Library of the University of Pennsylvania as well as the Annenberg School's head librarian, Dr. Susan Williamson. Throughout the process, Deborah Stinnett pursued sources, footnotes, and elusive facts tenaciously and with good humor; Maxine Beiderman transformed my scratching and scrawling into script; and Lorraine Hannon organized my days into hours, onto notecards, and into red folders. To them and to the others on the Annenberg School staff— Deborah Porter, Debra Williams, Mary Brennan, Pam Robinson, Ellen Reynolds, and David Graper—who make this a sane if sometimes silly place to teach and work, my thanks.

Appreciation too to Elena Cappella for Law cits, Joe Cappella for bluefishing and protections from exogeneity, Bob Hornik for Jewish texts, Judy Turow for Nasalcrom, Carolyn Marvin for clippings, and Larry Gross for books unfindable in libraries.

I am particularly grateful to Michele Strano, Deborah Stinnett, Nancy Weiner, Doug Rivlin, Robin Nabi, and Susan Stone for the interviews they conducted and the texts they tracked down. The book is informed by interviews they and I conducted at the 1993 Women Presidents' Summit in Washington D.C., June 8–11, 1993, at the National Womens' Political Caucus Convention in Los Angeles July 8–11, 1993, and at the Summer Institute for Women in Higher Education Administration in Bryn Mawr, Pennsylvania, June 27– July 23, 1993. Focus groups were organized by Anita Hall in Minneapolis, Pat Petrello in Detroit, Karen Anne Jones in Atlanta, and Nancy Patrick in Dallas.

This will be the fifth Oxford book that Laura Brown and Rosemary Wellner have sweated through with me. Each is as much a sister as an editor. I benefited as well from the help of Oxford's David Roll, now en route to becoming a doctor, Kathy Banks, a sensitive structural editor, and Gioia Stevens, whose taste in restaurants is among her many virtues. I am grateful as well to Carroll Smith-Rosenberg, Ellen Willis, Carrie Ricky, Michael Schudson, Rod Hart, and Shelley Fisher Fishkin for their helpful critiques of early drafts.

I remain intellectually and personally indebted to Karlyn Kohrs Campbell whose friendship and support brought me into book writing. And to the men in my life—Bob, Rob, and Patrick—who take out the trash and change the cat box, although not in that order, when they are home and, when they are not, phone to assure me that they have not been devoured by bears.

Finally my thanks to art-collector, bass-fisher, publisher, and philanthropist Walter Annenberg—the person to whom this book is affectionately dedicated. The United Negro College Fund, the Annenberg CPB project, public elementary and secondary education, and the Annenberg School for Communication of the University of Pennsylvania are among his beneficiaries. On a hundred-degree day in the desert five and a half years ago, he insisted that I would not regret accepting the offer of the Annenberg faculty and the University of Pennsylvania and becoming its dean. I haven't.

For that I have the Annenberg faculty—Carolyn, Charlie, Joe T, JNC, Bob, Oscar, Larry, Paul, Klaus, Elihu, Pan, Roberta, Louise—and the almost 87-year-old only brother of seven sisters to thank.

Contents

Beyond the Double Bind

Myth

Long afterward, Oedipus, old and blinded, walked the
roads.   He smelled a familiar smell.   It was
the Sphinx.   Oedipus said, "I want to ask one question.
Why didn't I recognize my mother?"   "You gave the
wrong answer," said the Sphinx.   "But that was what
made everything possible," said Oedipus.   "No," she said.
"When I asked, what walks on four legs in the morning,
two at noon, and three in the evening, you answered,
Man.   You didn't say anything about woman."
"When you say Man," said Oedipus, "you include woman
too. Everyone knows that."   She said, "That's what
you think."

"Myth" by Muriel Rukeyser
from *Out of Silence*, 1992,
TriQuarterly Books, Evanston, IL.
© William L. Rukeyser.

1

The Binds That Tie

IN 1631, in *Cautio Criminalis*, Julius Friedrich Spee identified one no-win situation in which prosecutors placed women accused of witch-craft. The suspected witch was submerged in a pond. If she drowned, she deserved to; if she didn't, she was a witch. In the first case, God was revealing her nature; in the second, the devil. Under torture, women either did or did not admit to complicity with Satan. If they did, they were executed for their crime. If they didn't, their silence was attributed to solidarity with Satan and they too were marched off to the stake.[1]

Although he didn't know it, Spee had identified a trap set for women throughout history. When our foremothers overstepped prescribed boundaries, they confronted situations constructed to ensure that they were guilty until proven guilty.

More often than not, the accused was vulnerable because she was un-protected by a father, husband, or son and had increased her susceptibility to suspicion by asserting her right to influence other women. These "witches" were likely to be older, unmarried, childless women who prac-ticed healing, transmitted advice about contraception, or used speech in socially disapproved ways.

Unprotected by men and outside a woman's "natural" sphere, these women were presumed susceptible to the wiles of Satan. It was with him that they coupled at night. In his name, they slaughtered infants and tormented townspeople. Once suspected of witchcraft, a woman was pro-pelled into the no-win situations described by Spee. Viewing the inevitable penalty attached to public speech and private healing, women were en-joined to stay where, the argument went, nature and nature's God re-quired.

Three and a half centuries later, the penalties are disdain and financial loss, not death, and the sanctions social, not theological, but it can still be hazardous for a woman to venture out of her "proper sphere." Some advocates say that female military personnel, for example, do not report sexual harassment for fear of being tagged lesbians and, under the ban on gays in the military, driven from the service. Alternatively, they will submit to unwanted heterosexual activity simply to establish that they are not gay. Their behavior is circumscribed in more subtle ways as well. "You have to project an image that is feminine enough that you won't be called a dyke," reports a former female captain in the air defense artillery who was investigated and "cleared" of the charge of being a lesbian, "but not so feminine that you won't be taken seriously."[2]

Nor does a woman have to venture so far into what were once all-male preserves. Businesswomen and mothers, Democrats and Republicans, young and older—a broad spectrum of contemporary women describe themselves as caught in situations in which they too are damned if they do and damned if they don't. "I had learned from more than a decade of political life that I was going to be criticized no matter what I did," wrote former First Lady Rosalyn Carter in her autobiography, "so I might as well be criticized for something I wanted to do. If I had spent all day 'pouring tea,' I would have been criticized for that too."[3]

"I don't think I was as bad, or as extreme in my power or my weakness, as I was depicted—especially during the first year, when people thought I was overly concerned with trivialities, and the final year, when some of the same people were convinced I was running the show," wrote Nancy Reagan in her autobiography a decade later.[4]

Dilemmas like these burden youngsters as well as adults. As an adult, Barbara McClintock would win the Nobel Prize for identifying the "jumping" genes on corn. As a child, she played catcher on her block's baseball team. Embarrassed by having a girl in such a position, the boys on the team refused to let her play in an important game but the opposing team welcomed her on its side. "On the way home," reports a biographer, "her neighborhood buddies accused her of being a traitor. 'So you couldn't win,' Barbara realized, concluding reasonably enough that 'you had to be alone. You couldn't be in a society you didn't belong to. You were only tolerated by the boys . . . I knew I couldn't win—and that's a dreadful feeling as a child.'"[5]

The history of Western culture is riddled with evidence of traps for women that have forcefully curtailed their options. This is not to say that, in some circumstances, men haven't been ensnared by such constructs. Masculinity has its own constellation of double binds, including the assumption that decisiveness and competence are masculine traits, so that a man considered effeminate is open to questions about his ability. At the other end of the spectrum, drafting men but not women for military service constitutes a double standard, with men subjected to all the risks and women escaping them simply because they are women. Indeed, psychologist Gregory Bateson, who formulated the concept of the double bind in an examination of schizophrenia, assumed that it was primarily mothers deploying double binds who induced schizophrenia in their sons.[6]

These facts aside, the double bind is a strategy perennially used by those with power against those without. The overwhelming evidence shows that, historically, women are usually the quarry.

Binds draw their power from their capacity to simplify complexity. Faced with a complicated situation or behavior, the human tendency is to split apart and dichotomize its elements. So we contrast good and bad, strong and weak, for and against, true and false, and in so doing assume that a person can't be both at once—or somewhere in between. Such distinctions are often useful. But when this tendency drives us to see life's options or the choices available to women as polarities and irreconcilable opposites, those differences become troublesome.

Business theorist Rosabeth Moss Kantor calls them "self-defeating traps."[7] Author Joseph Heller termed them Catch-22s. Twentieth-century psychologists label them *double binds*. The notion has become a catchphrase to describe the dilemmas confronting contemporary women and is a commonplace in feminist scholarship. "This double yardstick of gender appropriateness and managerial effectiveness often leaves women in an unbreakable, untenable double bind," writes Nancy Nichols in the July–August 1993 issue of the *Harvard Business Review*. "Women who attempt to fit themselves into a managerial role by acting like men . . . are forced to behave in a sexually dissonant way. They risk being characterized as 'too aggressive,' or worse, just plain 'bitchy.' Yet women who act like ladies, speaking indirectly and showing concern for others, risk being seen as 'ineffective.'"[8]

"Women were again in a double bind," writes Jane Ussher in *Women's Madness: Misogyny or Mental Illness?*, "for the association of femininity with sexual innocence and purity or conversely, with insatiable lust, could be used to categorize women as mad. Thus the rigid image of 'woman' or 'femininity' could be used to punish, to convict, to control—women out of control were clearly sexual and dangerous, and mad."[9]

Even a sense of humor "catches women in a double bind," observes American Studies scholar Nancy Walker. "While they are not supposed to be creators of humor, inasmuch as this role would ascribe to them power and intellectual qualities denied them by the majority culture, they are supposed to applaud the humor of that majority culture and, above all, not take themselves too seriously."[10]

Others describe the dilemma without attaching "double bind" to it. "In my experience as a trial attorney," writes Carrie Menkel-Meadow, "I observed that some women had difficulty with the 'macho' ethic of the courtroom battle. Even those who did successfully adapt to the male model often confronted a dilemma because women were less likely to be perceived as behaving properly when engaged in strong adversarial conduct. It is important to be 'strong' in the courtroom according to the stereotypic conception of appropriate trial behavior. The woman who conforms to the female stereotype by being 'soft' or 'weak' is a bad trial lawyer; but if a woman is 'tough' or 'strong' in the courtroom, she is seen as acting inappropriately for a woman."[11]

Double binds are frequently involved in descriptions of women's speech. "In 1972, to be a woman in politics was almost a masochistic experience, a series of setbacks without a lot of rewards," writes Senator Barbara Boxer. "If I was strong in my expression of the issues, I was strident; if I expressed any emotion as I spoke about the environment or the problems of the mentally ill, I was soft; if I spoke about economics I had to be perfect, and then I ran the risk of being 'too much like a man.'"[12]

Catch-22 is a synonym for double bind. "Mothers are caught in a perfect Catch-22," writes psychologist Paula Caplan. "They are supposed to be concerned with emotions and closeness in relationships, but because autonomy has been designated by the white male middle class in North America as the pinnacle of emotional health, both mothers and those offspring who remain close to them are treated as immature or even sick."[13]

Progress that women have made in the profession "will not ensure that more women or minorities will study or stay in science," writes Marguerite Holloway in *Scientific American*. "Many scientists describe the situation as a catch-22: more women will enter the field only when there are more women in it. And, they say, the only way out of the conundrum is to change society's attitudes toward women—and men."[14]

Attitudes aren't all that matter, however. Money talks too. "[M]oney is unlikely to flow into a campaign unless credentials are in place, along with a record of voter confidence expressed through successful elections," write Witt, Paget, and Matthews. "This catch-22 has been particularly maddening to women candidates: You need money to be credible, but you have to be credible to obtain money."[15]

The concept also has taken hold in the press. Here and elsewhere "double bind" along with its alternative expressions has become, in many instances, a protean concept and, as such, of limited use. At worst, it can hobble women who see double binds as indestructible, tensile constraints that will lash back just as women believe they have broken free of them.

My take on the history of women differs from that of those who argue that liberation for women is an endless war in which small battles are won only to be met with violent repression. In *Backlash*, Susan Faludi argues that gains for women have been subverted by a backlash engineered by those in power. As one reviewer noted, her case is "basically a conspiracy theory. . . . that a cabal of villains has been at work successfully intimidating a large class of victims: women."[16]

Except for its epilogue, pessimism pervades Faludi's text. "A backlash against women's rights is nothing new in American history," she writes in Chapter 3. "Indeed, it's a recurring phenomenon: It returns every time women begin to make headway toward equality, a seemingly inevitable early frost to the culture's brief flowering of feminism."[17] She cites the feminist movement of the mid-nineteenth century launched at Seneca Falls. "By the end of the century, a cultural counteraction crushed women's appeals for justice. Women fell back before a barrage of warnings virtually identical to today's."[18] "The malaise and enervation that women are feeling today aren't induced by the speed of liberation," she asserts, "but by its stagnation. The feminist revolution has petered out, leaving so many women discouraged and paralyzed by the knowledge that, once again, the possibility for real progress has been foreclosed."[19]

Is this really the case? Is the history of women one of Sisyphean struggle against odds that remain constant and overwhelming? Not quite. Women's progress has been thwarted by double binds that, when surmounted, have in fact been replaced by other double binds, as I will show here. But as women have conquered the no-win situations confronting them, they have marshaled resources and refined aptitudes that have made them more and more capable of facing the next challenge, the next opportunity. At the same time, they have systematically exposed the fallacious constructs tradi- tionally used against them, and changed and enlarged the frame through which women are viewed. Although the result is not a steady move toward equitable treatment of women, it is a world in which progress is certainly sufficient to justify optimism.[20]

Enthusiastic reviews of Faludi's book by women's rights advocates sug- gest that it did touch a chord. The concept of backlash has been percolating through the women's movement for quite a while. "[T]here is a real back- lash against the quality and personhood of women—in America, as in Islam and the Vatican," wrote Betty Friedan in 1981.[21]

Faludi tapped into a widespread feeling that the Reagan-Bush adminis- tration's challenges to equal opportunity and abortion rights placed women's rights advocates under siege. "The backlash years drove us, or certainly drove me, into feeling embattled and needing to huddle together against the inhospitable forces out there," reports Naomi Wolf. "And this created a subculture within feminism. We have to end that."[22]

In Faludi's view, "Millions of individual women, each in her own way, spent the last decade kicking against the backlash barricades. But much of that effort proved futile."[23]

As women's rights in education illustrate, however, even in the Reagan years, setbacks weren't permanent. Take, for example, the evolution of Title IX, an important component of the Education Amendments of 1972. It outlawed sex discrimination in elementary, secondary, and post- secondary schools receiving federal funds. That victory was solidified in 1974 by the Women's Educational Equity Act, which authorized funds for research in this area. In 1979, the Court added teeth by holding that individuals could file private suits against institutions over enforcement of the law.[24]

In 1984, three years into the Reagan administration, the Supreme Court narrowed the mandate of Title IX significantly. In *Grove City College v.*

Bell,[25] it held that only those programs within an institution receiving federal funding—and not the institution as a whole—were bound by Title IX. So, since only the financial aid office was receiving federal aid at Grove City College, only its programs were subject to Title IX. But the rollback didn't hold. In 1985 the Civil Rights Restoration Bill was introduced. It passed in January 1988, while Reagan was still in office.

Faludi, however, finds little consolation in history. "This pattern of women's hopes raised only to be dashed is peculiar neither to American history nor to modern times," she says.[26] Yet at the close of the supposed backlash decade, women saw progress. In March 1992, 82% of the American women polled by CNN/Time reported that they "have more freedom than their mothers did." By 50% to 22% the same women also reported that they enjoyed life more than their mothers had.[27]

I do not mean to suggest that discrimination against women no longer exists in American society. Instead, my point is that even in the face of an administration unsympathetic to many of the goals of the women's movement, there was progress—slow and insufficient, but progress nonetheless. To deny it, and in fact to close our eyes to the evolution of the struggle for women's rights, invites despair. To perceive gains for women as inevitably thwarted by crippling backlashes is to take on ourselves the role of the permanently shackled.

The thesis of *Beyond the Double Bind* is a simple one. Historically, women have faced and transcended double binds. Until recently, however, as one was overcome, another, often a ghost of the one surmounted, took its place. Meanwhile, vestiges of the surmounted bind lingered in the language through which women were invited to view their new challenges. But, over centuries, the lessons learned in breaking the binds persevered as well. These lessons open options for women that are either closed or less accessible to men. By exercising these options, women deny others the power to define and as a result confine them in false options. Put simply, over time women have learned to turn potatoes into vichyssoise.

A Bundle of Binds

The social binds that have undercut aspiring female scholars are illustrative. Women who sought to learn and teach faced a network of interlinked binds—binds that go farther back than those faced by Spee's witches in 1631. The New Testament's most often quoted statement about women demands silence, submission, and procreation. Indeed, it is reproduction that will enable a woman to expiate her personal guilt for the sin of Eve. "I am not giving permission for a woman to teach or tell a man what to do," says I Timothy 2:12–15. "A woman ought not to speak, because Adam was formed first and Eve afterwards, and it was not Adam who was led astray but the woman who was led astray and fell into sin. Nevertheless, she will be saved by childbearing."

Some theologians in fact argued that the first woman's abuse of the power of speech and of teaching wholly accounted for the Fall. Since their maternal ancestor had misused her power of speech, argued some, Eve's daughters should be grateful that they retained the capacity to bear children. "The woman taught once, and ruined all," wrote St. John Chrysostom in 500 A.D. "On this account . . . let her not teach. But what is it to other women that she suffered this? It certainly concerns them; for the sex is weak and fickle, . . . [T]he whole female race transgressed . . . Let her not however grieve. God hath given her no small consolation, that of childbearing." Indeed, it is "by means of children" that women will be saved.[28]

Even as some theologies began to emphasize the "purity" of womanhood instead of woman's original sin, the bind remained in place. A loose tongue bespoke loose morals. Women who did not remain silent might be burned at the stake as witches or heretics under the fundamentalist impulse; in other times and places, the woman who would not be enjoined to silence was condemned as unwomanly. So in 1617 the pseudonymous Constantia Munda responded to a misogynistic diatribe by pointing out the trap it embodied:

> [Y]ou surmised that, inveighing against poor illiterate women, we might fret and bite the lip at you, we might repine to see ourselves baited and tossed in a blanket, but never durst in open view of the vulgar either disclose your blasphemous and derogative slanders or maintain the untainted purity of our glorious sex. Nay, you'll put gags in our mouths and conjure us all to silence; you will first abuse us; then bind us to the peace. We must be tongue-tied, lest in starting up to find fault we prove ourselves guilty of those horrible accusations.[29]

Denied access to literacy and the learning that comes with schooling, women were condemned, as Munda pointed out, for their illiteracy, their inability to think profound thoughts or contribute to the great intellectual debates of their day. And since public speech was considered immodest, women were instructed that defending themselves or their work was unladylike. Those who did learn to read and write found that communicating their learning to others was difficult. Among other constraints, they were denied access to Latin, the language of scholarship. "If we be taught to read," wrote the pseudonymous Mary Tattlewell in 1640, "they then confine us within the compass of our Mother Tongue."[30]

As the pseudonymity of Munda and Tattlewell suggests, some female writers escaped strictures by hiding their identities. Adopting a *male* pseudonym was the best route to publication, but this strategy presented other problems. When women resorted to it, evidence needed to rebut the claim that women have produced no great scholars often disappeared with their deaths. And when a masculine pseudonym was unmasked, the woman writer was ridiculed and her sexuality questioned. Indeed, the use of masculine pseudonyms led, by the sometimes extraordinary mechanics of the double bind, to some strange assertions. "The notorious sexologist of Vienna, Otto Weininger, maintained in 1903 that all the great women of history (Queen Christina of Sweden, Catherine the Great of Russia, math-

ematician Sofia Kovalevskaia, artist Rosa Bonheur) had been either homosexual or bisexual. Why else would intellectual women (George Sand and
George Eliot, for example) take masculine names?," he asked.[31]

In books, libraries, and classrooms, men preserved the knowledge they
developed; meanwhile, women were expected to hide their learning and
were denied access to the lectern. "Being a Woman [I] Cannot . . .
Publicly . . . Preach, Teach, Declare or Explane [my works] by Words of
Mouth, as most of the Famous Philosophers have done, who thereby made
their Philosophical Opinions more Famous, than I fear Mine will ever be,"
wrote Margaret Cavendish, Duchess of Newcastle in 1663.[32]

Women also were denied access to the professorships that facilitated
dissemination of their ideas. So, for example, "[i]t took three attempts,
four years, a German revolution, and the intercession of Albert Einstein to
get [future Nobel Prize winner] Dr. Emmy Noether the most junior faculty position possible—without pay."[33]

Noether was told that she could not be a faculty member but could serve
as a faculty assistant. "Thus, the 1916–1917 university catalog read,
'Mathematical-physical seminar. Theory of invariants: Prof. Hilbert, with
the assistance of Frl. Dr. E. Noether, Mondays 4–6 p.m. free of charge.'
She could not even earn the usual student lecture fees."[34]

Once women gained the power of the pen, they faced a set of assumptions that disparaged their actions and impugned their motives. The sanity
of female authors was called into question, even as late as the nineteenth
century. In *A Room of One's Own*, Virginia Woolf quotes the reaction of a
young woman to a new work by a female author. "Sure the poor woman is
a little distracted," observes the reader, "she could never be so ridiculous
else as to venture at writing books, and in verse, too."[35]

Absorbing the Limits

Some learned women accepted their exclusion as in their own best interest.
For many, the price of authorship was anonymity. A noteworthy instance
was eighteenth-century scientist Thiroux d'Arconville, an associate of
Voltaire. Conditioned by centuries of strictures on public presentation by
women, she believed that "'intellectual women garner only ridicule; if their
work is good, they are ignored; if it is bad, they are hissed at.' As a
consequence, she worked within the confines of her own home and veiled
her science in a cloak of anonymity."[36] Things did not change much when
the century turned. In 1811, the educator and playwright Madame de
Genlis advised: "If a woman does write books, she should avoid all pub-
licity; she should show a great respect for religion and austere morals; she
should not respond to critics of her work for fear that in the response she
might transgress feminine delicacy, modesty, and softness."[37] But in an
academic world built on the notion that truth emerges in the clash of ideas,
the author unwilling to defend her work damned it.

Indeed, the work of women particularly needed defending. Women who wrote learned prose or elegant poetry were dismissed as derivative or the by-product of creative happenstance. "If what I do prove well, it won't advance," wrote Anne Bradstreet in the 1650 edition of *The Tenth Muse*. "They'll say it's stol'n, or else it was by chance."[38]

The Absence of a Legacy

When women published under their own names, academic institutions did not usually perpetuate or their libraries preserve their work. As a result, great women scientists, philosophers, historians, and authors vanished from history. And without the traces of their lives and work, new generations of women could be subjected to the assumption that if women were capable of such intellectual activities, they would have produced the very forms of evidence that had been suppressed.

Generation after generation was expected to rediscover the wheel. "In the preface to Erxleben's 1742 defense of women's right to higher education, the medical doctor Christian Leporin noted that Anna von Schurman had published a book on the education of women in the previous century, but that 'despite all my efforts, it was not to be had.'" With Schurman's work misplaced by tradition, Dorothea Erxleben—Leporin's daughter—never had the opportunity to sharpen her young mind on Schurman's mature thought. Erxleben could not have known that the same fate would befall her own work. Some fifty years later, Amalia Holst noted that Erxleben's *Inquiry into the Causes Preventing the Female Sex from Studying* was "no longer available." Holst could not procure a copy, nor could Erxleben's stepson—a professor.[39]

The stolen legacy of our foremothers denied generation after generation the role models that testified to the possibility of female learning. Generation after generation assumed that it was the first to surmount the biases; as one generation succeeded, its legacy was lost to the next. Cumulatively, that vacuum made it possible for misogynists to reiterate their claim that women had not demonstrated a capacity for original thought because it was not in their nature to do so.

A similar collection of arguments was used to justify excluding work by women from the canon. Critics rationalized the omission, or what Joanna Russ terms "suppression," with a familiar litany:

She didn't write it.
She wrote it, but she shouldn't have.
She wrote it, but look what she wrote about.
She wrote it, but "she" isn't really an artist. . . .
She wrote it, but she wrote only one of it.
She wrote it, but it's only interesting/included in the canon for one, limited reason.
She wrote it but there are very few of her.[40]

The suppression of scholarly work by women was the product of a bind I call silence/shame. Other binds came into play in the education of women, among them the proposition that admitting women to the classroom would harm men.

When women entered the classroom in respectable numbers at the turn of the century, the guardians of the diplomas hypothesized that men would leave disciplines dominated by women; if that proved untrue, then the presence of women in the classroom would undercut men's education by distracting them. "Women, charged with sex repulsion and sex attraction, both of which interfered with the holy process of educating the future leaders (males) of the country, simply could not win," writes historian Barbara Miller Solomon.[41]

Admitting women into the masculine sanctums of higher education would dilute the manly virtues of the institution, if not of the country itself. So in 1925 *The Nation* parodied opponents of female admission to Columbia's School of Law. "If women were admitted to Columbia Law School, the faculty said, then the choicer, more manly and red-blooded graduates of our great universities would turn away from Columbia and rush off to the Harvard Law School!"[42]

Nevertheless, women persisted. They gained access to learning through the libraries of fathers, brothers, and husbands. They gained access to schooling by arguing, among other things, that to properly fulfill their roles as mothers and educate the next generation, they required the knowledge books and classrooms could provide.

By the time the Education Amendments of 1972 put Title IX on the books, a lot of ground had been covered. Armed with education and gaining increasing access to the public sphere, women had claimed the tools that tore down barriers against them, stripping bare the pseudosciences that snared them in double binds. So, for example, the work of female legal scholars framed the Supreme Court ruling that outlawed fetal protection policies in the workplace, and the testimony of a female psychology professor lay the ground for the Supreme Court's decision, in *Hopkins v. Price Waterhouse*, that promotional decisions may not be determined by imposing gender stereotypes on women in the workplace.

Moving from education and public speech into action, women fought for and finally won the right to vote. The ballot became another major weapon in the war against binds. When exercised in force, it could be used to threaten the political survival of those who opposed women's rights. It also had the potential to put women in positions of influence, ready to create social change, when the opportunity arose. So, for example, after Congressman Howard Smith in a move designed to scuttle the entire proposition, amended Title VII of the Civil Rights Act to include the word "sex," Congresswoman Martha Griffiths led the battle so that it stayed there. Once the legislation was passed, she policed the Equal Employment Opportunity Commission to ensure that sex discrimination was taken seriously. Lindy Boggs did the same for equal credit protection.

And it was Congessswoman Patsy Mink who shepherded Title IX through the House in 1972. By barring discrimination in educational institutions that recevied federal support, Title IX opened graduate programs and tenure-track jobs once closed to women. I am of the generation that benefited—my first university job was created under the affirmative-action pressure induced by the impending passage of Title IX.

By persisting, women of past generations proved the truth of Sarah Grimke's observation in the nineteenth century that educational reformer Horace Mann would "not help the cause of women greatly, but his efforts to educate her will do a greater work than he anticipates. Prepare woman for duty and usefulness, and she will laugh at any boundaries man may set for her."[43]

The Construct of the Double Bind

Grimke might not be surprised at women in the U.S. Congress or women justices of the Supreme Court, but neither would she fail to recognize the sorts of constraints women still face. They are the vestiges or ghosts of the same double binds that have confronted women throughout history.

As described by Gregory Bateson in the mid-1950s, double binds involve a powerful and a powerless individual, or, in the cases of interest to us, social and institutional norms and a vulnerable class—women. For Bateson and his associates, a double bind occurs if two or more persons, one of them the victim, undergo a repeated experience in which one "primary negative injunction" conflicts with a second, both "enforced by punishments or signals which threaten survival," and from which the victim has no means of escape.[44]

In her 1975 book *Language and Woman's Place*,[45] linguist Robin Lakoff applied Bateson's notion to society's requirements of women. "Now the command that society gives to the young of both sexes might be phrased something like: 'Gain respect by speaking like other members of your sex,'" writes Lakoff. "For the boy, as we have seen, that order, constraining as it is, is not paradoxical: if he speaks (and generally behaves) as men in his culture are supposed to, he generally gains people's respect. But whichever course the woman takes—to speak women's language or not to—she will not be respected. So she cannot carry out the order."

Bateson and his colleagues concluded that responses to a double bind include "helplessness, fear, exasperation, and rage."[46] They theorized that schizophrenia could be induced by repeatedly subjecting vulnerable individuals to double binds. Other scientists found that in so-called "normal individuals," double binds increase expressed anxiety as well.[47]

Rhetoric, as critic Kenneth Burke notes, is a reflection as well as a selection and a deflection. Rhetoric makes sense of otherwise inchoate experiences. It structures. It orders. It focuses. *It attempts to limit our angle of vision to that of the writer or speaker.* A double bind is a rhetorical construct that posits two and only two alternatives, one or both penalizing the person being offered them. In the history of humans, such choices have been constructed to deny women access to power and, where individuals manage to slip past their constraints, to undermine their exercise of whatever power they achieve. The strategy defines something "fundamental" to women as incompatible with something the woman seeks—be it education, the ballot, or access to the workplace.

When a bind casts one alternative as loathsome, it points to the other as a woman's only appropriate choice. So, childbearing is expected to be chosen over intellectual pursuits, silence over shame, and invisibility over acknowledgment of aging. When both alternatives in the construct carry clear penalties, as equality and difference do when they are defined by male norms, the woman is encouraged to abandon whatever goal has aroused the equality or difference debate. Finally, when a bind casts two supposedly desirable states as mutually exclusive, the woman is invited to believe that she is incapable of attaining success.

The intended result is stalemate for those whose sex chromosomes are XX rather than XY. You might say that this book is a study of the binds that tie.

Double binds are constructions derived from theology, biology, and the law, and rhetoric's fashioning of each. In some cases, the external constraint is invariant. Men cannot bear children; women can. Men cannot breastfeed; women can. In other cases, the constraint is either hard and fast or fabricated, depending on one's belief system.

Theology

For a fundamentalist, the story of Adam and Eve reflects the way it really was. The sin of Eve condemns women to childbearing; in the New Testament Paul bars them from preaching on that account. Many avowed Christians deny this construction as a simplistic and archaic view; other people treat the whole story of Eve and Eden as the conjuring of those wanting to justify an absolute patriarchy. For others, the scriptural text is absolute, and the dictates of religion are as constraining as the laws of gravitation and matter. They describe the natural order of life—sin against them and be damned. Only through silence, submission, and reproduction could women be redeemed. Since silence and motherhood were twinned, a corollary assumption was formed of the alliance: Public speech by a woman is the outward sign of suspect sexuality.

Biology, created by God, enforced this "natural order." Since women could bear and nurture infants, it followed that they must. Once a child is born, the assumption that the mother has primary responsibility for raising offspring kicks in. Nothing inherent in a woman's physiological makeup uniquely equips her for child*rearing*. The suggestion that childrearing is a woman's natural role is not a statement of natural law, but an assertion. Yet it was granted for millennia.

The scientific community of earlier centuries perpetrated the idea that abandoning her natural sphere carried physiological penalties, among other punishments. "Female illnesses" were the outward signs of an unsubmissive soul "[T]he most significant cause of a woman's menopausal disease, virtually every [nineteenth-century] doctor believed, lay in her violation of the physiological and social laws dictated by her ovarian system," writes historian Carroll Smith-Rosenberg. "Education, attempts at birth control or abortion, undue sexual indulgence, a too fashionable life style, failure to devote herself fully to the needs of husband and children—even the advocacy of woman's suffrage—all might guarantee a disease-ridden menopause."[48]

Such assumptions prompted the infamous "rest cure" prescribed for female patients in the nineteenth century, and bedeviled the development of safe contraceptive drugs and devices.

The Law

For women to venture beyond the confines of rearing children in the home would destroy the natural order of things and in fact subvert the state. So scientific advances toward safe means of birth control were met with legal constraints. Moreover, women who married literally disappeared under the law of coverture.

"By marriage, the husband and wife are one person in law," wrote William Blackstone in 1765, "that is, the very being or legal existence of the woman is suspended during the marriage, or at least is incorporated and consolidated into that of the husband; under whose wing, protection, and *cover,* she performs everything."[49] Under the law of coverture, married women could not sue, sell, or contract without first getting their husband's permission. They were, in the eyes of the courts, represented in and through the man they had married.

"Now if a woman holding public office were to marry, two possibilities would follow," observed the German philosopher Johann Fichte in *The Science of Rights* (1798). "First, she might not subject herself to her husband in matters regarding her official duties which would be utterly against female dignity . . . secondly, she might subject herself utterly to her husband, as nature and morality require. But in that case she would cease to be the official and he would become it. The office would become his by marriage, like the rest of his wife's property and rights."

From the invariant facts of a woman's physiology—her ability to bear and nurse children—came a host of assumptions rooted in "nature and morality" that cast women in double binds. Codifying these assumptions were the powerful institutions of the church, the scientific community, and the state. Enmeshed as they were in theology, biology, and law, the binds were seemingly non-negotiable—until women began slipping the knots.

Beyond the Double Bind will identify five binds, their archaic origins, and the ways their vestiges continue to shape contemporary culture. These binds include the following constructs:

- Women can exercise their wombs or their brains, but not both.
- Women who speak out are immodest and will be shamed, while women who are silent will be ignored or dismissed.
- Women are subordinate whether they claim to be different from men or the same.
- Women who are considered feminine will be judged incompetent, and women who are competent, unfeminine.
- As men age, they gain wisdom and power; as women age, they wrinkle and become superfluous.

And, in a latter-day bind, women who succeed in politics and public life will be scrutinized under a different lens from that applied to successful men, and for longer periods of time.

At their base, these binds concern power and place. Across Western history a metaphor has emerged to express each. The first—who is in charge?—is expressed as a contest over who will wear the "breeches" and operates on the zero-sum notion that there is one pair of pants per couple. The second—place—is manifest in the claim that in their proper place, women nurture. That notion is often symbolized by the assumption that women should stay in the kitchen.

In the *Canterbury Tales*, the story of the Wife of Bath is about "the fight for the breeches." Centuries later the question remained current. Asked who wears the pants in his family, Prime Minister Margaret Thatcher's spouse Denis responded, "I do, and I also wash and iron them."[50]

The figurative enjoinder not to wear the pants in the family was incarnated in requirements that women in public wear appropriately "feminine" dress. Feminist activist Flo Kennedy recalls the era when women first began to wear slacks. "I can remember—I was still practicing law at the time—going to court in pants and the judge's remarking that I wasn't properly dressed, that the next time I came to court I should be dressed like a lawyer. He's sitting there in a long black dress gathered at the yoke, and I said, 'Judge, if you won't talk about what I'm wearing, I won't talk about what you're wearing.'"[51]

Historically, the place for women is in the private sphere of the home—centered, metaphorically, not in the bedroom or the parlor, but in the nursery and the kitchen. "A man is in general better pleased when he has a good dinner upon his table, than when his wife talks Greek," observed Samuel Johnson in the mid-eighteenth century.[52] Two centuries later, the spouse of the Democratic nominee for president, Hillary Clinton, was ensnared in a dispute over the meaning of her remark that she chose not to stay home and bake cookies and have teas. In such instances, food is taken as a symbol for home and home for "a woman's proper place."

Some activists stepped out of that noose by granting that a woman's place is in the home, while enlarging our notion of what constitutes home. "Home is where the heart is, where your loved ones are," argued Carry

Nation in her efforts to save men from the saloon. "If my son is in a drinking place, my place is there."[53]

The binds that constrain women's power and women's place are not, in fact, discrete. Their relationship is prismatic—one magnifies another. Each of the chapters on binds will trace the origins of the double bind, track the social sanctions put in place to preserve it, and chart the progress women have made in overcoming it—in the process drawing power and advantage from what were moves originally designed to disable. A final chapter will summarize the strategies available to surmount the residues of the binds that in the past have tied.

Underlying the binds are specific constructs: The no-choice-choice; the self-fulfilling prophecy; the no-win situation; the unrealizable expectation, and the double standard. Each circumscribes choice.

The no-choice-choice is the focus of Chapter 3, "Double Bind Number One: Womb/Brain." This bind casts the world as either/or, with one option set as desirable, the other loathsome—hence a no-choice-choice. Women could use their brains only at the expense of their uteruses; if they did, they risked their essential womanhood. Exercise of the uterus was associated with the private sphere, exercise of the brain with the public. Here was a question of a woman's proper place: Those who chose to exercise their intellects in public life upended the natural order, endangered the family, and called into question whether they were really women. Women broke the bind by gaining access to education and using the tools of scholarship to establish that childbearing does not destroy intellect, and vice versa, and then gaining access to forms of contraception and birth control that made reproductive choice and timing possible.

Once women ventured into public space, they confronted binds designed to deny them power and to undercut what power they could attain. In Chapter 2, I explore how Hillary Clinton was caught in a derivative of the womb/brain bind—the social assumption that someone who valued career must despise those who elect full-time homemaking. As the *New York Times* put it during the 1992 campaign, "She is a lightning rod for the mixed emotions we have about work and motherhood, dreams and accommodation, smart women and men's worlds." Scores of other media sources agree. "[T]he squirming over Hillary Clinton," said the *Los Angeles Times,* "isn't so much about a First Lady as about ambivalence over women, power, work and marriage." Clinton became a national test case, subjected in fact to *all* the binds traditionally deployed against women.

Self-fulfilling prophecies are the subject of Chapter 4, "Double Bind Number Two: Silence/Shame." Sociologist Robert K. Merton defined a self-fulfilling prophecy as "a *false* definition of the situation evoking a new behavior which makes the originally false conception come *true*."[54] The silence or shame bind condemns women for failing to do something they are forbidden to do. So, for example, women were forbidden to speak and then condemned for failing to produce great oratory. The first condition becomes a self-fulfilling prophecy of the second. The bind was overcome by women who weathered the social sanctions imposed on women who spoke, in the process demonstrating a capacity to speak. Through the exercise of public speech they were able to access other resources such as the courts and the ballot.

No-win situations are the subject of Chapter 5, "Double Bind Number Three: Sameness/Difference." In the no-win situation, by winning, you lose. If the no-choice choice is "either-or," this bind is "neither-nor." In it, women are judged against a masculine standard, and by that standard they lose, whether they claim difference or similarity. The bind is broken by positing a form of equality not solely based on a male norm.

Unrealizable expectations are a corollary of the no-win situation. Treated in Chapter 6 addressing the bind of femininity/competence, unrealizable expectations are also designed to undercut women's exercise of power. By requiring both femininity and competence of women in the public sphere, and then defining femininity in a way that excludes competence, the bind creates unrealizable expectations. By this standard, women are bound to fail. The power of the bind is rooted in a woman's willingness to grant someone else the right both to *define* and *impose* the requirement of femininity. The presence of this bind has led those studying women and leadership to conclude that women "have to reconcile contradictory expectations to succeed, contradictions not imposed on men."[55]

Denying others the power to define appropriate behavior breaks the bind. Being feminine as femininity was traditionally defined may be incompatible with being competent, but being a woman is not. When that view was embraced by the Supreme Court, women's rights advocates gained important ground in their fight to alter persistent attitudes that did what the law once had done—keep women in their assigned place.

The *double standard* is a construct that reinforces the competence/femininity bind. Blossoming as women gain power, it functions by ensuring that they cannot successfully exercise it. Women's sexuality is treated differently, their actions judged differently, their competence tested differently and for a longer period. Our expectations of women are more difficult to meet. At the core of this bind is the assumption that woman is other and defective.

Chapter 7, "Double Bind Number Five: Aging/Invisibility," examines the double standard holding that as men age they acquire wisdom and power, while women gain wrinkles and hot flashes. In pervasive stereotypes about female aging, we see the residues of the womb/brain and femininity/competence binds as well. These binds are reinforced throughout the culture. Chapter 8, "Newsbinds," lays out examples from news.

The final chapter contrasts stories that presuppose backlash and those that assume bind-breaking and argues that it is the latter, not the former, that best capture the history of the struggle for women's rights. It also charts the ways and means now available to women and men bent on bind-breaking.

That struggle has not, once and for all, been won. Neither have many other struggles over the rights of groups who have been history's victims, yet few would argue that we have not progressed since the days that the ballot was routinely denied to women as well as to African Americans and other minorities. The vestiges of ancient prejudice die hard, and the ways in which they are fought are matters of disagreement among those who fight them. The civil rights movement has included advocates of non-violence and Panthers, integrationists and isolationists. So the women's movement has found itself split into opposing camps more than once over time.

One problem with the backlash hypothesis is that it assumes that the movement has been a homogenous whole, moving steadily toward agreed-on goals. That view is not born out by history. Once suffrage was gained, for example, the movement split, with one flank advocating and the other opposing the Equal Rights Amendment. Feminist opponents of divorce reform laws in the early twentieth century turned into advocates,[56] and when an attained goal proved counterproductive—as did special protections in the workplace and different ages of sexual consent for men and women—advocates changed directions.

What can be construed as a backlash often was a serious disagreement among women's advocates about means and ends. When the Supreme Court faced the issue of a minimum wage for women in *Adkins v. Children's Hospital* in 1923, for example, some women's groups—including Alice Paul's National Women's party—favored the court's position, which ruled the law unconstitutional. Others supported what they saw as protection for women.[57] A similar divide separates those who favor and oppose bans on pornography.

Faludi's argument for backlash is, moreover, selective, and the resulting spin on events invites despair. A blue-collar female union member loses her job under conditions she considers discriminatory, sues, and loses. "Desperate for work to support her two children, King cleaned houses, then took a job as a waitress. She lost all her benefits. 'Today I cleaned the venetian blinds at work,' she says. 'I make $2.01 an hour and that's it, top pay. It's demeaning, degrading. It makes you feel like you are not worthwhile.'"[58] Of note is the fact that the woman's case was decided at the District Court level not on grounds of sex discrimination but on a question of the applicability of OSHA regulations. The first opportunity that the Supreme Court—with three Reagan justices on it—had to hear a comparable case framed as sex discrimination, it held for the plaintiffs, 9–0.[59]

Although feminist historians lined up on both sides of the case, Faludi also treats the loss of the class-action suit by women presumably denied commission-sales positions at Sears as a major defeat for women. Ignored entirely is Ann Hopkins' vindication in 1989, at both the District and Supreme Court levels, in her suit against Price Waterhouse for sex discrimination.

Selectivity is at work in Faludi's economic claims as well. While women continue to earn less on average than men, such economic factors as part-time work, time out of the labor force for childbearing and rearing, and a shorter work week account for some of that difference. So, for example, Census Bureau data indicate that in 1992 married women with pre-schoolers were less likely to work year-round and more likely to work part-time than mothers whose children were six years old or older.[60] And in key places, Faludi is simply wrong. The difference between the earnings of men and women, which Faludi says hasn't improved much since 1955,[61] in fact has changed. And change occurred during the supposed "backlash" decade that is Faludi's focus. In 1980, women earned 64 cents for every dollar earned by men; in 1990, that figure had jumped to 72 cents.[62] But more important is that in her presumed backlash decade of the 1980s, the gap narrowed more dramatically than it had in the 1960s or 1970s.

None of this is to suggest that the wage gap and other forms of discrimination do not persist and urgently require change. They do. My point is that even in the 1980s advocates won some major victories.

There have been periods in which the women's movement was largely quiescent, as it was from the early 1920s until the early 1960s. Other times are characterized by steady, ongoing activity, and with it progress. Beginning in the early 1970s, the Court has overturned laws that enshrined unequal access, unequal opportunity, and unequal obligations based on sex in estate administration,[63] as well as unequal access to fringe benefits in the military,[64] employment benefits,[65] and alimony.[66]

Viewed in broad perspective, progress has been clear, whether in the areas of employment rights, reproductive rights, rights to credit, or protections from sexual harassment and gender stereotyping. In denying such advances, and their cumulative effects, we risk seeing ourselves as perpetual victims.

Not all contradictions fit Bateson's definition of the double bind. Paradoxes pervade literary, philosophical, and theological discourse throughout the culture. Where they are not grounded in fallacies, they may be used to elicit higher levels of awareness. In some Eastern religions, for example, contradictions are the focus of meditation. The goal is to free the contemplative from the material world in which the contradictions seem to inhere. Even in physics, the idea that light is both a particle and a wave is seen as a paradox: As the question was framed by classical science, it had to be one or the other; it could not be both. The poet Pat Parker incorporates another contradiction, this one regarding human behavior, in her poem "For the white person who wants to know how to be my friend": "The first thing you must do is to forget that i'm Black. Second, you must never forget i'm Black."[67] In all three cases, enlightenment results from paradox.

Alternatively, as Rosalyn Carter suggested in her autobiography and as other leaders such as Gloria Steinem have pointed out, recognition that you will be condemned no matter what you do can liberate a person to do whatever she wants.

The double bind is durable, but not indestructible. Examined as rhetorical frames, double binds can be understood, manipulated, dismantled. The bind that crops up after one has been vanquished is often a pale ghost of the more vigorous form that preceded it. *Backlash* is a more insidious idea. Seen as an external and invariable constraint, it does real damage when women accept and *internalize* its assessment of social and political reality. "The most effective backlash against feminism almost always comes from within, as women either despair of achieving equality or retreat from its demands," writes Wendy Kaminer.[68] I agree. As a rhetorical construct, *backlash* can be as constraining a frame as the constructs that form double binds. The notion that moments of progress for women are met by an inevitable and inevitably successful backlash turns the political and social activity of women into Sisyphean gestures. *Backlash* invites women and their allies to give up.

A more inclusive view of the history of women shows them surmounting, sometimes one by one, a series of double binds whose roots are deeply embedded in the past. Women who unmasked one dilemma faced the next and challenged it, bumped into a third and pirouetted around it, confronted another, and denied it its power. In the process they enlarged the scope of science, changed laws, altered behaviors, and changed the political complexion of this country. If they do not disable themselves with the rhetoric of disempowerment and victimization, they will enter the twenty-first century able to stand, speak, dance, and redefine the world as the need arises.

2

Hillary Clinton as Rorschach Test

WHAT is sauce for the gander has until recently been poison for the goose. We see this in press treatment of and citizen reaction to Hillary Clinton, which manifest the binds that tie women in the public sphere. "Coverage of Hillary Clinton is a massive Rorschach test of the evolution of women in our society,"[1] observed Betty Friedan.

The mirror that segments of society hold up reflects some dark and ugly sentiments as well as some nobler ones. "What do you call a cross between a draft-dodger and a dyke?" asks a caller on Kevin McCarthy's call-in radio show in Dallas on March 25, 1994. "Chelsea." "What do you call the meanest woman in the world?" "Tonya Rodham Bobbitt."

At issue in public and private discussions of Hillary Clinton first in the campaign and then in the White House were fundamental and to some extent unresolved relationships between concepts taken as antithetical for women by those of our grandmothers' generation: women versus power, work versus marriage, childrearing versus career.

Hillary Clinton did not live in a world of either-or, however, but of both-and. The Yale-educated wife of a powerful man, she had earned the respect of the nation's lawyers while raising a child and managing a career. The fact that she had to earn her husband's respect as well is a revealing reflection on the constraints still facing women. In mid-summer of the 1992 campaign, Bill Clinton implied that his confidence in his spouse's abilities as a lawyer did not extend to her capacity to combine a career with marriage and motherhood. "I am most proud of how she's raised Chelsea," he told *People* magazine. "I say that because from the first time I met her, I knew she would be a great lawyer. She's achieved a lot of things I'm proud of, from the time we were on the mock trials together in law school to the Watergate committee to leading educational reforms here in Arkansas. I always knew she could do that." What then did he doubt that she could do? "But the fact that she was able to have this incredibly full professional and public life and still be a wonderfully successful person as a good mother and wife, and grow over the years is, I think, her greatest achievement."[2]

In additon to having to prove even to her spouse that a successful lawyer could also be a good wife and mother, Hillary Clinton faced the assumption that as the wife of a presidential candidate she both held and sought illegitimate power—power unmerited by her credentials, exercised with the delicacy of a Lady Macbeth or Marie Antoinette.

Hillary Clinton became a surrogate on whom we projected our attitudes about attributes once thought incompatible, that women either exercised their minds or had children but not both, that women who were smart were unwomanly and sexually unfulfilled, that articulate women were dangerous. As a lawyer she had examined the nature of the double binds affecting female attorneys. In the campaign she learned the nature of double binds as they play out in life.[3]

Midway through the primary season, nationally syndicated columnist Ellen Goodman identified some of the no-win situations in which Hillary Rodham Clinton was cast:

> When she held onto her own name in 1980, she was blamed for her husband's defeat. When she gave it up, she was criticized for self-defeat. When she stuck up for her husband and marriage on "60 Minutes," she had to prove that she was no Tammy Wynette. When she proved it, the Tammy Wynette fans wanted to know what's wrong with standing by your man. At campaign rallies, when she speaks with a strong political voice, someone invariably asks, "Why don't you run?" If she were as quiet as [Republican candidate Pat Buchanan's spouse] Shelley Buchanan someone would undoubtedly ask if her husband was an impossible chauvinist.[4]

"Perhaps this [criticism of Hillary Clinton] signals the ambivalence society has toward changing roles in an era of backlash against women," editorialized *El Nuevo Herald*. "Men may be projecting on Mrs. Clinton their hostility toward feminism, while women, distressed by their multiple responsibilities, may be projecting their frustrations."[5]

Of interest here is not so much the campaign strategy that ultimately transformed a lawyer in the Rose law firm in Arkansas into a First Lady, but rather the complex interplay between Hillary Rodham Clinton and the labels through which she was viewed by reporters, columnists, supporters, and antagonists. In them, we see the residues of the complex and sometimes contradictory expectations we carry into our encounters with women. Understanding how 51% of us are shaped by and shape a particular set of dilemmas that I term double binds is the task of this book.[6]

What pundits called "the Hillary factor" was brought to the fore by two of the more often replayed sound bites of the 1992 presidential campaign. The first occurred in the hour after the 1992 Superbowl, in a special *60*

Minutes interview, when she responded to a charge that her husband had engaged in a twelve-year affair with a nightclub singer. There Mrs. Clinton averred that she was "not sitting here, some little woman standing by my man like Tammy Wynette."

The second incident was her response to charges that she capitalized on her relationship with the Arkansas governor to draw state business to her law firm or, alternatively, that he had funneled funds to the firm. In response, Hillary Rodham Clinton noted that she "could have stayed home and baked cookies and had teas," but instead chose to follow the career begun before her marriage.

Over time, even otherwise careful scholars conflated the two. "[A]s late as 1989," write journalist Linda Witt, political scientist Karen Paget, and historian Glenna Matthews in *Running as a Woman: Gender and Power in American Politics,* "according to the editor in chief of the *New York Times,* the all-male-is-normal paradigm—except at tea parties, of course—was the *Times's* editorial philosophy, and the only women who might be news would be those wives who inappropriately exercised their husbands' authority. This may explain why Hillary Clinton's statement—'I'm not just a little woman who can stay home and bake cookies and have teas'—so startled the nation's news editors. It virtually ended weeks of front-page explorations into her husband's alleged extramarital affiars."[7]

"Stand by Your Man"

The contexts of the original statements were quickly lost in press accounts and the featured portions taken as controversial comments on a woman's role. The generative moment for the "stand by your man" comment occurred when interviewer Steve Kroft implied on the *60 Minutes* segment that the Clintons had "reached some sort of an understanding, an arrangement" in remaining married. "You're looking at two people who love each other," replied Bill Clinton with a tone of irritation. "This is not an arrangement or an understanding. This is a marriage. That's a different thing." Added Hillary, "Now, I'm not sitting here some little woman standing by my man like Tammy Wynette. . . . I'm, I'm sitting here because I love him and I respect him and I honor what he's been through and what we've been through together. And you know, if that's not enough for people, then heck, don't vote for him."[8]

"Determined not to appear like a victim, Hillary Clinton said coolly that loving her husband throughout the trouble spots in their marriage did not mean she was 'sitting here like some little woman standing by my man like Tammy Wynette,' recalled the *Los Angeles Times.*"[9]

"Mrs. Clinton responded with a loyal wife's indignation, an ambitious politician's fervor, and a practiced litigator's skill to the long-expected bedmail. . . . She is nobody's 'little woman'" observed *New York Times's*[10] columnist William Safire.

From a woman rated one of the nation's top one hundred lawyers, supporters heard a defense of her husband and their relationship made out of respect and love, not from the viscerally felt obligation of a wife to condone whatever her husband had done. On the other hand, in "cookies and tea" her champions heard the right of a woman to choose both marriage and a career. Critics perceived an indictment of women "loyal" to their husbands in the first case, and in the second an indictment of both motherhood and traditional homemakers.

The country and western singer whose song enjoined women to stand by their men publicized her response to Mrs. Clinton. "Mrs. Clinton," she wrote, "you have offended every woman and man who love[s] that song—several million in number. I believe you have offended every true country music fan and every person who has 'made it on their [sic] own' with no one to take them to the White House."[11]

"Cookies and Tea"

"Cookies and tea" was transmuted as well. When reporters simmered Hillary Rodham Clinton's fragmentary response to charges by Jerry Brown down to a sound bite, they did what she had not explicitly done: pitted traditional homemaking against career, diminishing in the process the legitimacy of the first. Lost in the reduction was the context in which Clinton had made the statement.

In a televised debate on the eve of the Illinois and Michigan primaries, former governor Jerry Brown accused Bill Clinton of funneling business to his wife's firm. Implicitly, Brown also accused Hillary Clinton of trafficking in her relationship with her husband to secure state business for her law firm. Hillary's response shifted the issue from one of possible conflict of interest in the public sphere to the more traditional question, Does a woman belong in the public sphere at all?[12]

Clinton aides George Stephanopolous, Paul Begala, and Richard Mintz staffed the pseudo-event that gave rise to the comment, a March 16 breakfast at a Chicago diner with subsequent handshaking with passengers entering the El. Pressed about Jerry Brown's charges, the Democratic contender told reporters to "Ask her. Ask her. You know what he said last night was absolutley false. He said I hustled business for her law firm. It was a typical thing to say of a person who respects the fact that women can be professional, can have their own work and do their own jobs."

Clinton expanded on his defense of his wife, attacked Brown for regularly reinventing himself, and noted "if he wants to go after my wife I'm gonna hit him just like I did last night." Here as in the debate, Clinton cast himself as the aggrieved husband defending his assaulted wife.

In a statement outside his hotel earlier that morning, the Arkansas governor translated what Brown had said into the language of the back alley brawl. "I went back over what we dealt with last night," he said. "And I think we know I have some old-fashioned values. If somebody jumps on my wife, I'm gonna jump em back. . . . And he jumped on her and I jumped him back and I still feel good about it this morning. I think most people will identify with that." By noon a different set of old-fashioned values would be at issue.

Twenty minutes later at the Busy Bee restaurant he returned to the theme. "[7:51 A.M. EST] I've got a lot of new ideas but I have some old-fashioned ones too and if he goes after her again, I'll hit him again." That comment set old and new as the boundaries for the day's discussion. A literary critic would call it presentiment.

Stand by your wife was the organizing principle of Clinton's statements. "[7:52] You know I don't mind what he says about me and I've never said anything about him or Senator Tsongas except where we differed on the issues. If he wants to go after my wife I'm gonna hit him just like I did last night." The rules of attack differ when wives are involved.

Hillary, meanwhile, had literally been standing by her man. At 7:51, by her husband's side, she entered the discussion stating that she had not "shared in one dollar of state funds that has ever gone to my firm," noting that she didn't know "what else I could have done," and defending the fact that her firm represented banks.

At 7:54 Bill Clinton again assumed the role of "Rocky." "I don't mind hitting him personally after what he said about Hillary last night."

That posture had strategic advantages. As Republican consultant Roger Ailes told the audience of the *Today Show* on March 17, "Jerry Brown helped Clinton because a man defending his wife gains points." Brown's attack, however, pinned Hillary Clinton in a double bind. If she marshaled the rebuttal, she would seem to be acting as a candidate in her own right, the debate now between her and Brown. If she didn't, she raised questions about her competence as a lawyer, about her ability to defend against Brown's charges, and about the truth of her claims to *60 Minutes* that theirs was a partnership.

Indeed, as the more hide-bound traditionalists saw it, Bill as champion of Hillary was inconsistent with Hillary's rejection of the "little woman" role in the *60 Minutes* interview. A woman in need of such chivalrous conduct ought to be at home tending the hearth and nursing the children—not dodging bullets with her partner at the OK Corral. "Stop trying to have it both ways"; columnist William Safire advised, "you cannot be gallant about a feminist."[13]

At 7:57, reporters broached the question to Hillary herself. What was her response to Brown's charges? Her fragmented syntax suggests that she is groping for an answer. By stressing that she is being attacked for trying to have "an independent life" and having her "own life," Mrs. Clinton underscores her husband's assertion that she did not benefit financially from his governorship. But by introducing both "an independent life" and "motherhood" into her rambling answer, Hillary made it possible to read into "cookies and tea" another idea entirely. "I thought number one it [Brown's attack] was pathetic and desperate," she said, "and also thought it was interesting because this is the sort of thing that happens to the sort of women who have their own careers and their own lives. And I think it's a shame but I guess it's something that we're going to have to live with. Those of us who have tried and have a career, tried to have an independent life and to make a difference and certainly like myself who has children but other issues, uh you know I've done the best I can to lead my life but I suppose it'll be subject to attack but it's not true and I don't know what else to say except it's sad to me."

Asked whether there wasn't a way to avoid the appearance of conflict, she responded "I wish that were true. You know I suppose I could have stayed home and baked cookies and had teas but what I decided to do was fulfill my profession which I entered before my husband was in public life. And I've tried very, very hard to be as careful as possible and that's all I can tell you."

As she uttered the "cookies and tea" remark (at 7:58), Clinton aides, in the words of one, "felt the air go out of the room," and noting "that look" in NBC's Andrea Mitchell's eye, ended the press opportunity at 7:59. "That's the sound bite of the day," noted one reporter moving outside to use his cellular phone to file. As he was doing so, the aides drew Mrs. Clinton aside to recommend that she recontextualize the statement.

The extent to which Hillary Clinton was surveying the terrain only after she had walked it is apparent in the fact that she did not realize the "cookies and tea" remark could be heard not as a rejection of the role of full-time hostess for a governor, but as an indictment of stay-at-home motherhood. One reporter who overheard the discussion between Mrs. Clinton and the campaign aides recalls that she initially argued that her statement could not be interpreted that way.

The aides persuaded her otherwise. Twenty minutes later, the Clinton staff encouraged reporters to move from covering Bill, who was shaking hands with passersby, to Hillary. "You know," she told them, "the work that I've done as a professional, as a public advocate, has been aimed in part to assure that women can make the choices that they should make—whether it's full-time career, full-time motherhood, some combination, depending upon what stage of life they are at—and I think that is still difficult for people to understand right now, that it is a generational change." As she made this statement, not all the cameras had moved to her. When they arrived, she reworked the remark. This accounts for the disparity betwen the versions recorded by the *Washington Post* and NBC News.

Hillary Clinton made her last statement at 8:24 Central Standard Time. Thirteen minutes later, the story would be on the AP wire. For those who wonder at the comparative power of television and print to set agendas, the subsequent treatment of "cookies and tea" is instructive. At 8:37 Chicago time, the AP wire service transmitted its story, dropping those portions of the original statement that distracted from the piece's focus: conflict of interest. "His wife, Hillary Rodham Clinton, said she always made sure her work as a partner at the prestigious Rose law firm in Little Rock, Arkansas, did not create any appearance of conflict. 'I've done the best I can to lead my life,' she said as she campaigned with her husband. 'I suppose I could have stayed home and baked cookies and had teas. I've tried very, very hard to be as careful as possible."[14]

It was the broadcast story, however, that worried the campaign staff. "CNN started running 'cookies and tea' almost immediately," recalled a staff member. Within hours CNN was broadcasting Hillary Clinton saying, "I suppose I could have stayed home and baked cookies and had teas, but I—what I decided to do was to fulfill my profession, which I entered before my husband was in public life. And I've tried very, very hard to be as careful as possible." In mid-afternoon, CNN's *Inside Politics* led with the statement. But in closing the show, the link between the statement and the sentence about the law firm was dropped. Gone was the claim that she'd tried to be careful. The sound bite was beginning to migrate from a statement about conflict of interest to a claim about homemakers and careerwomen.[15]

On NBC, Andrea Mitchell, whose nonverbal reaction to the original statement had worried Clinton staffers, dropped the "careful" statement and introduced the quote with the words "But in trying to rebut him [Brown], she [Hillary Clinton] may have offended a lot of women voters who work at home." Mitchell also noted that "Worried campaign aides urged her to soften her message right away," and then quoted Mrs. Clinton's statement from the second news encounter: "To be a full-time mother and homemaker and to be a full-time career person, to balance the two, to have those decisions at different stages of your life are very tough ones."

Other reporters saw the second sound bite as a simple extension of the first. "They took her away and she came back and tried to respond to it. I think it was a fuller explication of what she had tried to say [with cookies and tea . . . follow my profession]. I do think it was what she meant," recalls Gwen Ifill of the *New York Times*.[16]

As Clinton campaign aides and the press shifted their focus to the electoral effect of "cookies and tea," two facts were shunted aside. First, the majority of women, including married women, work outside the home.[17] It was possible that the statement would enhance Bill Clinton's prospects by increasing Hillary's identification with working women. Indeed, a *Houston Chronicle*-Hotline poll during the Republican convention found 50% of employed women supporting Clinton with 38% for Bush.[18] Second, women who stay at home are more likely to be both conservative and Republican and, as such, not targeted voters for a Democratic ticket.

Nonetheless, Mitchell's suggestion that the comment was a serious gaffe became the dominant news frame. That interpretation was given a helping hand by Republican consultant Roger Ailes, who told viewers of the *Today* Show the next day (March 17, 1992) that "She offended the Tammy Wynette vote again this week by saying she's not going to stay home and bake cookies. I mean, this is the only—the only reason I sort of still have some warm feeling toward Pat, is I think Pat Buchanan is a little warmer than she is."

Print reporters continued to focus on the original issue of conflict of interest. The next day, Dan Balz of the *Washington Post* offered two optics: conflict of interest and the appropriate public and private role of women. The article, titled "Clinton's Wife Finds She's Become Issue," reported both the "cookies and tea . . . profession" remark ("a comment that caused aides to shudder") and the statement about career choices. Separating them was the observation that "[m]oments later, she was the more politically correct voice of professional women."

There are telling generational differences in the ways in which the remark is reported. Writing for the *New York Times,* Gwen Ifill, 39, heard it simply as "a zingy comeback."[19]

By March 26, the press spin was clear. After excerpting "cookies and tea" without the law firm reference on *Nightline,* Jackie Judd observes, "Never mind that Clinton went on to say feminism means the right to choose work or home or both; the damage had been done. She'd been tagged an elitist and an ultrafeminist." Nowhere in the remarks of either Clinton was feminism mentioned or at issue.

A month after the initial remark, NBC's Lisa Myers would report that "although she helps her husband among some constituencies, she also hurts him. Many women still are fuming over this remark." Hillary Clinton is then shown saying "You know, I suppose I could have stayed home and baked cookies and had teas, but I—what I decided to do was to fulfill my profession."[20] Gone is the context—conflict of interest; absent is the information that she had begun her career before he entered public life.

The next piece to disappear was fulfilling her profession. Gradually, the sound bite was winnowed to staying home baking cookies and having teas.

"You certainly did, however, touch more than a few nerves out on the campaign trail with this now infamous line 'I suppose I could have stayed home and baked cookies and had teas,'" noted Katie Couric in a *Today Show* interview with Hillary on April 2. "I know you have since said that was not meant as a slight to homemakers or women who choose to stay at home and work, but it did sound like a put down." In that interivew, Hillary opines that "I wish what I'd said before and said after had all been part of the sound bite but I'm learning that that's not always what you could expect."

In September a poll confirmed that "cookies and tea" was still resonating with the electorate. When the *Today Show* reported the poll, it was a working woman, Katie Couric, who questioned Hillary's backing away from the implications of the statement.

KATIE COURIC: ". . . when you compared working women and stay-at-home moms, it seemed apparent to me that her comment about, 'could have stayed home and made cookies,' really has—has some residual effects in terms of—of turning stay-at-home moms off."

TIM RUSSERT: "That's one in five. That's pretty striking, particularly for someone—for a first lady. And it's also her comment about, 'Stand by your man.' And there's a certain, I would say, reference point where people began to look at Hillary and said, 'She's not like me. She is different.' And for better or for worse, the word feminist can have some negative overtones in America even in 1992. I think she's working very hard to change her image in that regard and try to become per—become more perceived as Bill Clinton's wife, the mother of Chelsea, who, yes has a career, but she understands her own value as a woman, including that of being a mother and a wife."

COURIC: ". . . [which] is conversely turning off some working women who say, 'Hey, what's—so bad about being an ambitious, career-minded woman? Why does she have to pretend, or why does she have to take on this new role?'"[21]

"Cookies and tea" and "stand by your man/Tammy Wynette" entered the public vocabulary as telegraphic references that neither activists nor focus group participants felt a need to define or explain. Like "It's not over until the fat lady sings" and "Here's looking at you, kid," its context and meaning were now assumed. But, in fact, the meanings heard in the phrases varied widely.

On March 26, conservative columnist William Safire weighed in with the charge that cookies and tea statement betokened "elitism in action."[22]

In mid-summer, conservative activist Phyllis Schlafly stoked the coals by treating the "cookies and tea and Tammy Wynette comments" as ideological markers of feminism, an affirmation that the prospective First Lady saw homemakers as second-class citizens. At the same time, she implicitly framed womb/brain, competence/femininity, and equality/difference binds. She condemned Clinton for dismissing wives and homemakers and expressing disdain for those who merely stand by their man. She also averred that Clinton stood for sameness, not difference, in wanting women to be treated just like men.

"Hillary Clinton's view about marriage, and the feminist view is that wives are a servant class [sic] a dependency relationship, a second-class citizen, and that is what they believe, and it's pretty obvious from many statements that Hillary has made, for example, when she looked down her nose at the homemaker who stays home and baked cookies, that was a typical feminist remark, when she went on 60 Minutes and said she wasn't going to be just a little wife [n.b.: Hillary Clinton had said 'little woman'] who stands by her man like Tammy Wynette. Now all of these things indicate that it is the feminist view, and that is a putdown of the homemaker, and people don't like that and I think that's perfectly fair game to talk about, because after all, the feminists do say they want to be treated like a man, and if she wants to be treated like a man. . . ."[23]

In mid-May, focus groups discussing Bill Clinton's electability took up the subject of the candidate's wife by concentrating on these same key phrases.

A 32-YEAR-OLD WHITE MALE ELECTRICIAN, MARRIED, THREE CHILDREN, WITH A SPOUSE IN THE LABOR FORCE: "What about Little Miss 'I won't stand by any [sic] man' Hillary?"

A 23-YEAR-OLD FEMALE CASHIER, SINGLE, NO CHILDREN: "She did. Didn't you see *60 Minutes*? You're just trying to for[?] an excuse to vote against [inaudible] and . . ."

THE ELECTRICIAN: "I don't need an excuse not [interrupted]"

A 27-YEAR-OLD FEMALE HOMEMAKER WITH TWO PRE-TEENAGE CHILDREN, WHOSE SPOUSE WORKS IN MANUFACTURING: "It comes down to whether she, I mean he, well both are against the family, 'cookies and tea.'"

THE CASHIER: "Why should she bake cookies and tea? All that stand by your man, Tammy Wynette. Why should she? She's a lawyer, for God's sake."

THE ELECTRICIAN: "I don't care what she is. I'm not going to vote for her. (laughter) OR for him."

AN UNIDENTIFIED MALE VOICE: "What about [voting for?] Tammy Wynette:" (laughter)

A 20-YEAR-OLD FEMALE COLLEGE STUDENT, SINGLE: "I'd vote for her, for Hillary [voice: "—not Tammy?"] No, for her. You can get cookies at the bakery."

Also in mid-May a group of young women in Atlanta, eighteen to thirty years old, moved from a discussion of the economy to the need for women to work and from there to a discussion of "cookies and tea":

FIRST WOMAN: "You know, I really identified with Hillary over 'cookies and tea.'"

SECOND WOMAN: "Juggling two careers is something [interrupted]. She shouldn't have to be an appendage for him to be president."

THIRD WOMAN: "Did you know she has a daughter?"

SECOND WOMAN: "Uh how old?"

THIRD WOMAN: "I'm not sure. Young."

SECOND WOMAN: "Who do you think takes care of her? Do you" [unintelligible]

THIRD WOMAN: "We do. [*laughter*] The same as the rest of us. You can bet it's not Bill."

FIRST WOMAN: "If he's president he can stay in the White House and make the cookies." [Laughter and someone says "and tea." Another voice adds, "Yeah, sure."]

Interestingly, among the patterns that emerged in the focus groups was the tendency of those who could vote for Bush or Perot to hold that after "cookies and tea" Bill silenced Hillary. By contrast, those who would ultimately support the Democratic ticket were more likely to conclude that she *voluntarily* refashioned her role.

Late the same summer, in a suburban group outside Detroit, a retired, married, white male salesperson, father of six grown children and self-identified Republican, declared: "You want somebody you're proud of as an American. . . . This isn't a guy that ran a thing like in Arkansas there that's—what is it? 50th in the country as far as advancement goes? It's a cesspool down there almost. I don't like the wife either. . . . And it isn't the cookie thing." Moderator: "What is it?" Retired salesperson: "I don't like the thing like children suing their parents and stuff like that. I'm not ready for this as a taxpaying American citizen." Moderator: "Where did you hear about children suing their parents? Salesperson: "Through the media . . . I tend to listen to all the different ones. I listen to ones I disagree with as well." Moderator: "Have you ever heard her responses to that or Clinton's response to that?" Retired salesman: "No. She has not responded. Well, they put the blinkers on her. After the cookie deal, they told her to shut up."

By contrast, the 46-year-old married male (a small business owner and father of two teenage daughters) who concludes this exchange voted for Clinton.

> FIRST WOMAN: "Uh huh. She became more, sort of the typical candidate's wife. Retiring, more quiet. More like standing behind her husband, less verbal."
> SECOND WOMAN: "It was pronounced."
> THIRD WOMAN: "Oh yeah. I mean, we were comparing a lot of what she said, what she was doing to all of a sudden. . . . I don't even recall seeing her in any of the film clips or. . . ."
> FIRST WOMAN: "Except for standing behind her husband."
> SECOND WOMAN: "Right. With the mouth shut."
> MALE: "My perception of it though is she was the one who made the call. That she herself looked back at what had happened during that period of time and looked at what was happening in the polls concurrently and determined that if she did not position herself otherwise, neither she nor he would be in the position that they wanted."

Lost to reporters' tendency to simplify, dramatize, and feature conflict was Hillary Clinton's tardy but nonetheless expressed recognition of the difficulty of the choices faced by women in both the private and public sphere. Telegraphed as "cookies and tea," the decontextualized statement projected a war between the uterus and the brain, motherhood at odds with career, private sphere as the antithesis of public.

In this worldview, the choices facing women were dichotomous and antithetical: motherhood and homemaking, symbolized by cookies and teas, or a career. With such surrogate symbolism one way of breaking the trap was to establish that a career women could bake cookies; Hillary

Clinton did just that, winning the *Family Circle* cookie bake-off against Barbara Bush with 55.2% of the vote.

The bake-off elicited one of the more bizarre journalistic moments of the campaign. On June 17, Paula Zahn and Connie Chung engaged in the following exchange on CBS.

> ZAHN: "[W]e're going to be talking about one of the defining moments of the political campaign here in 1992—Hillary Clinton on a woman's role in politics."
> HRC (on tape): "You know, I could have stayed home and baked cookies and had teas, but I—what I decided to do was to fulfill my profession, which I entered before my husband was in public life."
> CONNIE CHUNG, co-host: "Ouch."
> ZAHN: "Well, now it turns out Mrs. Clinton has, in fact, been in the kitchen."
> CHUNG: "Yeah."
> ZAHN: "And doing quite well, thank you."
> CHUNG: "Uh huh. You know why? Because Barbara Bush has, too. They've been allowed to—*Family Circle,* actually, has published their best chocolate chip cookies recipes. And I—we can try them, can't we. . . .
> ZAHN: "Oh. We—I want to."
> CHUNG: "Cause they are right here? You can even vote for your favorite with a postcard to the magazine."
> ZAHN: "I like those."
> CHUNG: "All right."
> ZAHN: "Clinton's chips."
> CHUNG: "All right."
> ZAHN: "Bush's batch. Charlie, you want a Republican cookie this morning or a Democratic one?"

Another means of arguing "both-and" was the traditional one employed by the Temperance League against liquor, by suffragists for the ballot, and by advocates of women's need for education. That too was dusted off as Hillary Clinton argued that her activities on the hustings protected the hearth.

Reading Hillary to say "cookies and tea are not for me," the press and the Republicans incarnated her sound bite in a war of the wives: Barbara Bush—mother, grandmother, and homemaker who had dropped out of college to marry George—and high-salaried Yale lawyer Hillary Rodham Clinton. "Offering a contrast between a 67-year-old grandmother of 12 who dropped out of college to marry and never again held a paying job and a 45-year-old attorney who earns six figures a year and has only one child, this 'race' has been said to represent a symbolic referendum on all America's conflicted feelings about feminism, family and child-rearing," wrote a reporter for the *Washington Post*.[24] "One sharp generational contrast many Republicans hope to draw is between Barbara Bush—who embodies the

'GI generation' of women who put their first priority on their families—
and the career-minded Hillary Clinton," wrote one reporter.[25] "After four
years of demonstrating to the country that she is *not* Nancy Reagan, the
President's wife tonight is staking out a new identity," reported the *Los
Angeles Times*. "Barbara Bush, mother of five, grandmother of 12, college
dropout, uncomplaining spouse of 47 years, is *not* Hillary Clinton."[26]
"The true generation gap looms like an abyss between Barbara Bush, the
67-year-old happy homemaker, and Hillary Clinton, the 44-year-old outspoken overachiever," noted Robin Abcarian in another piece in the *Los
Angeles Times*.[27] Here we have a doubled double bind. Outspoken condemns speech while overachiever questions competence and ambition.[28]
"Happy" functions as an antonym for "outspoken," "homemaker" for
"overachiever."

In some reports, the omissions were telling. "Barbara Bush, warm,
friendly and grandmother of 12 children, is considered by Republican
strategists the ideal of family values," noted *El Mundo,* "and a powerful
counterweight to the wife of Bill Clinton, a lawyer who fights for women's
rights."[29] As past head of The Childrens' Defense Fund, Clinton is best
known not as a women's rights but as a childrens' rights activist. Are we to
assume that because she is a warm, friendly grandmother of twelve, Barbara Bush is not a champion of women's rights? Why is Mrs. Bush described as warm and friendly but Mrs. Clinton bereft of adjectives? Is
Clinton, by implication, cold and unfriendly? And why in the catalog of
offspring is the Clintons' daughter Chelsea uncounted and unmentioned?

In an era in which polls are the nouns of political sentences, a pollster
asked in mid-March about the comparative favorable ratings of Barbara
Bush and Hillary Clinton. The *Washington Post*-ABC News poll "found
Barbara Bush with a 75 percent favorability score while Hillary Clinton
was seen favorably by only 28 percent."[30]

Marriage as a Zero-Sum Game

But in one respect Barbara Bush and Hillary Clinton were painted in the
same strokes. With marriage a zero-sum game, if one spouse was strong the
other must be weak. Directly or indirectly, attacks on the strengths of
the women simultaneously assailed the strength of their mates, and Barbara and George Bush came in for some of the same treatment.

"(Barbara) is described as Nancy Reagan with impeccable WASP manners, a backstage manipulator of a weak husband who makes Hillary Clinton look like a novice," wrote the *Los Angeles Times*.[31] The same assumption pervaded columnist William Safire's characterizations of Nancy
Reagan. "My point," he wrote in 1987, "is that Mr. Reagan is being made
to look wimpish and helpless by a wife who has crossed the line from
valuable confidante to behind-his-back political manipulator."[32] "Don't
sell Nancy Reagan short," he wrote in another column. "[S]he may be
tougher than she lets her husband appear to be."[33]

Former president Richard Nixon took the same tack. The campaign should make judicious use of Hillary Clinton, he observed. "If the wife comes through as being too strong and too intelligent, it makes the husband look like a wimp."[34] "Mr. Nixon," said the report in the *New York Times,* "praised Barbara Bush as a model of a wife who has her own opinions without upstaging her husband, and suggested that many Americans are still put off by a male politician who does not seem to be as strong as his wife. The former President allowed that, unfortunately, some voters agree with Cardinal de Richelieu, who said, 'Intellect in a woman is unbecoming.'"[35]

The notion is not uniquely Western. In Japan, the engagement of diplomat Masako Owada to Crown Prince Naruhito was greeted with predictions "that given the new crown princess's background and character, she would dominate her husband and fundamentally change things inside the palace." "She's Japan's Hillary Clinton," observers told a U.S. reporter.[36]

"There's never been a candidate's wife quite like Hillary Clinton," noted Jackie Judd on *Nightline,* "outspoken, independent, smart, but her strengths have been used to make Bill Clinton look like a wimp, even by a president who used to be accused of wimpiness himself." Judd then shows Bush saying, "And then there's Clinton, a very formidable candidate, but would Mario Cuomo run as Hillary's vice president?"[37]

On the same program, Ted Koppel asks a guest, "[T]o what degree are we still such a retrogressive society that when we see a smart, tough woman up there, we almost infer that it reflects badly on the guy, that maybe he can't handle it on his own, maybe she's the one who wears the pants in the family, you know, one of those old cliches?"

The claim that Hillary was in charge had also been reflected in comparisons between Bill and Hillary that advantaged Hillary. The suggestion was that they could not both be comparably effective. In the zero-sum game, one has to be better. Marriage can't be a win-win partnership.

"She was in control," noted Steve Kroft who interviewed the Clintons about the Gennifer Flowers accusations for *60 Minutes.* "Hillary is tougher and more disciplined than Bill is. And she's analytical. Among his faults, he has a tendency not to think of the consequences of things he says. I think she knows," he said, not divining the emergence of cookies and tea. "She's got a ten-second delay. If something comes to her mind she doesn't think will play right, she cuts it off before anybody knows she's thinking it."[38]

"Hillary Clinton seemed almost eager to put to rest the [Gennifer Flowers] issue and the rumors. Bill Clinton looked as though, on the whole, he'd rather be in Philadelphia. . . . She managed to have a sense of humor about it; he appeared less relaxed," said the *Washington Post.*[39] Bill Clinton "looked like a scared kid . . . Hillary Clinton, on the other hand, appeared impressively impervious, suggesting perhaps that the wrong Clinton is running for office."[40] "Indeed, many who watched the couple's appearance . . . thought that, in many ways, Mrs. Clinton handled the questions better than her husband did," observed the *New York*

Times.[41] The notion persisted after the election. "Hillary Clinton's reputation in Arkansas was as the disciplined half of the Clinton-and-Clinton governing duo, the one who perhaps lacked her husband's rhetorical skills, but did her homework and didn't promise what she couldn't deliver."[42]

In late September 1993, CNN and *USA Today* actually asked a national sample "Who do you think is smarter, Hillary Clinton or Bill Clinton?"[43] The replies: 40% answered Hillary, 22%, Bill, and 17% said "Both are the same."

In its benign form, this reportorial frame suggested that one complemented the other. "An Arkansas political columnist once melded their names into 'Billary Clinton' with good reason," noted a reporter for the *Los Angeles Times*. "They are complementary, a political and marital team; if they were a law firm, she would be the litigator, he the mediator."[44] In its malign form, "Billary" said that B[ill] was whatever [H]illary made of him.

Along the same lines, conservative columnist William Safire described Bill Clinton as "looking like a hanger-on hungering for home-baked cookies." When Hillary delivered a lengthy introduction after his win in Illinois, Safire charged her with "usurpation of the candidate's moment."[45] In a revealing slip, Hillary Clinton's biographer Judith Warner misquotes this passage to say "Safire accused Hillary of 'usurpation of a candidate's strength.'"[46]

The contrasts between the Clintons also pointed out who made more money and by how much. Hillary reportedly earned three to four times her husband's $35,000 salary."[47] After noting the salary disparity, a piece in the *Los Angles Times* added, "He earns less, as governor, than a successful Park Avenue dog walker."[48]

In a marriage based on the partnership model, one might assume that it doesn't matter who contributes what to the common pool on which the family draws for its needs. But after the election, the country learned that who was supporting whom was in fact an issue for one member of the first household. Asked on January 13 whether Hillary would have a job outside the White House, the president-elect replied, "No. I don't want her to have a job outside the White House. I want her to help me. You know, besides that, I want to support her. She's been supporting me for 15 years. This is going to be the first time—we've been married since 1975, and the only year where I made more money than she did is the first year we were married. So I've got a lot of catching up to do. I haven't supported her very well all these years, so I want her to work there."[49]

Until the Republican convention bewildered voters with its mean-spirited misrepresentations of Clinton's political positions, the Democratic nominee's spouse was more often than not cast as the *defective* other in her pair-offs with Mrs. Bush. She was not simply a feminist, but a *brash* feminist. "The campaign has pitted two strong women on each side against one another," noted the *Washington Post*. "Barbara Bush, the matriarch; Hillary Clinton, the brash feminist."[50] Her cookies and tea remark was that of a "defiant feminist."[51] A pollster for *Time* went so far as to ask respondents whether they agreed that Hillary Clinton "doesn't pay enough attention to her family."[52]

When Hillary Clinton is described as tough, the word is most often a pejorative that exists a whisper away from the condemnatory "hard" and "hard-edged." "She may not want to be the candidate, but Hillary Clinton is a practiced politician, one who is not afraid of a fight, not in Arkansas politics. . . . Friends say Hillary Clinton is smart, tough, the next generation of political spouse. Critics say she is coming off too tough," noted a reporter on the *NBC Today Show*.[53]

The words "tough" and "soft" have strange usages in American public address. In politics, "tough" doesn't mean difficult or hard to chew. A candidate tough enough to do the job is, in a word, competent. A candidate tough on crime and communism is interventionist and aggressive. Similarly, hard and soft form a related set that expresses a distinction between feature stories and straightforward reporting about real-world events. Indeed, "[o]ne irreverent woman journalist once commented that it came as no surprise to her that 'men made a distinction between 'soft' and 'hard' news and found the 'hard' kind much more desirable.'"[54] Simply put, *tough* embodies testosterone.

Madison Avenue ties "softness" to toilet tissue, the skin of babies and women, and low-key selling techniques. Saying that something is soft can mean vulnerable (as in Churchill's description of the soft underbelly of Europe). In politics, it seems to be an antonym of tough and hard or hard-edged.

Soft is also implied by other indictments of masculinity. Bush went to great lengths, as a result, to dispatch his "wimp" image and when Republican consultant Roger Ailes wanted to express disdain for Senator Paul Simon, who was opposed by an Ailes' client, Ailes described Simon as a "weenie."

Among other uses, these presumed tough/soft antonyms and their progeny signal whether a male candidate is taking stands on issues in what is presumed to be a gender appropriate way. At issue in 1988 was whether Dukakis was "tough" on crime, in 1972 whether McGovern was "soft" on Communism. When red-blooded, adrenaline-charged candidates went so far as to snort that their opponents were "squishy soft" on communism, the masculinity of the other party's candidate was on the line.

In the political arena, where toughness ostensibly means discipline and courage, the real subtext continues to be masculinity—just as it is in war and contact sports.

If men and women are seen as yin and yang, animus and anima, bipolar opposites, complementary parts of a whole, then it follows that an attribute necessary to one is not needed in the other. Moreover, what is *positive* in one is *negative* in the other. The assumption that men and women are inevitably and naturally ordained opposites is one of the sources of the double binds women historically have faced. Breaking those binds requires that we recognize fundamental human similarities as well as differences between women and men.

Recasting Hillary Rodham Clinton

Where "soft" condemns a male candidate, it rains praise when sprinkled on the wife of the Democratic nominee. Until she "softened" her appearance, behavior, and speech. Mrs. Clinton's activities in the public sphere were enwrapped in adjectives such as "hard," "tough," "aggressive," and "feminist." The reports of the Republican convention and the convention itself used the word "feminist" as an epithet, an assault with a lethal part of speech. By contrast, Hillary Clinton's efforts to re-identify with the role of wife and mother were swathed by the press in softness. Characterization and condemnation of her speech would be tied to the former role, not the latter. During the primaries, after the votes for Super Tuesday indicated a clear victory for her husband, Hillary had transformed a speech introducing him into a speech in her own right. "Not just an introduction, this is a speech by Mrs. Clinton," observed NBC's Tom Brokaw.[55] She also spoke after Clinton's win in Illinois. Again network anchors expressed chagrin. In her wife-and-mother incarnation, Hillary was seen as silent or speaking softly.

"Hillary Clinton defined as the essence of hard-edged feminism," noted the *Washington Post*.[56] "They don't like Hillary because she's a working mom with a hard edge," reported the *Chicago Tribune*.[57] "Mrs. Clinton is trying to balance her image as a tough-minded working spouse and her new, softer look," observed the *New York Times*.[58]

Slipping to the surface were hints that the "harder" side of Hillary Rodham Clinton was genuine, the other a charade, as if a woman could not be both tough and soft, or tough in some environments, soft in others. "Despite the kinder, gentler image Clinton has been working hard to project, her harder edge showed through in a few unscripted moments. When Brown delegates interrrupted her address, she shot back, 'I've never known Jerry not to speak.'"[59] "Now the critics snicker that, with polls telling her that her old assertive self hurt—especially, alas with women— she's softening her image," noted the *New York Times*.[60] "But I've noticed a definite change in her style, from being openly combative at the start of the race to being far more judicious now," one pundit, Dr. Myra G. Gutin, told the *New York Times*. "She's cloaking her aggression in velvet."[61]

When framing her choices in caricature, the reporters parodied one pole of the womb/brain choice but not the other. Implicitly her status as mother was treated as evidence that Hillary was all right after all. "Hillary Clinton's campaign to get her husband of 16 years elected has taken an unacknowledged mid-course change in emphasis," noted a profile in the *Los Angeles Times,* "to put forward the kinder, gentler Hillary Clinton, to round off some of the sharper edges, to convince voters that she is not an ambitious, hectoring manipulator but one more working mom juggling through hectic days—a new American traditionalist, as down-home likable as she is intellectually admirable."[62]

"Hillary Clinton, who was portrayed in a political cartoon as the rubber-clad, whip-wielding vixen, gave up the role this week domesticating her image with ladylike teas and cookies recipes and motherly appearances with her 12-year-old daughter, Chelsea,"[63] said the *New York Times,* apparently believing that Clinton previously played the role of vixen. How else could she "give it up"?

The *Philadelphia Inquirer* captured the antitheses this way: Was she a "cold, mouthy . . . career-crazed . . . feminist who commandeers microphones, a gaffe-prone, power-starved liability," or a "warm, down-to-earth mother"?[64] Hillary Clinton was compared to Lady Macbeth, Eva Peron, and tagged the Winnie Mandela of American politics. Was she, reporters asked, a supportive spouse or "the overbearing yuppie wife from hell"?[65]

Meanwhile, Republicans, said the press, painted her "as an unwifely feminist with undue influence on her husband's policy-making—Gloria Steinem with the claws of Madame Nhu."[66] "[I]n portions of the George Bush generation," noted one profile, "she has come across to some as a bossy, humorless Valkyrie with a briefcase, the arrogant advance scout of an unknown, unnerving shift of generation and class."[67] "Claws of Madame Nhu" and "bossy, humorless Valkyrie" were, of course, the dramatic riffs of the reporters on far more pedestrian Republican prose.

Some articles hinted that for the Clintons career and marriage formed an unholy alliance. The *Los Angeles Times* reported "Now in the general election, the brassy, career-minded wife who alienated some homemakers is being tested at every turn by Clinton's opponents."[68] "People said they were driven, this couple," observed the *New York Times,* "hanging on to each other for appearance, or self-protection, tied by a thread of ambition."[69]

Adaptations by Mrs. Clinton and the Democratic campaign created a climate in which attack would backfire. Those alterations in image and tone were prompted in part by an April memo from Clinton pollster Stan Greenberg reporting that public disapproval of Hillary and her relationship to the Democratic contender opened the Democrats to a Republican attack on "family values." The public believed, implied the memo, that Hillary Clinton wanted power for herself.[70]

Clinton modified her approach. Where her speechmaking had provoked comment in the primaries, she was now more often seen applauding from the sidelines. Noted at the Democratic convention was the fact that she had, moreover, "softened" her hair style and her clothing. "Though some voters say they like her precisely because she is a modern role model, so many others have been put off by her assertiveness that she has begun favoring her softer side," wrote the *New York Times.*[71] "En route from Arkansas to New York the wardrobe had been softened to favor pastels," commented the *Chicago Tribune.*[72]

In the process of "softening" her appearance and her tone, the press observed, Mrs. Clinton also had begun to more clearly focus on her dedication to the cause of children. And the voice in which she spoke, noted the reporters, was now the voice traditionally identified with women who move from private to public spheres in order to defend the virtues of the home.

"A high school student . . . asked Hillary Clinton a question that has bedeviled her husband's Presidential campaign for months. He wanted to know what her role would be in a Clinton Administration. . . . 'I want to be a voice for children in the White House,' she said softly. . . . The label is well-suited to Mrs. Clinton's experience as a longtime advocate of children's rights, but is also carefully tailored to match voters' expectations of what is appropriate work for the President's wife."[73]

"Hillary Clinton is talking softly and gently these days about her interest in children," observed the *Chicago Tribune.* "Her primary goal, she says, is to make life better for children."[74]

Reporters saw "soft and silent" as Hillary Clinton's new watchwords. So too did the Democratic campaign. Where Barbara Bush would speak from the podium of her party's convention, Hillary Clinton would not. With deft stage management, the Democrats used their convention in July to swaddle Hillary Clinton in protective layers that for decades have shielded women in the public sphere from attack. Daughter Chelsea emerged from the background to be displayed on the cover of *People* with her loving parents. Mrs. Clinton no longer spoke in her own right. With these moves came press descriptions that put him back in charge of the relationship. "Mr. Clinton has done what it takes, from restyling his hair to confessing his sins to muting his wife."[75]

She now spoke when invited by him or when his failing voice dictated that someone speak in his stead. And when called on to play this role she "echoed" her husband's views, said the press. Gone was the assumption that she spoke on her own behalf.

As Lynn Sheer of ABC pointed out, in her first incarnation Mrs. Clinton had violated what once was a sacrosanct rule for spouses. "Rules to Wives of Candidates" issued by Republicans back in 1972, advised them: "When your husband is speaking, watch him proudly. Never appear bored. Never detract from him. Steer clear of controversial statements."[76]

No longer was she usurping; instead she was "pitching in" and spoke words one might easily associate with a mother.

ANDREA MITCHELL: "Clinton has campaigned so much in Michigan that he jokes about taking up residence. His voice almost gone, Hillary Clinton has been pitching in."

MRS. HILLARY CLINTON: "We'll have to be real quiet to be able to hear him."

GOVERNOR CLINTON: "If you will be my voice tomorrow, I'll be yours for four years."[77]

And in Illinois, "Hillary Clinton urged Illinois delegates on Tuesday to vote for her chocolate-chip recipe in a bake-off with Barbara Bush," reported the *Chicago Tribune*. "The kinder, gentler, more pastel Hillary heaped praise on the mayor and Chicago's unique political history."[78]

The Republicans, however, hadn't noticed.

Politics speaks in the language of veneration and vilification. In 1992 the Democratic and Republican exchanges over Hillary Rodham Clinton mirror the extremes available to characterize powerful women. Literature sold in the Republican "Spirit of America" hall next to the arena at the Republican convention called Mrs. Clinton a "femi nazi."[79] Republican National Committee Chair Richard Bond described her ideology as "radical feminism."[80] From the podium she was denounced as "a militant feminist lawyer who equates marriage with slavery,"[81] the "ideological leader of a 'Clinton-Clinton' administration that would push a radical-feminist agenda,"[82] "a feminist extremist,"[83] "a champion of 'radical feminism' who believes that 12-year-olds should have the right to sue their parents,"[84] and as a "radical feminist," who along with her husband pushes an agenda that includes "abortion on demand, a litmus test for the Supreme Court, homosexual rights, discrimination against religious schools and 'women in combat.'"[85] A *New York Times* article put the pieces together and saw them portraying "an unwifely feminist."[86] Outside the Republican convention, a live elephant crushed a Hillary doll. *Spy* magazine pushed such rhetoric to its absurdist extreme in February 1993 by depicting the First Lady as a leather-clad dominatrix with a whip.

According to incumbent President George Bush, Hillary Clinton was not simply an aggressive lawyer but "a very aggressive lawyer" who had "inject[ed] herself into the issue business" as part of a Clinton campaign that included the claim "two for one" and had an "activist past." As a result, he felt justified in "going after the wife."[87] The Democrats responded by questioning George Bush's testosterone level. Of the Republican attacks on Hillary Clinton, the Democratic nominee observed, "Nobody ever talked about a co-presidency. There wasn't a co-governorship in Arkansas. You'd think George Bush was running for first lady instead of for president half the time."[88]

Press response to the Republican attacks was disapproving. Dan Rather labeled Buchanan's speech "raw meat," the *Washington Post* suggested that Buchanan "sprayed his targets" and "assailed" Hillary Clinton,[89] and issued "blistering opening attacks."[90] The *Post* had expected the convention "to be a festival of Clinton-bashing" but noted, as if it were unexpected, "As it turned out, the target has been not only the candidate, but also his wife, Hillary."[91] "Pat Buchanan can be mean and intolerant, as he demonstrated on Monday night, when he . . . misrepresented Hillary Clinton's views on children," opined Colbert King in an op-ed.[92] The *New York Times* declared: "They bashed Hillary and Bill Clinton. Day after day some Bush surrogate spoke of Governor Clinton's sex life. But probably nastiest of all were the references to Mrs. Clinton."[93]

"Bashing" was the word of choice to describe the attacks, which also were tagged "unnecessarily divisive." Military descriptions escalated. "In his unnecessarily divisive attack on homosexuals, Hillary Clinton and abortion rights, Buchanan hurled his polemicist's spear," reported the *Los Angeles Times*.[94]

Employing the rhetoric of war to describe the Republican rhetoric recontextualized Hillary. The Republican attacks positioned her the same way that her husband had in the Busy Bee cafe in the hours before she undercut that view with the "cookies and tea" observation. But this time, the "softer and more silent" spouse fit the role of vulnerable victim when attacked by the Republicans—and protected wife when defended by her man.

The new role, of course, created problems of its own. Where before she was criticized for being outspoken, now Hillary was chided for failing to be herself. "We ought to have reached the stage where we accept women for what they are, not try to put them in some cookie cutter," noted feminist leader and former Congresswoman Bella Abzug.[95]

While drawing attention to this apparently calculated "makeover," the press nonetheless shifted adjectives in response to the Republican forays. No longer was Hillary Clinton portrayed as tough, hard-edged, and Machiavellian. She was cast instead in a role rich with empathy, the unjustified object of unfair attack. Characterizing the Republican attack as an assault with a blunt instrument, the press created, by implication, a vulnerable heroine. And they implicitly invoked the cultural rule: Men don't attack women, they protect and defend them.

The Republicans, for their part, played straight into their hands. Where Hillary had been read as a champion of the liberated woman, Marilyn Quayle was seen as the defender of traditional homemaking. But in her speech at the Republican National Convention, the vice president's wife seemed more explicitly intolerant of the choice of career than Hillary Clinton had ever seemed about traditional homemaking. "Most women do not wish to be liberated from their essential natures as women," she proclaimed. "Most of us love being mothers or wives, which gives our lives a richness that few men or women get from professional accomplishments alone." And in what was widely seen as a jab at Hillary, she added, "Not everyone believed that the family was so oppressive that women could only thrive apart from it."

Where Hillary Clinton had stumbled into "cookies and tea" in a crowded restaurant in Chicago, here was a carefully scripted, nationally televised address from the podium of the Republican National Convention. Women in our focus groups, Republican as well as Democratic, including many who had expressed serious reservations about Hillary Clinton's views of marriage, family, and career, now ricocheted into her corner. "I could not believe it," said a 28-year-old homemaker and mother of three in Dallas, who earlier had stated she would probably vote Republican. "I said to my mother, I don't know where she gets off speaking for women. She makes the other one, Hillary, seem warm and cuddly for comparison." "My essential nature says that I am not going to vote for her essential nature," added a 56-year-old grandmother who works as an insurance broker. "They are trying to attract to, the right wingers who think women, those at home, housewives, belong in caves," said a 43-year-old male mechanic and father of two. "My response was to defend Hillary who I don't even like. You just have to say, it has nothing to do with the president. It's politics." "I can't get over the hypocrisy of it at all," said a 24-year-old male high school teacher and father of one. "[Marilyn] Quayle is a lawyer . . . runs his office . . . chases around making speeches . . . is the brains of the two. She and Hillary could be twins. Where does she get off attacking working women?"

"The Republicans have described the enemy and she sounds like me," wrote Dianne Klein in a column in the *Los Angeles Times*.[96] The freedom to choose emerged as a responsive refrain. "We must also remember that it is not so important which road we choose to travel but that the decision to choose is ours to make," wrote a law student in an opinion piece for the *Los Angeles Times*.[97]

When Pat Buchanan and Marilyn Quayle suggested that Hillary Clinton would destroy the family as well as traditional roles for women, reporters responded, in effect, that it was the Republicans who had violated tradition by "bashing" a woman.

The tide suddenly turned. As the *New York Times* noted prophetically, "Hillary-Taunting" had its perils "if the electorate comes to see it as intolerance of working women in general."[98] Conservative commentator George Will expressed the same sentiment on ABC after George Bush's acceptance speech. "Tonight was a sustained innuendo against the Democratic Party—that they don't like marriages, families, women in the kitchen, or children. . . . They're saying Bill Clinton—and generically Democrats—are bad people. . . . I think there are an awful lot of Democrats who are going to be profoundly offended by the innuendo."

What the evolving press analysis of Hillary Clinton as working woman, working mother, working wife ultimately produced was commentary about the stereotypes that had governed the perceptions and coverage of Hillary Rodham Clinton. "The irony is that if Mrs. Clinton were up on that podium as a candidate, she would be golden, with her Yale Law Degree, her board positions, her smarts and her looks," noted the *New York Times*.[99] "But Hillary Rodham Clinton is running for First Lady, an anachronistic title for an amorphous position. The job description is a stereotype that no real woman has ever fit except perhaps June Cleaver on her good days."

The ongoing scrutiny provided the country with an extended opportunity to examine its changing understanding of women in power. In the end, Clinton defused voter's fears that she would function as the power behind the throne by adopting the guise women used historically to gain power: She would be the advocate of children, she said—softly.

Forming the backdrop to the instinct to condemn Hillary were the twin assumptions that working women are feminists, and feminists devalue full-time homemaking and homemakers. "Women who choose not to work . . . don't sit home and bake cookies," observed Lee Hart, spouse of 1984 and 1988 presidential contender Gary Hart. "One of the problems in the early years of the feminist movement was the real put-down of women who chose to stay at home, take care of the children and be a mother."[100] Hart was echoing the comments of historian Jean Bethke Elshtain: "In many early feminist accounts mothering was portrayed as a condition of terminal psychological and social decay, total self-abnegation, physical deterioration, and absence of self-respect," Elshtain maintained. "Mothers were demeaned under the guise of 'liberating' them."[101]

Whether a fair characterization of early feminism or not, this idea was fueled by clashes such as the argument between Betty Friedan and Simone de Beauvoir over whether mothers should be compensated for staying at home caring for their children. In June 1975 Friedan argued that "There could be a voucher system which a woman who chooses to continue her profession or her education and have little children could use to pay for child care. But if she chooses to take care of her own children full time, she would earn the money herself." "No woman should be authorized to stay at home to raise her children," replied de Beauvoir. "Society should be totally different. Women should not have that choice, precisely because if there is such a choice, too many women will make that one."[102]

Confirming a Woman's Right to Choose

But an important result of the 1992 campaign controversy was in fact the affirmation to the contrary that it produced. Nearly lost in all the fuss was the fact that women as different as Barbara Bush, Hillary Clinton, and Marilyn Quayle would finally come to recite the same premises: that the truly liberated woman was free to choose full-time homemaking, full-time career, or some combination.

Quayle, in an effort to calm the negative reaction to her convention address, said, "We don't have to reject the prospect of marriage and children to succeed. We don't have to reject our essential natures as women to prosper in what was once the domain of men. It is no longer an either-or situation."[103] And Hillary Clinton added a stock line to her repertoire. At the hint of an opportunity, she reiterates that each woman should choose what is best for her. She carried the same message into the White House. When asked by Larry King whether she had changed the pattern for First Ladies, Clinton responded, "Larry, I don't think there should be a pattern.

I really think that each individual ought to be free to do what she thinks is best for herself and her husband and her country. I have a lot of respect for all the women who have been in this position and I think every one of them made a significant contribution, but they may have done it in a different way. . . . Everybody should be permitted to be who they are."[104] Barbara Bush took the same position when told initially of the "cookies and tea" remark. "Everybody's different and that's a great thing," she said.[105]

After the media had put its spin on Hillary Clinton's "cookies and teas" remark, Clinton aides revised it to fit the new media frame, and the media accepted those revisions, in the process sharpening the focus on women's roles. More pointedly, the media began to examine the frame itself.

By November 20, 1992, a caption on the photo of Barbara Bush greeting Hillary Clinton at the White House read, "Tough Political Insider Meets Warm Friend and Devoted Mother." The crash of stereotypes hitting the ground in shards could be heard in the next sentence. "Both descriptions fit Hillary Clinton and both fit Barbara Bush, who met at the White House to search for other common ground."[106]

Gwen Ifill, who wrote the story for the *Times,* recalls, "The story came before the picture. The photographer came and showed me the picture. They took my lead and made it the cut line. When I was writing that story I decided to play to stereotypes that applied to both women. Barbara Bush had been portrayed as a warm, fuzzy grandmother. Hillary Clinton had been portrayed as a hard careerist. It didn't take much to see that the opposite had been said about both. They both had been stereotyped."[107]

The original context in which "cookies and tea" had been uttered—charges of conflict of interest surrounding Mrs. Clinton's Arkansas law practice—would re-emerge during the Clinton administration in the form of allegations about investments that came to be known simply as Whitewater. But continued press coverage and argument over appropriate roles for women had exposed stereotypical assumptions to the harsh light of day, and they did not weather it well. In a real sense, voters had chosen to send both Mr. and Mrs. Clinton to the White House. Once there, Hillary Rodham Clinton reclaimed her birth name and became the point person in the most important proposed piece of domestic legislation of her husband's first term.

The break from the past was seismic. A decade and a half earlier, First Lady Rosalynn Carter had stirred controversy by merely attending cabinet meetings. "When I got into trouble with the press," she recalled, "it was about going to Cabinet meetings. . . . I just sat in a chair by the door and listened. I never entered into any conversation or discussion."[108]

Hillary Clinton's competence was acknowledged by those who heard her Congressional testimony. "Hillary Clinton, on Capitol Hill, Wins Raves, If Not a Health Plan," read the front-page headline in the *New York Times.*[109] In the opening paragraph where she "captivated and dominated two usually grumpy House Committees," Clinton was credited with the traditionally masculine (she "dominated") and well as feminine (she "capti-

This picture, which appeared on the front page of the *New York Times* on November 20, 1992, was captioned, "Tough Political Insider Meets Warm Friend and Devoted Mother. Both descriptions fit Hillary Clinton and both fit Barbara Bush, who met at the White House to search for other common ground. Mrs. Bush gave Mrs. Clinton a tour a day after their husbands met there." (Jose R. Lopez/NYT Pictures.)

vated") virtues. She was tough ("trolling for votes") and caring ("offering . . . reassurance"). "So beyond restating the health plan's principles and offering every bit of reassurance she could muster, she was there trolling for votes."[110] In another line "[h]er voice [was] momentarily snappish," a phrase unlikely to be applied to a man; at another point she garnered a more typically male adjective when she gave "Republicans a steely reception at both committees when they expressed doubt about the financing of the system."[111]

"I was aware of the pairings ("captivated . . . dominated") when I wrote the lead," recalls Adam Clymer. "I wanted to convey that she was a very special witness who had both the knowledge and the authority that came with her position and was an attractive, smart woman who affected Congess for that reason as well."[112]

Transcending the Traditional Constraints

The descriptions of Mrs. Clinton's presentation suggested that for the moment, at least, she had overcome two of the no-win situations haunting women. She had transcended the narrow range of attributes traditionally assumed to constrain a woman speaking in public—the ancient restraints of silence or shame. And she established her competence without sacrific-ing her femininity. Typically antithetical adjectives, adverbs, and verbs were now paired to laud Hillary Clinton's success. "But no previous First Lady has occupied center stage so *aggressively* or *disarmed* her critics more effectively (emphasis mine)," noted another article in the *New York Times*.[113]

Reporters and columnists informed their readers that with her perfor-mance, Hillary Clinton had widened the range of options open to future presidential spouses. "As the debate over the Administration health care plan she helped to create comes to the fore," noted a *New York Times* caption, "Hillary Rodham Clinton is solidifying her position as the power beside, rather than behind, the throne."[114] "The national consciousness has shifted, slightly but perceptibly," noted Jane R. Eisner, the *Philadelphia Inquirer*'s deputy editor of the editorial page. Hillary Clinton "proved the early critics wrong, and for that I believe, many women in this country are privately grateful."[115]

When a national sample was asked late in September 1993 to agree or disagree with the statement "She [Hillary Clinton] is a good role model overall for American women," 76% agreed. To the statement: "She is a good role model for mothers," 61% agreed; 46% found her "too pushy"; 41% said that she "is too strong a feminist."[116]

Queried about his changed attitude toward Mrs. Clinton, Republican Senator Robert Dole, who is married to Red Cross President Elizabeth Dole, noted, "I'm used to smart women. At least Hillary doesn't ask for blood."[117] "She has established her qualifications," observed Representa-tive John Dingell. Concern about her role has "all abated."[118] "[S]he has earned the respect of everyone (except the wackos) with her handling of the health care issue. Indeed, she has gotten everyone (except the wackos) to agree that we need health care for everyone," commented talk show host and columinst Larry King.[119] On September 28, 1993, a *USA Today* poll found 60% approving and 29% disapproving of Hillary Clinton's han-dling of health-care policy.[120]

The change was reflected in our focus groups, In Minneapolis, in mid-October 1993, a group of white, middle-aged, middle-class individuals who had met throughout the 1992 campaign reassembled. Midway through the discussion, attention turned to Hillary Clinton's testimony before the Congressional committees.

FEMALE NEWSLETTER EDITOR: "[T]he only experience I have of Hill-ary Clinton is through the papers and I can only speak to that. But it seemed as if everyone was saying, 'This woman knows her stuff' and that even the Republican House and Senate have a great deal of respect for her and her knowledge and that any question that they had for her, she was able to answer, but of course, I am getting this from the newspaper. I didn't watch any hearings."

MALE GRADUATE STUDENT: "I did."

MODERATOR: "What was your reaction?"

MALE GRADUATE STUDENT: "She sounded so confident. That she handled the questions from people so straightforward with each person. That before they got to the newscasters who did the analysis of it, you could feel the shift in the room of people really paying attention and respecting how she handled herself . . . [she] was so genuine. It was a sharp contrast to him, I thought."

MODERATOR: "To whom?"

MALE GRADUATE STUDENT: "To Bill Clinton. The contrast being that I never understood where the slick stuff came from until I heard her speak. When I heard her speak, she sounded like she had some real centering, some real balance to what she was saying when she would answer questions. She could do the back and forth, but she had a kind of professional tone to her. She didn't backslap with it and she literally answered everybody. It was sort of amazing. So I was really impressed."

MODERATOR: "Did the fact that she was a woman make any difference in your response?"

MALE GRADUATE STUDENT: "As impressed if she were a man? No, maybe not. I think my impression was probably because I had been slanted toward thinking . . . that she was sort of like pushy from behind and that she was kind of like trying to angle stuff."

FEMALE ENGINEER: "More manipulative than really knowledgeable."

MALE GRADUATE STUDENT: "Yeah, that is the word, than knowledgeable and I just picked up such brief snippets of things from the newspapers, but when I actually heard her . . . it was a really long time before anybody broke in. It must have been an hour's worth of questions back and forth to her presenting and people really . . . I don't know who the Senator was, but one guy was really trying to get to her and it was just like water off of a duck's back. She had her answers. She didn't take offense to the kind of remarks that were being made and yeah, so I am guilty of being overly impressed because she seemed much more than I thought she was."

FEMALE HOMEMAKER: "Part of it was the shock, you know the surprise as you said because you have been led to believe something different about her."

FEMALE PHYSICIAN: ". . . she is an excellent role model for women. Someone who knows her stuff and presents herself very well she doesn't have to look at her notes. . . . You know, she just handles it all and all the press and even the medical papers have been pretty favorable towards her. Even though physicians in general are having a hard time with this health care reform stuff. So I am very impressed and my respect for her has grown tremendously."

Even Richard Nixon had come around. No longer was he concerned that a strong woman makes her husband look like a wimp. Instead he told the *Today Show*'s audience on February 17, 1993, that "if I wouldn't criticize Bill Clinton I certainly wouldn't take on Hillary, because she is a very intelligent, very strong, very effective First Lady. I think it's very appropriate for her to do what she believes is the right thing to do. . . . [A]s far as Hillary Clinton is concerned, with her great abilities, her intelligence and her strong beliefs, she can be a very effective help to her husband, the president, and I think the American people will like that. For example, if she can come up with a solution on health care, then I say we're all for it, because we certainly need one. The national health care is a scandal."[121]

Questions about career and motherhood, however, lingered. Larry King asked Clinton, for example, why she and Bill had not had a second child and learned that they had tried. And long-lived assumptions about who is responsible for child care continued to surface. A widely circulated story, unconfirmed by the White House, reports that when Chelsea sought an aspirin at school, the nurse insisted that she obtain parental permission. "Call my Dad," says Chelsea. "My mom's away."

In November, our Minnesota focus group struck up a dialogue about Hillary Clinton's role as a mother. Joyce, who stayed at home with her three young children, was disturbed by press comments that Hillary had "put to rest" any questions that a woman could be a good mother and also have a "tremendous" impact on national affairs. "I am sure it has impacted her relationship with her child and that to say that it doesn't is misleading." There followed a lengthy discussion in which both male and female respondents suggested, variously, that Bill Clinton's responsibilities might also complicate his role as a father; that having a mother who was prominent and successful might have a positive impact on a child; and that a 13 year old might not need as much mothering as younger children. The majority of respondents defended Hillary's both-and role, with the emphasis on her function as a role model for other women. The discussion largely centered on what were considered overwhelming demands placed on women who had to juggle homemaking and career. What was resisted was the possible re-emergence of the "super-Mom" idea as it was embodied in Hillary Clinton. Respondents suggested that necessary compromises between home and career had become acceptable, and questioned whether the female partner should accept the full burden of household responsibilities.

While this discussion explored and attempted to untangle the various vestiges of the womb/brain bind, in other dialogues other binds remained fully operative. When the Minneapolis focus group was shown a clip of an interview of Hillary Clinton by Katie Couric, the group's response subtly undercuts Clinton's claim to equality by focusing at length on facets of her *appearance* that signal difference. It is hard to imagine a male eliciting a comparable response.

> FEMALE: "She looks like June Cleaver in the clip there."
> [inaudible]
> FEMALE: "Yeah, where are the pearls?"
> FEMALE: "It doesn't seem to be appropriate hair for her. . . ."

The discussion of the hair style continues. Then:

> FEMALE: "I wonder if this hair style doesn't have something to do with having to sell her health plan to the House or Senate."
> MALE: "Well it certainly is not one to make you think of her as at all appealing. My first reaction to it was, like, 'Yeuhh.' I certainly wouldn't want to run my fingers through that hair."

After more discussion of the hair, this:

> MALE: "I have heard rumors that she had her face lifted."
> MODERATOR: "Anyone else?"
> MALE: "Me."
> FEMALE: "Not I."
> MODERATOR: "Has anyone else heard it?"
> MALE: "I did. I heard that she had a facelift and it was very noticeable if you looked and I don't see enough of her to know."
> FEMALE: "You heard that rumor also?"
> MODERATOR: "What did you hear?"
> MALE: "It was a bad job. . . ."

In a similar vein, publication of a series of portraits of Hillary Clinton in *Vogue* elicited a rush of stereotyped notions from the media that pitted femininity against competence. If Hillary Clinton's "softer" approach had helped turn the tide of public opinion in her favor, it had not resolved an old dilemma. One can, say the canards, be feminine, expressed photographically in soft focus and flowing silk, or masculine—expressed *not* in appearance but in substance, coded here as "an individual," and "tough"— the hoary cover for "competent" but not both.

"It just seems kind of odd," said Michael Deaver, once President Ronald Reagan's image maestro. "Now that she's proven herself as an individual and been tough, why then get back to an image strictly based on femininity?" In other words, appearing "feminine" risks perception of competence, toughness, individuality. Either/or, not both/and.[122]

Barely veiled in *New York Times* reporter Maureen Dowd's summary of reaction is the assumption by supporters and critics that power and motherhood, testifying and baking are incompatible. If they co-exist, one must be more real than the other. "Her admirers say that she has adroitly helped the country adjust to the notion of a First Lady sharing power with her husband by periodically *doling out* softer, more traditional images like hostess, mother and wife," notes Dowd. "But some complain that her cascading images—changing hairstyles so many times, testifying on the Hill about health care one minute and chatting happily about Christmas baking the next, now adding Rodham to her name, now posing for fashion layouts—are dizzying and unsettling, and suggest that the 46-year-old First Lady still wakes up every day struggling to create a persona. 'It doesn't feel genuine,' said Sheila Tate, who worked as an aide to Nancy Reagan and George Bush" (emphasis added).[123]

Although it is invigorating to believe that a visible, talented, powerful woman can once and for all dispatch the stereotypes into which others would box her, that's just not the way the world works. Despite her "disarming" but "aggressive" performance on the Hill, residues of the silence/shame bind pocked the surface of an exchange in a segment of the *McLaughlin Group* after the hearing.

"Was she a vixen?" McLaughlin wondered, in an exchange laden with double meaning. "No," responded an admiring Mort Kondracke, "She was *good*." But if not a vixen, Fred Barnes points out, neither was she a virginal saint. The canine interlude from Jack Germond is an unexpected metaphorical fillip.

MORT KONDRACKE: "She did it with skill and with flair."

JOHN McLAUGHLIN: "You mean like a vixen?"

KONDRACKE: "No, not like a vixen. She's good. She is very good. . . ."

McLAUGHLIN: "Is she tough?"

KONDRACKE: "She's tough."

McLAUGHLIN: "Is she resilient?"

KONDRACKE: "She's resilient."

McLAUGHLIN: "Is she brilliant?"

KONDRACKE: "Yes, I'd say she's brilliant."

JACK GERMOND: "I don't know why we're all marveling at her performance just because she's a First Lady and we've never seen this before. We've never seen a dog play first base, either, but you know it's going to happen one of these days."

FRED BARNES: "The comments by people like Rostenkowski (D. Ill.) and most of the other members of these committees were patronizing. Of course she did a good job testifying. Armey [who had claimed earlier that the Clinton plan was a "Dr. Kevorkian prescription that will kill 3.1 million American jobs"] was one of the few people willing to treat her as a smart, tough person, and not as Mother Teresa."[124]

If her testimony on health-care reform seemed to confirm that a first lady could lead, indeed that the Clintons were operating a partnership on health reform, discussion of the demise of the plan presupposed her responsibility and his accountability. The questions raised focused not on her gender but on the hazards of entrusting a liberal family member with power. Would a better plan have been produced by someone more accountable to the electorate or more likely to be fired if she failed? Had Hillary overridden Bill's best instincts and drawn the White House to a plan more liberal than the country would accept?

The experiences of Hillary Rodham Clinton as spouse of a Democratic contender and as First Partner are an ongoing demonstration of the power of cultural binds that enjoin women in general, and of the ways in which women continue to surmount or maneuver around them. These binds are not new conspiracies, but invocations of ancient constraints—traps that emerge in different guises over time, even as they continue to lose their teeth. What those basic constraints are, how they have been modulated to fit contemporary issues, how they play into each other, and how women continue to expose and loosen them, form the subject of this book.

3

Double Bind Number One: Womb/Brain

THROUGHOUT history, women have been identified as bodies not minds, wombs not brains. The distinction is captured in the cliches of our culture. Where men think, women feel. The man is the head of the family, the woman the heart. Boys never make passes at girls who wear glasses.

Indeed, many women of my generation recall being praised for defying the constraints of our sex and "thinking like men." The results were unsettling. When a high school teacher encouraged me into varsity debate on the assumption that I had "a masculine mind," my "masculine mind" wondered whether it was having an out-of-body experience.

In ways that defy logic, women are treated as if they are governed by their bodies and men as if they are ruled by their minds. So, although it was President John Kennedy who took mood-altering steroids to control his Addison's disease, Senator Hubert Humphrey's physician Edgar Berman ruled out the possibility of a female president on the grounds that the mood-altering effects of her "raging hormones" would disqualify her.

At a session of the Democratic party's Committee on National Priorities in April 1970, Representative Patsy Mink of Hawaii argued that the cause of women's rights deserved "the highest priority." Physician and committee member Edgar Berman responded that "raging hormonal influences" caused by the menstrual cycle and menopause should exclude women from executive responsibility. In a subsequent exchange, Berman attributed Mink's anger at his claims to her "raging hormonal imbalance."

"There are just physical and psychological inhibitants that limit a female's potential," Berman told a reporter. "So, I reiterate, all things being equal, I would still rather have had a male J.F.K. make the Cuban missile crisis decisions than a female of similar age who could possibly be subject to the curious mental aberrations of that age group."[1]

In earlier times, women's bodies were in fact controlled by someone else. Only recently could a woman charge that she had been raped by her husband. Unless separated from his control by a divorce decree, her body was literally his. So, indeed, was her mind.

Even today, if a woman shows expertise, she may find someone assuming that it was gotten from a man. Or so Republican Congresswoman Nancy Johnson discovered recently. Johnson is one of a handful of congressional experts on health-care reform and the only woman on the house subcommittee responsible for health-care legislation. On March 16, 1994, midway through an intricate discussion of cost controls, the committee chair, Rep. Pete Stark, observed that "The gentlelady got her medical degree through pillow talk and the gentleman from Washington (Congressman Jim McDermott, a psychiatrist) got his through going to school." Stark was imposing on Johnson, who is married to a doctor, the residues of a long-lived tradition that reduced women to the role of wives and in the process questioned their competence in any other domain. Later, Stark apologized. His original remark is a contemporary reminder of the womb/brain bind. How we got from that bind to apologies for assuming it is the subject of this chapter. There were no apologies to its early victims.[2]

Before women entered colleges in majority proportions, the remnants of these brain/body distinctions existed in the assumption that educating the brain makes women less desirable to the opposite sex. Eleanor Roosevelt's daughter recalled, for example, that her grandmother believed that "Girls who went to college were very apt to be 'old maids' and become 'bookworms' . . . a dire threat to any girls' chance of attracting a husband."[3]

A more contemporary form assumed initially that one could have either career or marriage and motherhood, but not both. "Sometimes I envy some of my female colleagues, those who managed to have both a career and a family," notes Betty Boothroyd, current Speaker of the British House of Commons. "I'm sure I could never have been Speaker if I'd been married."[4] When that bind fell, a second replaced it. One could have both career and family, but not at the same time. Now women are confronted with the corollary assumption: They can have both at the same time, but only at the cost of cheating one or the other.

Three hundred years ago, the penalities for pursuing the life of the intellect were more dire than those cited by Eleanor Roosevelt's daughter. For her belief that the indwelling spirit relieved sinners of their sins, Anne Hutchinson, a seventeenth-century midwife, healer, and mother of fifteen, was tried for heresy in Massachusetts Bay in 1637. Her prosecutor was Governor John Winthrop. Her penalty: banishment. She and her family were forced to move, and resettled in Rhode Island.

In his account of the trial, Winthrop claimed that Hutchinson was "a woman of a haughty and fierce carriage, of a nimble wit and active spirit, and a very voluble tongue, more bold than a man, though in understanding and judgment, inferiour to many women."[5] When she miscarried soon after arriving in Rhode Island, her tormentor read the event as a sign from God. "Mistres Hutchinson being big with child," wrote Winthrop, "and growing towards the time of her labour . . . she brought forth not one . . . but . . . 30, monstrous births or thereabouts, at once, some of them bigger, some lesser, some of one shape, some of another; few of any perfect shape, none at all of them (as farre as I could ever learne) of human shape. . . . And see how the wisdom of God fitted this judgement to her sinne everyway, for looke as she had vented mishapen opinions, so she must bring forth deformed monsters; and as about 30."[6]

Throughout the nineteenth century, theologians and scientists argued that when women followed the dictates of nature, they found happiness in this life and salvation in the next. The natural order saw woman's role as reproduction, childrearing, and maintaining the home. As punishment for Eve's abuse of the power of speech, women were denied the option to teach or preach. In a society in which silence and childbearing were marks of a good woman, Winthrop's account of Hutchinson's 30 monsters was plausible.

The brain and the womb both required energy—so went the conventional wisdom of the day. A woman's intellectual activity cheated her uterus of the wherewithal to sustain her health and her ability to reproduce. Those who violated the natural order—by engaging in public or intellectual activites or by working in industry—supposedly paid a price. Their uteruses shriveled, their ability to conceive constricted, and, if they reproduced at all, they risked bearing monsters.

The resulting dilemma was apparent. Since you cannot exercise both your brain and your uterus, you must choose one over the other. Select childbearing and sacrifice the satisfactions of the intellect. Favor the brain and forego the pleasures of motherhood. Here was a classic no-choice choice: choose marriage and motherhood and society approved. Choose the life of the mind and be punished by man *and* God.

The prominent British psychiatrist Henry Maudsley (1835–1918) went so far as to claim that "female education threatened to produce a lowering of the racial stock and even an eventual disappearance of the species."[7] His more moderate colleagues argued instead that education would produce in women "infertility or defective offspring" or "monstrosities of Nature."[8] An underlying assumption was that "Every body of the least experience must be sensible of the influence of menstruation on the operations of the mind."[9]

When a woman in puberty chooses books over knitting, went the medical logic, her ovaries may remain undeveloped or, if already developed, shrivel. "There have been instances, and I have seen such, of females in whom the special mechanism we are speaking of remained germinal— undeveloped," wrote one nineteenth-century doctor. "They graduated

The professionalization of medicine edged women from their traditional role as midwives. Males with forceps in hand replaced women with herbal remedies; the hospital room rather than the bedroom became the place in which childbirth occurred. (Man-Midwifery is an illustration from S. W. Fores, *Man-midwifery dissected; or, The obstetric family-instructor, London, 1793.* Used with permission of the National Library of Medicine.)

from school or college excellent scholars, but with undeveloped ovaries. Later they married and were sterile."[10]

In the wake of the industrial revolution, the explanatory metaphors spoke of mechanisms and engines. "Schools and colleges, as we have seen, require girls to work their brains with full force and sustained power at the time when their organization periodically requires a portion of their force for the performance of a periodical function, and a portion of their power for the building up of a peculiar, complicated, and important mechanism—

the engine within an engine," noted retired nineteenth-century Harvard Medical School professor Edward Clarke in 1873.[11] Unasked was: If God and nature ordained motherhood as woman's role, why in the struggle between brain and uterus did the uterus always lose? Why wasn't brain-strain the result instead?

If an intellectual woman did conceive a healthy child, her ongoing public activities would certainly jeopardize the child's well-being. "Already it is well known," declared an enterprising doctor, "that a sad physical deterioration of the children is the penalty of a mother exhausting her nervous power in public life."[12]

The dispute about womb and brain was not only about female sexuality but also about homemaking. Indeed, some thought that sexual desire correlated inversely with domesticity: the better the wife and mother, the less passionate the lover. In the nineteenth-century classic *The Functions and Disorders of the Reproductive Organs,* Dr. William Acton asserted that good wives and mothers felt more passionately about their domestic role than about their husbands. "I am ready to maintain," he wrote, "that there are many females who never feel any sexual excitement whatever. . . . Many of the best mothers, wives, and managers of households, know little of or are careless about sexual indulgences. Love of home, of children, and of domestic duties, are the only passion they feel." And, he added, "As a general rule, a modest woman seldom desires any sexual gratification for herself. She submits to her husband's embraces, but principally to gratify him; and, were it not for the desire of maternity, would far rather be relieved from his attentions."[13]

Even reading was risky. *The Ladies' Guide of Des Moines* informed its readers in 1882 that "Reading of a character to stimulate the emotions and rouse the passions may produce or increase a tendency to uterine congestion."[14]

Although many argued that women's smaller brain signaled their inferiority, no one seriously contended that women lacked brains altogether. Instead an elaborate theory explained that a woman's mind is different in ways uniquely suiting her for life in the private sphere. The Darwinians, for example, argued that the brain of the woman had evolved consistent both with her nature and her social role. "A woman's brain," claimed Social Darwinist Herbert Spencer, "evolves emotion rather than intellect. . . . The best wife and mother and sister would make the worst legislator, judge and policy."[15]

Working outside the home also was thought to jeopardize a woman's reproductive capacity, or at least her womanliness. Even some educated women embraced the latter assumption. "[T]he truth is that mentally and emotionally I really had three categories of gender—men, women who are men, and women who stayed home," noted twentieth-century newspaper editor Deborah Howell.[16]

The supposition that a brilliant woman must have sacrificed her sexuality is evident in a eulogy delivered after the death of Nobel Prize winner

Emmy Noether in 1935. "She was a one-sided being who was thrown out of balance by the overweight of her mathematical talent," said her former colleague. "Essential aspects of human life remained undeveloped in her, among them, I suppose, the erotic."[17]

Stereotypes aside, education did produce social differences. Educated women were less likely to marry. When they did, they had fewer children.[18] What many of these college-educated women were in fact doing was opting for the "loathsome" alternative in the no-choice choice between womb and brain. They stayed single and pursued careers.

Where the nineteenth-century theorists focused on the claim that intellectual activity damaged a woman's reproductive and sexual capacity, the twentieth century has concentrated on the corollary: Woman's sexuality interfered with her intellect. As I noted earlier, the twentieth-century version of the belief that biology disqualified women for leadership took form in the claim that a woman's "raging female hormones" barred her from the presidency.[19] Those raging hormones—presumably evident in premenstrual syndrome and menopause—disabled women in leadership. Her sexuality is assumed to affect a woman's job performance: "It's that time of month," say co-workers, or "All she needs is a good lay."

Fighting the Womb/Brain Bind

There were formidable barriers to securing access to education for women, given what society believed to be at stake. Women made initial inroads by framing their arguments in their opponents' terms: Education, they said, would make them better wives and mothers. From Mary Wollstonecraft to Catherine Beecher, women's advocates saw their problem as educational discrimination. To attain a seat in the classroom, advocates argued that women were responsible for educating the young. To educate they required education.

Once in the door, women were able to establish that exercising their brains did not destroy their health. "I am convinced," observed University of Michigan president and psychologist James Angell, near the turn of the century, "that a young woman, coming here in fair health, is likely to remain in health as good as that which she would have had had she stayed at home."

At the same time, women were able to show that their aptitudes were not limited to those areas traditionally defined as "woman's work." "There is no branch of study pursued in any of our schools in which some women have not done superior work," concluded Angell.[20]

Starting in the late nineteenth century, women began to account for an increasing portion of the total enrollment in institutions of higher education. In each decade from 1870 to 1920, the percent increased, from 21% in 1870 to 47% in 1920. The percentages then drop to a low of 30% in 1950, explained in part by the dramatic increase in college attendance by

World War II veterans on the GI Bill. The rise after 1950 is steady. In 1980 women passed the 50% mark.[21]

Armed with an education, female scholars turned the scientific method against the assumptions that had perpetuated the womb/brain bind. In 1903, psychologist Helen Thompson demonstrated both that men and women show comparable proficiency at basic intellectual and motor tasks, and that social scientific proponents of the womb/brain bind had interpreted their results to confirm their stereotypes. The powerful pull of social expectations is evident, however, in the fact that when Helen Thompson became a wife and mother, she left the classroom.[22]

One of the pioneers in the effort to dispatch stereotypes with science was Leta Hollingworth, who 80 years ago established that women's performance did not decline during menstruation. She also challenged the notion of a maternal instinct, arguing instead that society used its resources to engender in women an expectation that childbearing is "normal."

Female economists also entered the debate over the effects of industrial work on women. "If it could be shown that the women of to-day were growing beards, were changing as to pelvic bones, were developing bass voices, or that in their new activities they were manifesting the destructive energy, the brutal combative instinct, or the intense sex-vanity of the male then there would be cause for alarm," wrote Charlotte Perkins Gilman in *Women and Economics,* published in 1898. "But the one thing that has been shown in what study we have been able to make of women in industry is that they are women still, and this seems to be a surprise to many worthy souls."[23]

But establishing that the brain-drain hypothesis was specious was only half the battle. Having overcome the socially and legally constructed assumptions that the brain and uterus existed in a perpetual tug-of-war, women faced the biological fact that they were the species' childbearers.

Reproductive Rights and the Double Bind

Without information about contraception and access to reliable birth control, celibacy was the only sure way our great-grandmothers could avoid motherhood. Before women secured these means, and access to abortion when birth control failed, regular sexual intercourse and childbearing bore an all-but-inevitable correlation. And that's the way some wanted it to stay. "One of the principal objections to contraception by medical and lay persons in the nineteenth century," writes Pulitzer Prize-winning historian Carl Degler, "was that it seemed to call into the question the idea that women were meant to bear children."[24]

Measured by the average number of children delivered by the typical white woman before menopause, the nineteenth century marked a dramatic change in woman's autonomy: from 7.04 in 1800 to 3.56 in 1900.

Degler terms this drop "the single most important fact about women and the family in American history."[25]

The smaller family size occurred during a century in which there were no important innovations in birth control methods themselves.[26] When not just saying no, women and their cooperative husbands were probably practicing coitus interruptus or— employing means that did not require spousal cooperation—aborting.[27] In 1821, Connecticut was the first state to outlaw abortion; most states had followed by the early 1860s.[28]

Those wishing to control the size of their families were inadvertently aided by medical practitioners who counseled in the euphemisms of the day that "excessive connection" could induce ill health in both husband and wife. "My own opinion," wrote Dr. Acton in one of the more widely read books on reproduction in the nineteenth century, "is that, *taking hard-working intellectual married men residing in London as the type,* sexual congress had better not take place more frequently than once in seven or ten days. I could point to the case of many a married man suffering from derangement of health solely, or at all events mainly, attributable to unsuspected sexual excesses, the best proof of which is that the health becomes restored as soon as the excesses are left off."[29]

Interestingly, however, Acton also blames the women's movement for encouraging women to turn their husband away. "During the last few years," writes Acton in a section titled "Sexual Suffering in the Married," "and since the rights of women have been so much insisted upon, and practically carried out by the 'strongest-minded of the sex,' numerous husbands have complained to me of the hardships under which they suffer by being married to women who regard themselves as martyrs when called upon to fulfill the duties of wives."[30] Clearly Acton's focus was on the health of the male, who needed just enough—but not too much—sex. By indicting certain wives as "strong-minded" women who affected martyrdom in the bedroom, he once again underscored the opposition of brain and uterus. And since in the same work he described the "best wives and mothers" as "modest women" who only "submitted" to their husbands' desires, in Acton's world it would appear to be easier for a woman to squeeze through the eye of a needle than to assume a proper wifely role in bed.

Until comparatively recently, the law and biological reality colluded to deny women the reproductive choice that would increase their predictable control over their condition in the productive workplace. Not until the availability of the birth control pill in the 1960s was a reliable contraceptive available. The level of control women began to exert over their reproductive lives at that point is evident in the fact that the average woman in the mid-1960s would bear no additional children after she reached the age of 28.[31]

Before the advent of the Pill, for most of the population sterilization and abstinence were the only sure and acceptable means of avoiding pregnancy. From 1873 until 1938, the Comstock Law forbade transmission of contraceptive information and devices in the U.S. mails. That law was invalidated in *U.S. v. Nicholas* in 1938.[32] Until *Griswold v. Connecticut*[33] in 1965, state law could still ban the use of contraceptives by married couples. And state restriction on distribution of contraceptives to the unmarried was legal until 1972.[34] A year later, in *Roe v. Wade,* the Supreme Court protected a woman's right to choose abortion.[35]

Congress entered an important additional guarantee for women with the Pregnancy Discrimination Act in 1978. Beginning October 31, 1978, employers could no longer discriminate because of pregnancy, childbirth, or related medical conditions.[36]

The battle over the right to contraception and abortion is the ground on which the womb/brain bind is now being fought. Supreme Court Justices O'Connor, Kennedy, and Souter explicitly recognized this fact when in June 1992 in *Planned Parenthood v. Casey* they defended *Roe v. Wade* on the grounds that "For two decades of economic and social developments, people have organized intimate relationships and made choices that define their views of themselves and their places in society, in reliance on the availability of abortion in the event that contraception should fail. The ability of women to participate equally in the economic and social life of the Nation has been facilitated by their ability to control their reproductive lives."[37] At issue is whether women will be able to decide whether and when to conceive and carry to term. When women have that option, enshrined in law, and accessible without financial risk or social stigma, the hold of this bind will have been broken.

With access to reliable forms of birth control and confirmation that women can think without triggering biological melt-down, women have before them three ways to respond to the biological and social constraints of childbearing and rearing. Some either live the life of a traditional wife and mother or, whether single or married, elect to be childless career women. Some choose a route that sequences career and childbearing. Some, through choice or economic need, live lives in which career and childrearing co-exist.

Both/And: Co-existence

The double bind faced by the working mother lies in society's persistence in linking a woman's identity to a man and to the role of mother and homemaker. When Congresswoman Pat Schroeder was asked how she could be both a member of Congress and a mother at the same time, she challenged the assumption that the two were mutually exclusive by commenting, "Because I have a uterus and a brain, and I intend to use them both."[38]

A man is not defined by his relationship to a woman or by fatherhood. When the minister intones "I now pronounce you man and wife," he is confirming that the ceremony has transformed the woman into a wife and left the man's identity unaltered.

Carol Browner, Clinton's Administrator of the Environmental Protection Agency, has said, "When I talk to teen-age girls, they want to know if they can have children and a career and will someone still want to marry them. For an awful lot of women it's still very difficult. I guess it's progress when the husband says, 'I'm going to baby-sit the kids.'"[39] The fact that fathers often see their role as "babysitter"—surrogate caretaker, not equal partner— illuminates the persistence of traditional gender roles. They persist too among women who feel guilty about time devoted to career rather than children, especially the many who were raised to believe that childbearing is primarily the mother's responsibility. And they raise questions in child custody cases where, for much of the history of the country, fathers were given custody and from the turn of the century to the 1970s, mothers received preference. Now the courts are ruling in favor of the "primary caregiver," a standard that raises the question: Are men rewarded for exceeding social expectations and showing an interest in the primary parenting role and women punished because devotion to a career appears unmaternal? Reliable child care and family leave can ease the time constraints, but not the competing pressures.

The double task of raising children while pursuing a career has earned its own tag—"juggling." The word expresses the assumption that women are mainly in charge of child care—men are generally not expected to juggle. On average, men are now performing more household chores, but they are still picking up only about 20% of such daily tasks as laundry, cooking, and cleaning.[40] And while the amount of time fathers spent with their children increased from the mid-1970s to the early 1980s, the total still is a third or less of the time put in by mothers.[41]

This places female executives in a bind not faced by their husbands. Male corporate executives "place every working woman on a continuum that runs from total dedication to career at one end to a balance between career and family at the other," writes Felice Schwartz in the *Harvard Business Review*. "Male corporate culture sees both extremes as unacceptable. Women who want the flexibility to balance their families and their careers are not adequately committed to the organization."[42]

Women inevitably try to conform to these social and professional expectations. Senator Dianne Feinstein notes that as a single mother, she hesitated to tell her supervisor that she had to leave work to attend to her child. "I still think women have to be very careful. One of the reasons they are denied promotions is because they are looked at in the office as the ones responsible for the family."[43]

The view that a working woman purchases success in a career at a cost in childraising is what led to Nannygate. Attorney General nominee Zoe Baird was asked how many hours she spent at home with her children, and who took care of them in her absence. No one asked these questions of her male counterparts, all confirmed for positions in the Clinton cabinet. Baird withdrew because of disclosures that she and her husband had employed an illegal alien as a live-in. Since the Justice department is supposed to enforce the law in question, the concern was understandable. More prob-lematic is the fact that these questions were never addressed to any male nominee nor had they been asked if they had covered the Social Security of their housekeepers.

Next up for the post of Attorney General was Federal Court Judge Kimba Wood. Here the plot thickens. Before being officially nominated, Wood withdrew because she had employed an illegal as a nanny *before* the law barred doing so. How does one account for the fact that Baird and Wood were judged differently from their male peers? Perhaps mothers forced into the labor market by economic need are considered dutiful mothers when they obtain child care. But high-salaried women in posi-tions of power are presumed to be working for self-satisfaction or luxuries, not to meet basic family needs. And when they purchase child care, they are, as a result, regarded as negligent mothers. The low-wage mother, in other words, is assumed to be working for her children; the high-wage mother is presumed to be working for herself.

This is an aspect of the double bind that says a working mother is shortchanging either career or children. Its flip side is the unrealizable expectation that a woman can perform two demanding jobs simul-taneously without exhausting herself. The father, meanwhile, gets off scot-free, and the need for safe, cost-effective day care for children and family leave for their parents can be sidestepped.

The pressures on women created by combining career and motherhood have resulted in calls for a more equitable distribution of child-care respon-sibilities between mother and father, reliable, affordable day care, viable family leave, and a work environment that makes it possible for men and women to meet both their family and professional obligations well.

Passage of the Family and Medical Leave Act of 1993 was a dramatic move forward in this area. That act guarantees mothers and fathers 12 weeks of unpaid, job-protected leave to support the birth of a child. Here is an instance in which the law may be a bit ahead of the attitudes of em-ployers. A 1986 *Washington Post* survey asked 1500 human resource offi-cers at the nation's large companies what would constitute reasonable paternity leave; 63% replied "none."[44]

Sequencing

In earlier times, motherhood banished the poet and musician's muse. "Be-ing married in 1708, I bid adieu to the muses," wrote British poet, play-wright, and essayist Catherine Cockburn, "and so wholly gave myself up to the cares of family, and the education of my children, that I scarce knew, whether there was any such thing as books, plays, or poems stirring in Great Britain."[45] Before sequencing became a socially accepted choice, those women who were financially able retired from the public world at marriage or first pregnancy.

The uterine-drain hypothesis was uncontested by women who chose not to marry, including Susan B. Anthony, and women who sequenced work and childrearing. Elizabeth Cady Stanton was among those who "sequenced" their private and public lives by withdrawing from the work force to bear and raise children. Sequencing of childraising and work outside the home began in the Industrial Revolution and persists today.

In May 1927, sequencing gained the endorsement of Eleanor Roosevelt. In *Success Magazine,* ER wrote that engaging in political activity could "guard against the emptiness and loneliness that enter some women's lives after their children are grown." In 1956, sequencing obtained the blessing of sociologists Alva Myrdal and Viola Klein in their influential book *Women's Two Roles: Home and Work.*[46]

Sequencers argue that education, family, and work can be blended into a harmonious whole in one lifetime if each is given its own place in a chronological sequence. But sequencing carries a cost to one's career. Unless women can translate time spent rearing children into an asset in the public sphere, sequencing means that women of a given age will have less professional experience than men of the same age. Moreover, a woman's biological clock will draw her from the workforce at an age when management-level promotions are most likely.

"The problem of reconciling a scientific career with some semblance of a normal life is exacerbated by the tenure system," notes Shirley M. Tilghman, a professor of molecular biology at Princeton. "A woman is usually 30 years of age before assuming an assistant professorship at a university, which puts her tenure decision at age 35 to 36. Thus her critical scientific years, in which she is establishing her reputation, and her peak reproductive years coincide. This is a dirty trick. Many in my generation chose to forego child-bearing until the security of tenure had been granted, only to find that their biological clock had stopped ticking."[47]

The same is true of women in industry, the law, and academia in general. "The decade between age 25 and 35 is when all lawyers become partners in the good firms, when business managers make it onto the 'fast track,' when academics get tenure at good universities, and when blue collar workers find the training opportunities and the skills that will generate high earnings," writes economist Lester Thurow. "[This is also] precisely the decade when women are most apt to leave the labor force or become part-time workers to have children. When they do, the current system of promotion and skill acquisition will extract an enormous lifetime price."[48]

When sequencers step back from their careers while promotions are likely, their chances of competitive re-entry into the labor market are significantly reduced. The sequencer who leaves the workforce temporarily invites employers to argue that hiring women is more costly than hiring men. Whether or not they plan to return to the job, sequencers are assumed to be stepping out of the management or promotional track as soon as they decide to leave work to have a child.

On the other hand, delaying childbearing carries biological penalties—and not just in terms of decreased fertility. In fall 1993, Carol Higfoss

Rubin, an epidemiologist with the Centers for Disease Control and Prevention, released a study showing that women in a number of professional groups had a greater chance of dying from breast cancer than homemakers or other women in nonprofessional positions, in part because women in professional positions started their families later.[49] Earlier studies had also correlated delays in childbearing to an increased risk of breast cancer.

Associations of teachers—one of the groups at risk—responded with calls for increased use of mammography. Instead of trying to better detect cancer at an earlier stage, one might wonder why they are not calling for child-care facilities in universities and organizations that house high-risk professions. The assumption underlying the response was that "we" must not be doing something, in this case mammography, rather than that society and industry ought to be providing options that would make both early childbearing and professions compatible.

Mandated Sequencing

Before 1978, sequencing was not a choice but a mandate for many women. Women who landed professional positions were asked to leave them at the first hint of pregnancy. "Women voluntarily left the newsroom when they were pregnant—but often they were fired. Time was," says reporter Elsie Carper of the *Washington Post*, "that the day a woman even said she was getting married, much less having a child, was the day she was in effect resigning from newspaper work."[50] When Betsy Wade told her supervisors at the *Herald Tribune* that she was pregnant in 1953, she was fired.[51] Two decades later, she was among those filing a successful discrimination suit against the *New York Times*.

Even the prospect of pregnancy was once enough to deny a qualified woman a job. "One editor had refused to hire me several years before because he said I would probably just quit and have a baby," writes Lindsay Van Gelder. "When I replied that I was on the Pill—I cringe to recall that I thought this bit of information was his business—he still wasn't impressed. 'A pretty little thing like you ought to be home having a baby every year,' he gallantly insisted. And he was damned if he was going to help me thwart my biological destiny."[52]

In many jurisdictions, female teachers were dismissed when their nuptials were announced. As Gladys Borchers, the first female professor in the Department of Speech at the University of Wisconsin, told me, "The choice was put directly to women: teach or marry. One or the other. You can't have both. I chose to teach."

A National Educational Association survey completed in 1930-31 found that "77 percent of surveyed districts would not hire wives and 63 percent dismissed female teachers if they married."[53] In *Cleveland Board of Education v. LaFleur*[54] and *Cohen v.Chesterfield County School Board*,[55] the Court examined school policies that banished pregnant teachers from the classroom in the fourth or fifth month of pregnancy. The policies did not make exceptions for time of the year, the health of the teacher, or her wishes. "In *Cleveland Board of Education v. LaFleur*. . . [t]he Board had selected the employment cut-off date to coincide with the time most women 'begin to show,' both to save teachers from embarrassment at the expected giggles of schoolchildren, and to spare the children from the sight of pregnant women. The very fact that schoolchildren or male workers have been insulated from the sight of pregnant women has only helped to reinforce the mystery and embarrassment that justified the exclusions."[56] The Court held that the decision of a pregnant woman "to continue to work past any fixed time in her pregnancy is very much an individual matter."[57] Nor could the airlines fire or ground female flight attendants who married if no such action had been ever taken against a male.[58]

And when the Supreme Court held in 1976 that denying disability benefits to pregnant women doesn't constitute discrimination on the basis of sex,[59] Congress blunted the ruling with the Pregnancy Discrimination Act, which took effect in October 1978.

Remaining Childless

At the side of the early "sequencers" were professional women who, by choice, chance, or professional decree, remained childless. So, for example, a study by the American Chemical Society concluded that 37% of female chemists older than 50 years had no children; while only 9% of the men older than 50 were childless.[60]

The power of the double bind is heard in echoes of a descriptor that was still widespread in my youth: "old maid school teacher." Those who assumed heterosexuality as a given made little comment on the number of female scholars, including heads of major women's colleges, who "lived with" close female friends. Those who chose to exercise their brains but not their uteruses did not escape social sanction, however. They were cast as women too ugly or ill tempered to attract a husband, or—when the topic became a matter of public discussion—as lesbians. The pejorative "dyke" came to encompass both ideas.

The tag "frustrated spinster" was quickly tied to single women over a certain age. Susan B. Anthony was presumed to have channeled her twarted maternal instincts into the care of her friend Elizabeth Cady Stanton's children. Critics also assumed that it took the strength and wits of both the single Anthony and the married mother Stanton to make one complete woman.

Choosing a Traditional Role as Wife and Mother

We are quick to see the double binds hampering the mobility of those working outside the home. But the reason that the womb/brain is a double bind is because either choice carries penalties. The dilemma facing those aspiring to a traditional role was recognized by psychiatrist Bruno Bet-

The eloquent speeches of Elizabeth Cady Stanton and Susan B. Anthony helped shatter the assumption that there was no place for women in the public sphere. (Library of Congress.)

telheim in the early 1960s. At an early age a girl learns "that her main fulfillment will come with marriage and children," he wrote, "but her education has nevertheless been the same as that of boys who are expected to realize themselves mainly through work and achievement in society."[61] This was the bind explored so tellingly in *The Feminine Mystique*.

Where staying home was then the norm, a majority of the women in the population now work outside of it. That trend has magnified feelings that full-time housekeeping is less worthy than the "real work performed by the husband, a reality brought home when work wins out over time and love."[62] The phrase "just a housewife" acknowledges the low status some attach to the traditional role. The phrase "work inside the home" is an attempt to set such responsibilities on an equal plane with the alternative.

Whether a woman claims to manage public and private spheres competently or not, she will be subject to special forms of scrutiny. The residual power of the womb/brain bind is evident in recurrent references to a career woman's neglect of the private sphere, specifically her housekeeping chores.

So, for example, reporters told the public that Janet Reno did not keep a very neat house, but failed to inform us whether Ron Brown throws his underwear on the floor or Robert Reich washes the dishes after breakfast. Implicit in the observation about Reno are the notions that she ought to be responsible for cleaning her house and held accountable when it falls short of some standard. More important, such talk assumes that a high-powered woman in a public role will have to cheat her domestic responsibilities—in this case housekeeping. "I am not a good housekeeper," Reno admitted in a speech at the National Press Club, "I don't put much priority on housekeeping."[63]

The assumption that a woman should plagues women in public life. A profile on Democratic gubernatorial candidate Debbie Stabinow included the fact that her house could use a good dusting. "The story was focused on me as a woman and it talks about my kids, it talks about how my house was not as clean as it could be. A little tiny bit on issues," recalls Stabinow.

"She has only minimal (every-other-week) housekeeping assistance too," wrote the reporter, "although, truthfully, most days the modest little place on South Deerfield with tufted stucco ceilings and pink butterfly soaps in the bathroom dish *could* use serious commitment from someone wielding a broom and sponge."[64] Ms. Stabinow, it seems, lacks commitment to housekeeping.

The notion that a woman's place is in the home surfaces when one least expects it. When Virginia Senator Chuck Robb was asked by a reporter for the *Washington Post* what he would like to change about his spouse, Lynda, he responded with an answer uninformed by a reading of the *The Feminine Mystique:*" "I wouldn't say Lynda is a natural in terms of housework."

Lynda Bird Johnson Robb could deny that a woman ought to be *a natural* at housework. Instead she counterattacks. "Oh, he says that?"

Lynda asks later, her voice rising. "Well, I think it is true. But I'll just tell you, I've run the vacuum many more times than he has."[65]

Residues of the Womb/Brain Bind

Historically, the childless single woman was assumed to be defective, either an asexual spinster or a lesbian. The childless married woman was presumed to be so power-driven and selfish that she deliberately sacrificed her childbearing role for her profession. The bind remains, but shrouded by a thin veil. While no one can respectably claim today that women drain brain energy by depriving their wombs, opponents of women's rights can achieve the same effect when they suggest that women in public roles are sexually deviant.

By identifying a woman with her reproductive capacity and framing public discussion of women's rights on those terms, opponents of women's rights have legitimized a public focus on a woman's private behavior. Such a discussion has also legitimized a focus on female—but not male—sexuality.

The Focus on Female Sexuality

This focus on female sexuality was evident in filmwriter Penelope Gilliatt's belief that "It would be difficult for a woman to be, I should think, the production head of a studio or a manager without being called a bulldyke."[66] In a tone tinged with exasperation, a state legislator at a conference at which I spoke reported that when she told the male speaker of her own party that she was planning to introduce legislation codifying *Roe v. Wade*, he said, "You cunt!"[67] Female sexuality is at issue when male sexuality wouldn't be.

So, for example, the U.S. Bureau Chief of Britain's *The Guardian* wrote that "To meet Margaret Thatcher was to be aware of an extraordinary powerful sensuality, one that could inspire intense devotion and illuminate that great historical mystery: how the Virgin Queen, Elizabeth I, was able to bewitch sixteenth-century Europe. As France's President Mitterand once described Thatcher: She had the 'eyes of Caligula, and the lips of Marilyn Monroe.'"[68]

To many men, women's sexuality is always in play. *New York Times* reporter Nan Robertson recounts the following conversation overheard by a female *Times* reporter between two male editors:[69]

FIRST EDITOR: "Did you see _____ when Abe passed her that cake at lunch? You could almost see her nipples pucker for power."
SECOND EDITOR: "Yeah. These broads' nipples always pucker for power."

Some men equate desire for power with lust, and reach for such descriptions especially when they feel their own power is threatened.

The sexuality of women is fair game in a way that men's sexual habits are not. In the 1992 presidential campaign, Torrie Clarke was one of four female spokespersons for presidential candidates. At the Republican Convention, I attended a briefing at which Clarke, a Bush spokesperson, presented. Behind me were two male reporters for major newspapers. In the transition from Clarke's statement to the next in line, I overheard one reporter ask the other, "So, what did you think?" "Nice legs," came the reply. "A little thin for my taste," commented a third who had joined them during the session.

One 1992 focus group that included both men and women speculated about whether Clinton representative Dee Dee Myers had gotten her job by sleeping with Bill Clinton. Torrie Clarke was described as "available," "the sexiest woman on television," and a "real bitch when pushed." She and Bush advisor Mary Matalin were thought to "sleep around." No comparable comment was heard from men or women about Democrat James Carville, the man whom Matalin was involved with. Only Clinton's ad director Mandy Grunwald—single and childless—escaped sexual banter. She was perceived to be older than the others, and viewed as "a Jewish mother," immune, in this motherly incarnation, from any sexual commentary.

Where a man is presumed to be able to separate work from play, public from private relations, a woman is not. During the 1992 presidential campaign, for example, Clinton's adviser James Carville and Bush adviser Mary Matalin were having an affair. In the Bush camp, senior advisers speculated that the relationship compromised Matalin's effectiveness. No such speculation surfaced in the Clinton campaign about Carville. "[Bush's adviser] Teeter had actually said 'Maybe we better get Mary off this,'" recalls reporter Kit Seelye who covered the campaign. "Nobody in the Clinton campaign said the same thing about Carville. He was perceived as having this girlfriend on the side."[70]

Sexual innuendo also plagued Bernadine Healy, who served as director of the National Institutes of Health during the Bush presidency. When her marriage to a medical colleague on the Johns Hopkins faculty broke up, for example, the school's all male Pithotomy Club included a skit about the couple in its annual comedy revue—"a bonding ritual" for the male elite of American medicine.[71] In the skit, Healy's ex-husband was shown as widely suspicious that his ex-wife was sleeping with his colleagues and friends. "Healy, played by a man dressed in a blond wig, fishnet stockings and coconut-half brassieres, was depicted performing a variety of pornographic acts on other physicians until, at the end, she was discovered *in flagrante* by her ex-husband." The show ended with the song "Cardiology Girl," a bawdy takeoff of the Playboy centerfold-inspired hit song "Calendar Girl."[72]

Unprotected by a husband, Healy was fair game for sexualized attack. Such a skit would not have been performed about a doctor's wife.

Healy's experience was not aberrational. Congresswoman Pat Schroeder led congressional efforts to get to the bottom of a Navy and Marine scandal in which two dozen women were sexually assaulted at the 1991 convention of the Tailhook Association. In June 1992, at a Navy event known as the Tomcat Follies, participants invited Schroeder to perform fellatio in a rhyme that began "Hickory, dickory, dock . . . ."[73]

Intervening between the skit ridiculing Healy and that demeaning Schroeder was the testimony of Anita Hill before the Senate Judiciary Committee—and Tailhook itself. Consciousness had been raised. Where no punishment followed the humiliating portrayal of Healy, things were different for Schroeder. She was a member of Congress with some say over defense budgets. In July 1992, the Navy removed two officers from their command posts for not stopping the Tomcat Follies skit.

While women's real or persumed sexual activity elicits commentary that men would never expect to be directed at them, failure to produce the *progeny* of sexual union also draws comment. Women have to explain why they have no children.

"Those of us who remain on the sidelines childless also take our hits," writes Robin Young in *Newsweek* in 1990. "Too often, there's an assumption made that being childless is a calculated career move. Pretty dangerous assumption, if you ask me. For me, it has more to do with sorrow and disappointment when long-term relationships ended."[74]

Women without children are expected to explain why. The response that hints that chance, not choice, is responsible is politically smart. The idea that *a woman wants children, but hasn't been lucky enough to conceive* elicits sympathy instead of suspicion. In the boardroom, company men like assurances that company women have a "natural" desire for family, but have been thwarted by forces outside their control—quite conveniently, of course, for the company's demands. On the stump, the female politician can take a similar tack, drawing fatherly and sisterly sympathy, thwarting rumors about her frigidity or power-hungry sterility.

"When I was Local County Commissioner, which was a County Supervisor, a part-time job, I got questions about 'What did my husband think?' 'Did I have children?'" recalls Michigan legislator Debbie Stabinow. "I didn't have children at the time. People must have thought 'Poor dear, she can't have kids. Oh, how sad,' and I won."[75] And in the 1993 contest to determine who would become the next Prime Minister of Canada, the eventual victor—Kim Campbell, who was childless—was subjected to "comparisons with her opponent's attractive wife who, commentators noted, was the mother of young children."[76]

The vestiges of the womb/brain bind surface whenever women are accused of sublimating their sexuality to benefit their careers. The supposition that a nontraditional career substitutes for a satisfying heterosexual sex life survives in such statements as "She will have to get her ecstasy of orgasm some other way . . . maybe with a vibrator." That remark was made by Ernie Chambers, State Senator of Nebraska, about Nebraska Assistant Attorney General Sharon Lindgren.[77] A more subtle insinuation can be read in the labels used to identify New York City Comptroller Elizabeth Holtzman. She merits references in *The Macmillan Dictionary of Political Quotations* as "Ice Lady" and "Virgin Liz."[78] Because she is childless, the logic goes, she must be frigid or asexual.

Sexual Deviance

In the grab bag of caricatures of the female leader are images suggesting that she is asexual, a whore or dominatrix, or a lesbian. One of Anita Hill's champions noted, "If you are a single woman in a political environment, your sexual behavior will be subject to review. You will either be promiscuous or you'll be a lesbian. Those are the choices, you know. Women who run for office, whatever they are, in subtle or unsubtle ways get attacked. If they are married it is how strong is their marriage, or if they are divorced, who created the divorce?"[79]

An attack in *The Spectator* asked: "So Hill may be a bit nutty, and a bit slutty, but is she an outright liar?"[80] "Hill's obsessive, even perverse, desire for male attention is another sign that she may have been suffering from some kind of love-hate complex,"[81] speculated David Brock in that article. Charles A. Kothe, dean of the Law School at Oral Roberts University, who had hired Anita Hill at Clarence Thomas's recommendation, hypothesized that Hill's charges of sexual harassment were the by-product of "fantasy." He later recanted the charge.[82] To dismiss the scenarios being sculpted by Thomas's defenders, Senator Heflin asked her whether she was a scorned woman and whether she had a martyr complex.[83] With a reference to her "proclivities," Senator Simpson implied that she was a lesbian.[84]

Similarly, in the early 1990s, campaign literature of the Religious Right described board members of the Mainstream Voter's Project in ways reminiscent of nineteenth-century attacks on the suffragists: "One is a Democrat who wanted to be born a man but fate played a cruel hoax by giving her a womb. Another one is trying to convince people that she is not a bitter, manhating bitch. Someone was rumored to have no vagina. After months of comprehensive investigations involving background checks, public record research, DNA matching and a Pap smear or two, we have completed the report on the Mainstream Voter's Project."[85]

The charge that a powerful woman is a closeted lesbian is also predictable. So, for example, Republican Linda Chavez suggested in her attempt to unseat Senator Barbara Mikulski that the Maryland Senator was a "San-Francisco-Style Democrat" who was "anti-male" and ought to "come out of the closet."[86] Marriage and motherhood are not necessarily insulators. Conservative radio call-in host Rush Limbaugh characterized Democratic political consultant Ann Lewis as someone who "looks like she could be a dyke if she wanted to be."[87]

Similarly, while campaigning for Republican George Allen and against Democrat Mary Sue Terry during the race for the governship of Virginia,

Oliver North, of Iran-Contra fame, observed that the governor's mansion shouldn't be "a sterile building" but rather a home "where a man and his wife live, and with the laughter of their children."[88] Some audiences heard these as veiled illusions to certain rumors about the Democratic candidate. Behind the scenes, Allen's supporters were spreading the rumor that Terry was a lesbian. Before one of the debates, a reporter confronted Terry's media adviser Bill Knapp. After denying that his candidate was a lesbian, Knapp's partner Bob Squier urged the reporter to walk across the stage and ask Allen whether he was gay. The reporter refused.

In Kansas, Gloria O'Dell, who failed in her attempt to unseat Republican Senator Robert Dole, was called a lesbian by her primary campaign opponent. A divorced mother, O'Dell was subjected to signs at rallies referring to her as "Bull Dyke O'Dell." Finally, in exasperation, she called a press conference to "proclaim her heterosexuality."[89]

When Janet Reno was Dade County State Attorney, an opponent spread rumors that she was a lesbian. During her confirmation hearings, the Associated Press implicitly responded to comparable rumors with the claim that she had in fact been involved with men in the past. "She dates men, friends say, but hasn't found anyone she's wanted to marry," noted the reporter. "At age 54 and 6 feet, 2 inches tall, she described herself last week as 'an awkward old maid with a very great affection for men.'"[90]

Reno set aside the question of sexual preference by pitting the older stereotype of women in power against the newer one: the spinster against the lesbian. "I am just an awkward old maid with a very great attraction to men," she repeated to *People* magazine a few months later.[91]

Extending the notion that female leaders seek inappropriate power over men are cartoons and composite photos showing them in leather, wielding whips. In the nineteenth century female leaders were thought to sap their femininity and enervate their maternal capacities; in twentieth-century projection of them as dominatrixes, they emasculate men.

"The most striking theme that emerged in the media coverage" of Margaret Thatcher's victory as Conservative party leader in 1975, wrote a British critic, "was the idea of Mrs. Thatcher as a ruthless killer, triumphing over men, beating and humiliating them; and there were elements not only of fascination, but also of disturbance in this view. The image produced was suggestive of a sub-genre of pornography, where a woman holds the whip, an inversion of the normal conventions of sado-masochism where mastery and dominance are masculine and punishment and humiliation meted out by men."[92]

Where in the nineteenth century, women were expected to submit, in this view they now demanded submission. The dominatrix is an extension of an argument made in the nineteenth century claiming that what women sought was not equal rights but power over men—an allegation that activists in the women's movement met by declaring "women don't seek power over men but power over themselves."

In the treatment of Anita Hill, Margaret Thatcher, and Bernadine Healy we hear the echo of nineteenth-century ideas: they are whores, lesbians, dominatrixes, or castrators, whose intellectual activity endangered society and drove them into neuroses—specifically, in Hill's case, fantasy and erotomania. Between the lines of the press coverage of Janet Reno, we hear her sexual identity questioned and find her playing out an alternative stereotype—that of the awkward spinster so unattractive she had no choice but to seek out professional achievement.

The diminished power of the lesbian/whore whispers is evident in the success of the women who have weathered them. Barbara Mikulski was elected in Maryland. Janet Reno was confirmed and has proven a popular attorney general. Ann Lewis is a prominent political consultant. Although she is still plagued by questions about her role as first lady, Hillary Clinton is a respected member of the Clinton team. Bernadine Healy went on to head NIH and undertook a Republican senatorial campaign. Margaret Thatcher was, during her tenure, the most powerful female leader in the world. And a year after her testimony at his confirmation hearing, more people believed Anita Hill's story than that of the man now sitting as Supreme Court Justice. As women succeed despite them, insinuations of "deviant" sexuality are more and more likely to be viewed as signs of the desperation of those offering them.

At the same time, as we realize that our associates, neighbors, friends, and children include lesbians and gays, prejudice will predictably decline. Those who wonder, for example, why Phyllis Schlafly does not vilify gays found an explanation in the summer of 1992, when her own son came out of the closet.

The Law Assumes That Women Control Reproduction

The final frontier for the womb/brain bind was won by those who believe that unless pregnancy interferes with job performance what happens in a woman's uterus is her business. The courts have provided differing answers to the question, can an airline ground a pregnant flight attendant? At issue was whether the nausea and fatigue often present in the early months of pregnancy will impair performance, and whether a woman in her final months would be able in an emergency to safely open 100-pound exit doors, assist passengers down chutes, and drag them from danger. A number of cases permitted mandatory grounding.[93] One authorized service through the 26th week, with consultation by the woman's physician.[94] A third outlaws grounding in the first trimester, permits it if advised by a physician in the second, and bans flight attendance in the third.[95]

Conflicts also have arisen between the objectives of the Occupational Safety and Health Act (OSHA) and Title VII of the Civil Rights Act barring sex-based discrimination in employment. The reason: lead, polyvinyl chloride, radiation, anesthetic gases and organic solvents, among other chemicals and agents, pose a danger to fetuses. (They may, as well, pose hazards for men of reproductive age.)

Some employers—in many cases those who had once refused to hire women altogether—responded by minimizing their legal liability at a cost to female workers. The by-product was a no-win situation for women who were required to provide proof of sterilization or shift to other, often lower-paying jobs. That was the choice offered women of childbearing age at American Cyanamid.

Although she invites us to see history and the decade of the 1980s through the optic of "backlash," Susan Faludi recognizes that fetal protection policies posed what I call a double bind. "As these companies would have it," she writes, "women could choose to be procreators who stayed home—or workers who were sterilized. Take your pick, they told their female employees: Lose your job or lose your womb."[96]

American Cyanamid argued that it could not reduce the level of lead in its pigments department below what it saw as hazardous to fetuses. The resulting "fetus protection policy" confronted women of childbearing age with the no-win situation described above. Five female employees chose sterilization; two did not and were moved to lower-paying jobs in other parts of the company.

Backed by the Oil, Chemical and Atomic Workers Union, the women filed a sex discrimination suit under Title VII and a federal suit argued that the company had failed to comply with the OSHA requirement that workplaces should be "free from recognized hazards that are causing or are likely to cause death or serious physical harm" to its workers. They settled the sex discrimination suit out of court. What remained at issue was the OSHA requirement. In a ruling that played a role in his rejection for the Supreme Court, Robert Bork held that the women's situation was not covered under the applicable OSHA provision. Since the presenting issue was not argued under Title VII or under the protections of the Pregnancy Discrimination Act, this case was a poor test of "fetus protection" plans.

Had Bork been presented with the sex discrimination issue, he says that he would have decided the case for the women.[97] "American Cyanamid didn't present a sex discrimination issue," claims Bork. "That had been settled out of court. The issue in American Cyanamid was a narrow one under OSHA. If it had been framed as a sex discrimination suit, the women would have won in American Cyanamid."[98]

The same year, 1984, a hospital was found in violation of Title VII for dismissing a pregnant X-ray technician under a "fetus protection" policy.[99]

In 1991, the issue was laid decisively to rest by the Supreme Court in *UAW v. Johnson Controls, Inc.*[100] The parallels to the American Cyanamid case are striking. Johnson Controls made batteries that contained lead as a key ingredient. Like American Cyanamid, Johnson Controls required medical documentation of women employees' inability to bear children as a condition of employment in a battery-manufacturing job,

The Supreme Court's majority opinion is unequivocal. Johnson Controls "has chosen to treat all its female employees as potentially pregnant," said the Court. "[T]hat choice evinces discrimination on the basis of sex.

Decisions about the welfare of future children must be left to the parents who conceive, bear, support, and raise them rather than to the employers who hire those parents," said the majority decision. "Congress has mandated this choice through Title VII, as amended by the Pregnancy Discrimination Act. Johnson Controls has attempted to exclude women because of their reproductive capacity. Title VII and PDA simply do not allow a woman's dismissal because of her failure to submit to sterilization."

With reproductive control in her hands, and Title VII and the PDA mitigating the effects of the womb/mind bind in the workplace, a woman could choose both to produce and reproduce—and when.

4

Double Bind Number Two: Silence/Shame

REMBRANDT and Tintoretto were among thirty-three of the Old Masters[1] who translated the Biblical story of Susanna and the Elders to a lay audience.[2] From Paul Rebhun's play *Ein geistlich Spiel von der Gotfurchtigen und keuschen Frawen Susannan* in 1536 to Handel's *Susanna* in 1749 and Jean Gilbert's operetta *Die keusche Susanne* in 1910, the story has been a staple of popular culture. Susanna's tale is retold in dramas in French, German, Greek, and English.[3]

The narrative portrays a noblewoman blackmailed by two judges in ancient Israel. The tale is set in a private garden where the elders had hidden themselves to observe Susanna. Thinking herself alone, Susanna disrobed to bathe. Emerging from their hiding place, the elders offered a double bind. The noblewoman could either submit to sexual intercourse with them, thereby becoming an adulteress, or retain her virtue and be charged by them with adultery.

"Then Susanna sighed," says the Scripture, "and said, I am straitened on every side: for if I do this thing, it is death unto me and if I do not, I cannot escape your hands. It is better for me to fall into your hands, and not to do it, than to sin in the sight of the Lord (22-23)."[4]

When she refused their advances, they publicly accused her of betraying her husband. She was brought to trial where she wept and looked to Heaven (35). On the word of the elders, the tribunal convicted her of adultery. "No one bothers to ask Susanna if she has anything to say—not even her husband or her parents," notes Bible scholar Carey A. Moore. "The testimony of the two witnesses, whose age and rank put them above suspicion, is enough to convict her."[5]

Leviticus 20:10 specified death as the penalty. On hearing the sentence, Susanna cried out not to the tribunal, but to God. As she was being led to her death, God raised up a champion to protest her innocence. Using his verbal skills to outwit the elders, Daniel elicited confirmation of their plot. The elders were executed and Susanna freed.

In the privacy of the garden, Susanna had eloquently resisted the elders and defended her virtue. But, as Harvard Divinity School professor Margaret Miles observes, "She did not tell her husband—a man fully as powerful as the Elders—her side of the story. Nor did she speak at her impromptu trial, at which she arrived veiled, as if in shame." When she pleaded for divine intervention, a male advocate appeared to defend her virtue in public. "In the story," concludes Miles, "she is alone in a world of men, without voice. . . . [I]n the story the assumptions of the self-righteous community that was ready to execute Susanna without hearing *her* story are never questioned."[6]

As interesting is the fact that painted recreations of the story invite the viewer to see Susanna through the lustful eyes of the elders. The paintings locate her not at the trial but in the garden. And there, even as she wards off their advances, she is portrayed as sensual and alluring. "[T]he prevailing pictorial treatment of the theme typically included an erotically suggestive garden setting and a partly nude Susanna, whose body is prominent and alluring, and whose expressive range runs from protest of a largely rhetorical nature to the hint of outright acquiescence," writes art historian Mary D. Garrard.[7]

Only one of these descriptions invites us to empathize with Susanna. That rendering is the work of a female painter, Artemisia, who, as a teenager, had been raped by her art instructor, Agostino Tassi. When Artemesia's father, Orazio, took Tassi to court, the tutor argued that the young girl had not been a virgin. "He was subsequently acquitted, while Artemesia, whose testimony was put to the test of torture by thumbscrew, acquired a reputation as a licentious woman that has persisted to this day," writes an art historian.[8] And to compound the injustices against her, for centuries scholars attributed her best work to her father.

Artemisia's Susanna shows "an emotionally distressed young woman, whose vulnerability is emphasized on the awkward twisting of her body. The artist has also eliminated the sexually allusive garden setting, replacing the lush foliage, spurting fountain and sculptured satyr heads . . . with an austere rectilinear stone balustrade that subtly reinforces our sense of Susanna's discomfort. The expressive core of this picture is the heroine's plight, not the villian's anticipated pleasure."[9]

Susanna's contorted posture and tortured expression in Artemisia's picture suggest that she lacks the wherewithal to take control of the situation. Nor could she do so at her trial. Since public advocacy by a woman was seen as a sign of promiscuity, Susanna was trapped. Her public silence reflects her double bind. To establish that she was virtuous, she would have to engage in public behavior that confirmed she lacked virtue.

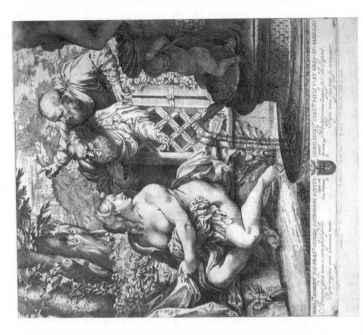

Susanna and the Elders by Artemesia Gentileschi, 1610. With permission of Graf von Schoenborn Collection, Pommersfelden, Germany.

Susanna and the Elders by Annibale Carracci, Bolognese, 1560–1609, etching and engraving plate. With permission of Ailsa Mellon Bruce Fund, National Gallery of Art, Washington, D.C.

In the portrayal by the female artist (*left*), Susanna is shown not as an object of desire but as the victim of uninvited advances. By contrast, the male artist (*right*) suggests that she is enticing and sensual.

Standing as a backdrop to the story of Susanna is that of Moses' sister, Miriam. In Numbers 12, Miriam and her brother Aaron object to the fact that Moses has married an Ethiopian woman. They argued as well that God had spoken not merely through Moses, but through them. God responded that he did indeed speak in a special way through Moses.

The story doesn't end there. Although both of Moses's siblings had questioned his role, only one—Miriam— is punished with leprosy. When Moses intercedes, God reduces the sentence to seven days. In justifying the punishment, Yahweh likens her to a shameful daughter (12:14). "The analogy God draws is quite astounding," writes biblical scholar Ilana Pardes. "Miriam's demand for greater expression seems to be synonymous with lewdness."[10] Miriam apparently learned her lesson. "[S]he dies shortly after this harsh incident without uttering an additional [recorded] word."[11]

The New Testament cast silence, salvation, and childbearing as confederates. Denied speech, women were unable to establish that their capacities were not exhausted by childbirth. The role of women was to be chaste, silent, subordinate to a male and fruitful. "Let a woman learn in silence with all submissiveness." wrote Paul.[12] "I permit no woman to teach or to have authority over men; she is to keep silent. For Adam was formed first, then Eve; and Adam was not deceived, but the woman was deceived and became a transgressor. Yet woman will be saved through bearing children, if she continues in faith and love and holiness with modesty."

Parallels exist in other times and cultures. Just as public speech by a woman invited inferences about promiscuity, so too her silence testified to her modesty. "For a silence and a chaste reserve is a woman's genuine praise and to remain quiet in the home,"[13] wrote Euripides. Aristotle too saw silence and modesty as conjoined virtues befitting a woman.[14]

The notion persisted throughout Western culture. Abbe Fenelon was expressing a societal commonplace when he wrote in 1687 that "The good woman spins, confines herself to her household, holds her tongue, believes and obeys."[15] "Silence is not only woman's greatest wisdom," wrote Soren Kierkegaard, "but also her highest beauty."[16] Similarly, in 1560, Thomas Wilson asked, "What becometh a woman best and first of all?" His answer was "Silence. What second? Silence. What third? Silence. Yea, if a man should ask me till doomsday, I would still cry silence, silence, without which no woman hath any good gift."[17]

As these statements suggest, silence was the outward sign of submission to appropiate authority. One of the lessons of the story of Susanna is that prayer and public silence are the sources of a woman's salvation. When the cautions implicit in such tales failed, various forms of force were marshaled. "If a woman speaks disrespectfully to a man," says an Urakaginan edict, "that woman's mouth" will be "crushed with a fired brick."[18] By falling silent or speaking submissively, women purchased protection.

In seventeenth-century colonial America, the ducking stool held a place of honor near the courthouse alongside the pillory and the stock. After being bound to the stool, the "scold," "nag," "brabling (*sic*)," or "unquiet" woman was submerged in the nearest body of water, where she could choose between silence and drowning. When the stool was raised, the drenched, breathless woman was offered the chance to renounce her verbal past. If she repentently promised to control her speech, the dunkings would cease. Her submergence and submission invited silence from women who might be disposed to disrupt the social order with speech. An English ditty and the laughter it invited carried the warning of the ducking stool to the pubs and playgrounds:

> If noisy dames should once begin
> To drive the house with horrid din
> Away, you cry, you'll grace the stool
> We'll teach you how your tongue to rule
> No brawling wives, no furious wenches
> No fire so hot but water quenches.[19]

Less deadly but comparably humiliating was the public gagging of women prone to disruptive speech. In Boston, in the century before the signing of the Declaration of Independence, "scolds" were gagged and publicly exhibited before their own doors. The humiliation hearkened back to branking, a Middle-Ages practice aimed at silencing blasphemers and raucous women. Also known as the scold's bridle, the branks consisted of a metal bit and muzzle that were strapped into the victim's mouth. As a warning to others, the bridled person was either tied to a post in the square or paraded through the town.[20]

Long after ducking stools and gossip's bridles had become curiosities in museums, the silence they enforced and the warnings they imposed continued to haunt women. "If women, in two different sexist stereotypes, are supposed to be chatterboxes or to remain silent," Robin Morgan writes to her lover, "does that make me a collaborator with one stereotype and you a collaborator with the other? Or are your silence and my speech both facets of our feminist rebellion? Or is your silence male identification, as a survival mode—and is my loquacity male identification, as a means of controlling the expression between us?"[21]

Whether caused by a stroke, a nervous breakdown, or a calculated desire to effect revenge, in 1993 Japanese Empress Michiko collapsed and stopped speaking altogether after a series of magazine articles described her as "domineering, extravagant and thoughtless . . . [and] boss[ing] the staff around."[22] But here silence elicited guilt on the part of those who ostensibly triggered it. "The magazines that criticized the Empress have spent the last month apologizing, and then apologizing some more," wrote the *New York Times*. "The author of one of the articles, Tadasu Ouchi, wrote that he 'nearly suffered a heart attack' when he learned that the Empress had fallen ill. He went on to say that everything he wrote was true, but that he would retract it anyway."[23]

As the Empress found, where few descriptors characterize the maleness

of male speech, the vocabulary exists to condemn female speech with variety, color, and dispatch. Being "bossy," however, is mild by comparision to
the alternatives. Women are "scolds," "nags," "schrews," "fishwives," "harpies," "viragos," "bitches," "harridans," "magpies," and "termagants." Unlike men, censored women "hector," "bitch," "boss," "scold," "shriek," and
are "strident" and "shrill." By condemning the expressive woman, these
names enjoin her sisters to silence. Be quiet and submissive, or suffer public
shaming.

The effects persist. Females report more difficulty than males in expressing themselves in public and gaining either a hearing or respect for their
ideas.[24] Consistent with social sanctions against aggressive speech by
women, they are also less likely than men to speak on controversial
topics.[25] And a study by the American Association of University Women
found that boys speak more in class, are more likely to be called on by
teachers, and are less frequently reprimanded than girls for not raising their
hands to speak and being "impolite."[26]

Even holding a position of authority does not ensure that one's ideas will
be heard. Anita Gottlieb, now a senior vice president at Defenders of
Wildlife in Washington, D.C., recalls that, as staff director of a congressional subcommittee, a male associate dismissed ideas raised by her but
praised them when they came from another man. "I'd say something [at
meetings] and he would ignore me or say it was a bad idea," says Gottlieb.
"Five minutes later, one of my male colleagues would bring it up and he'd
say it was a great idea."[27]

Girls benefit from same-sex education in part because in that environment their voice is the norm. The majority of women Nobel laureates, for
example, were educated in all-women schools. The same sort of advantage
seems to be gained through participation in high school and collegiate
debate. Among the features that Ann Richards and Janet Reno have in
common is such a background. When condemning a woman, however, the
term debater can serve as synonym for dogmatic and defiant.

Anita Hill was condemned for being "a relentless debater—among other
things." J. C. Alvarez, who had served for four years as Clarence Thomas's
special assistant, said of her: "On Friday [when she testified], she played
the role of a meek, innocent, shy Baptist girl from the South who was the
victim of this big bad man. . . . The Anita Hill that I knew and worked
with was nothing like that. She was a very hard, tough woman. She was
opinionated, she was arrogant, she was a relentless debater."[28]

As noted earlier, the pseudonymous Constantia Mundi recognized in
1617 that condemnatory labels create no-choice-choices.[29] She was responding to the pamphlet that fell into the same family as "The arraignment
of . . . women." Among other things, it claims that "As a sharp bit curbs a
forward horse, even so a curst woman must be roughly used, but if women
could hold their tongues, then many times men would their hands."[30]

The pseudonymous Ester Sowerman pilloried this tract and the man
who held such views. "You challenge women for untamed and unbridled

tongues," she wrote in 1617, "[but] there was never a woman was ever noted for so shameless, so brutish, so beastly a scold as you prove yourself in this base and odious Pamphlet. You blaspheme God, you rail at his Creation, you abuse and slander his Creatures. And what immodest or impudent scurrility is it which you do not express in this lewd and lying Pamphlet."[31]

Even after women were granted access to higher education, their right to present their own ideas in public was denied. Lucy Stone was admitted to Oberlin in 1837. Although permitted to attend classes with men, she was not allowed to read her own papers in class or participate in debates. When she fulfilled the requirements to graduate, she learned that she could write her own commencement address but not deliver it. As a leader of the woman's movement, Stone later helped establish women's competence as speakers.[32]

The Young Ladies' Academy of Philadelphia did permit its class leaders to speak their own words. In 1793 the salutatorian, Miss Priscilla Mason, used the opportunity to indict the self-fulfilling prophesies that constrained women. "Our high and mightly (sic) Lords (thanks to their arbitrary constitutions) have denied us the means of knowledge, and then reproached us for the want of it. . . . But supposing now that we possess'd all the talents of the orator, in the highest perfection; where shall we find a theatre for the display of them? The Church, the Bar, and the Senate are shut against us. Who shut them? *Man*; despotic man, first made us incapable of the duty, and then forbid us the exercise."[33]

As they established the right of women to speak on the topics of their choice and in the forums of their preference, early women's rights activists demonstrated that they had the intellectual capacity once denied in those of their kind. Ironies resulted. By forcefully arguing against woman's suffrage, anti-suffragist women established their competence to vote. "Once she had become skilled at public speaking, the anti-suffragist woman speaker implanted a smaller paradox inside the movement's larger paradox: for she revealed through her actions the competence of women in public affairs which in her speeches she vigorously denied," writes an historian of the British movement. "The Duchess of Atholl's was a rather similar story: she spoke so well for anti-suffragism that her audiences could not understand why, whatever she might say, such a woman should not vote."[34]

In the last half of the nineteenth century, feminist leaders began to identify the traps society had set for them. In her important speech "The Solitude of Self" (1894), Elizabeth Cady Stanton analogized the situation of women to that of a character in a Shakespearean play. "Titus and Andronicus contains a terrible satire on woman's position in the nineteenth century," said Stanton. "'Rude men seized the king's daughter, cut out her tongue, cut off her hands, and then bade her go call for water and wash her hands.' What a picture of woman's position! Robbed of her natural rights, handicapped by law and custom at every turn, yet compelled to fight her own battles, and in the emergencies of life fall back on herself for protection."[35]

Similarly, Sarah Grimke observed that man "has done all he could to debase and enslave her mind; and now he looks triumphantly on the ruin he has wrought, and says, the being thus deeply injured is his inferior."[36]

To hold the speech of women in check, the clergy, the courts, and the keepers of the medical profession devised labels discrediting "womanly" speech. "Heretics!" said the clergy. "Hysterics!" pronounced the doctors. "Witches!" decreed the judges. "Whores!" said a general chorus. "Harpies!" exclaimed those husbanding their power over women's names and property. "Manly," said those locked in the womb/brain dichotomy. These names invited the silence that in earlier times had been ensured by force.

In wielding such labels, opponents of women's rights exercised one of language's most powerful properties: the capacity to name. Each label identifies its object as a deviant to be shunned. Each deprives the speaker of those audiences disposed to listen to a woman, but disinclined to hear a witch, a whore, a heretic, an hysteric, or a woman masquerading as a man.

One way a woman could earn these labels was by engaging in forbidden or disapproved speech acts. In each instance some institutionalized male role was threatened. Where the clergy spoke in the name of the church, the judges in the name of the law, and the doctors in the name of science, those labeled whores, witches, heretics, and hysterics uttered individualized personal speech. The belief that an institution must credential public speech was challenged by the implicit personalized assumptions underlying the discourse of these condemned women. If God could speak through a woman, then the legitimacy of a male priesthood was suspect. By seeming to prefer speaking through female witches, the devil provided his fellow males with a consoling counterpoint.

Each indicting name specifies a nonrational genesis for female speech, and hence underscores the assumption that male speech is based in the intellect and is the superior form. The heretic's speech was spun either from the devil or from a demented mind. The speech of the hysteric arose out of her ungoverned emotions. The whore purchased her public speech at the price of her sexual propriety. The witch voiced the devil's sentiments. None, in other words, were credited with using their own brains, a concession that would have undercut the womb/brain bind.

In each case the threatened institution offered the threatening woman the opportunity to repent her public speech, an act reestablishing the power of male discourse. Only when women refused was their verbal defiance punished with the permanent silence of ostracism or death. Accordingly, both the heretic and the witch were invited to recant, a metaphor suggesting that they reclaim their speech from its public sphere and seal it within themselves. The hysteric was asked to return to the womb of the bed and, after disavowing her past patterns of speaking, to be reborn at male hands into the male-dominated household, a reformed woman. The whore was counseled to abjure her evil speech and public presentation of self, and return to the plain dress and modest demeanor of a submissive, homebound woman.

Speaker as Whore

On May 15, 1862, Major General Benjamin Butler decided that he and his union troops had endured enough from the Confederate women of New Orleans. The women he identified as "she-adders, more venomous than he-adders,"[37] had, among other actions, been spitting at Union soldiers. The result was General Order Number 28 which read:

> As the officers and Soldiers of the United States have been subjected to re-peated insults from the women (calling themselves ladies) of New Orleans, in return for the most scrupulous non-interference and courtesy on our part, it is ordered that hereafter when any female shall by word, gesture, or movement, insult or show contempt for any officer or private of the United States she shall be regarded and held liable to be treated as a woman of the town plying her vocation.[38]

The order was, according to Butler, self-executing. It succeeded, he argued, because it created a no-win situation for the women of New Orleans. "No arrests were ever made under it or because of it," recalled Butler. "All the ladies in New Orleans forebore to insult our troops because they didn't want ot be deemed common women, and all the common women forebore to insult our troops because they wanted to be deemed ladies, and of these two classes were all the women secessionists of the city."[39]

As Butler believed, the women's self-restraint may have been a product of their distaste for the label "whore." Alternatively, they may have read General Order Number 28 as a license to rape any woman who insulted a Union soldier. "I do not believe," wrote Butler piously in his autobiography, "any man of ordinary sense, of clear judgment, ever did misunderstood it or misinterpret *(sic)* how the order intended that such women should be dealt with in any other way than being put in the hands of the police."[40]

Literalists that they were, the British ministers protested the order to the U.S. State Department. Responding with characteristic subtlety, Butler described His Majesty's protestors as "the women-beaters and wife-whippers of England."[41]

Butler's order translated a social sanction into a legal one. For centuries, women who defied a male authority by taking stands in public had been branded prostitutes, at worst; immodest, at best.

As women edged through the classroom to the public forum, they employed speech to dispute the tie between modesty and silence. In her 1793 valedictory at the Young Ladies' Academy of Philadelphia, Eliza Laskey claimed the "liberty to speak on any subject which is suitable" and distinguished modesty from "dull and rigid silence," warning that silence must not be mistaken for virtue.[42]

Her speech took issue with a long-lived alliance between loquacity and lasciviousness. "[B]y their wanton laughter, loquacity, insolence and scurrilous behavior," Castiglione had noted in his treatise on the Courtier, women who were "Not unchaste" appeared to "be so."[43] Similarly, at the climax of Ben Jonson's play *The Silent Woman*, Morose links lewdness and female speech when he exclaims "[T]his is the worst of all worst worsts that hell could have devised! Marry a whore, and so much noise!"

These attitudes were translated into behavior when Abby Kelly, a nineteenth-century feminist, found herself the subject of a sermon delivered in church. "The Jezabel is come among us also." intoned the preacher. Using Scripture for their purpose, the congregation and those sympathic to its sentiments jeered Kelly, pelted her with rotten eggs when she rose to speak, and threw stones at her.[44]

The institution threatened by the "harlot" was the family. Public speaking took women from the home, where, their opponents assumed, they should be bearing and raising children.

Speakers as Heresiarchs

Since the times of myth and magic, men have chronicled their fears of the seductive powers and sinful purposes of women's speech. A popular seventeenth-century rhyme joined female preaching and fiendish glee: "When women preach and cobblers pray/ The fiends in hell make holiday."[45] Just as the songs of the Sirens lured men to physical death, Eve enticed Adam into spiritual demise. The sexuality simmering just below the surface in those seductions took explicit form. As the serpent spoke through Eve, so Satan spoke through women accused of witchcraft—specifically, according to prosecutors, through witches' vaginas.

One of the "heresies" rooted out by the Inquisition gloried in the proclamation that not only was the Holy Spirit a woman, but that she had been incarnated in Guglielma of Milan. In the thirteenth century, Guglielma proclaimed a new church headed by a female pope and female cardinals. By claiming to speak for and as God, she posed a profound threat to the papacy and clergy. Condemned post-mortem by the Inquisition, her remains were exhumed and burned.[46]

So concerned was Pope Boniface VIII about the rise of female prophetesses and heresiarchs that he issued the bull *Nuper ad audientiam* (1296) condemning women who proclaimed new dogma and revelled in immorality. The Guglielmites were not alone in claiming privileged speech and actions for women. Female Lollards also threatened Medieval Christendom. These women taught followers in their homes, reinterpreted scripture, and affronted the clergy with claims to being as learned as they. They too were burned.

As part of the case against these heretics, sexual excess was routinely alleged. When the heretic was female, these allegations discredited the speech acts. In fact, many heresiarchs were celibate. Still, the threat they posed to the male clergy was direct. By preaching and teaching, these women violated societal norms and usurped clerical functions.

By the seventeenth century, however, a few cracks were starting to appear in the patriarchal monolith. Religions that espoused spiritual equality permitted women to preach. While secular authorities were ducking and gagging outspoken women in the seventeenth century, female Quaker preachers carried their religious convictions not only to London and Dublin but to the New World as well.

Meanwhile, a number of women outside the Quaker community found prophesying a powerful means of self-expression and got away with it. Not only were the prophesies of Eleanor Davies, Mary Cary, and Anna Trapuel published, but they also delivered "oracular speeches" in such prominent places as Parliament.[47] Perhaps fearful that their own futures would be foretold and in the telling ordained, churchmen shied away from condemning the speech of those whose predictions came true. If it was God, not mammon, who spoke through a woman's mouth, which man would dare silence her?

But female preachers did face a double standard. Nineteenth-century African-American evangelist Julia Foote observed that "We are sometimes told that if a woman pretends to a Divine call . . . she will be believed when she shows credentials from heaven: that is, when she works a miracle."[48] Turning the tables, she responded, "If it be necessary to prove one's right to preach the Gospel, I ask of my brethren to show me their credentials, or I cannot believe in the propriety of their ministry."

Elizabeth Cady Stanton made a similar move when a clergyman criticized her book *The Women's Bible,* an early work of feminist biblical criticism. He proclaimed it "the work of women, and the devil." "This is a grave mistake," Stanton responded. "His Satanic Majesty was not invited to join the Revising Committee, which consists of women alone. Moreover, he has been so busy of late years attending Synods, General Assemblies and Conferences, to prevent the recognition of women delegates, that he has had no time to study the languages and 'higher criticism.'"[49]

As society became more civilized, ostracizing and institutionalizing replaced incinerating as a preferred response to the woman who claimed theologically suspect religious privileges. When Anne Hutchinson asserted leadership of the congregation of the Boston church in colonial America, she was banished. In 1860, a clergymen demonstrated his control over his "heretical" wife by having her committed to the Jacksonville State Hospital for the Insane in Illinois.

The dementia of Mrs. E. P. W. Packard was manifest in her disagreement with her clergyman husband on items of faith and religious observance. Among other things, Packard questioned the notion of original sin, argued that God approved of all religions, and contended that women had a right to defend their own beliefs.[50]

During her three-year stay in the Jacksonville State Hospital, Mrs. Packard kept a secret diary. In it, she observed that her ward was filled with married women. The superintendent "invariably kept these wives until they begged to be sent home. This led me to suspect that there was a secret understanding between the husband and the doctor; that the *subjection* of the wife was the cure the husband was seeking to effect under the specious plea of insanity."[51]

"Had I lived in the sixteenth instead of the nineteenth century," wrote Mrs. Packard, "my husband would have used the laws of the day to punish me as a heretic for this departure from the established creed—while under the influence of some intolerant spirit he now uses this autocratic institution as a means of torture to bring about the same result—namely a *recantation of my faith.*"[52]

After three years, while making plans to move the family to Massachusetts and her to an asylum there, the Rev. Mr. Packard removed her from the asylum and locked her in their home. Because such "house arrest" was illegal, her neighbors were able to obtain a writ of habeas corpus. The judge before whom she was brought concluded that she was sane.

On her release, the *Chicago Post* editorialized, "Mrs. Packard became liberal in her views, in fact avowed Universalist sentiments, and as her husband was unable to answer her arguments, he thought he could silence her tongue by calling her insane."[53]

Now free, Packard lobbied the Illinois state legislature for a law protecting anyone, man or woman, from involuntary commitment without a jury trial. The statutes she sought were passed in 1865 and 1867.[54]

Speakers as Witches

In Essex County, Massachusetts, more "witches" were convicted of "assaultive speech" than any other crime including "lying."[55] Encompassed in such assaults were "slander," "defamation," "filthy speeches," and "scandalous speeches." An inability to control one's tongue was a sign of witchcraft. Needless to say, the large majority of those accused and convicted of witchcraft were women. What the "witches" were saying was as menancing to the status quo as the fact that they were saying it. "You look at the things witches were accused of," notes medical historian John M. Riddle. "Most of them have to do with fertility. They're accused of causing sterility, babies born dead, causing impotence, miscarriages." Among the statements repeated during the Inquisition was "The devil works through herbs." The witches, in other words, may well have been midwives disseminating information about herbal means of contraception and abortion. Here is a link between the womb/brain and speech/shame binds.[56]

Because women presumably were not governed by reason, they were susceptible to the seditious advances of the devil. As men use language for their purposes, the devil uses the weak woman. When the devil did not speak through the woman's vagina, an act that explicitily tied female speech and sexuality, he spoke in his own voice through her mouth. The Rev. John Whiting reports that one victim "said she knew nothing of those things that were spoken by her, but that her tongue was improved to express what was never in her mind, which was a matter of great affliction to her."[57]

Consistent with social stereotypes and societal needs, the abrasive, contentious woman was more likely than her soft-spoken sister to be labeled a witch. The woman who had few children—or none—had another strike against her.

Not only did her speech threaten the established order but she further challenged the institutional structure by appropriating some of its skills. Where the heretic threatened the clergy by claiming knowledge of scripture and access to God, the witch was likely to have practiced "doctoring."[58]

Witches and heretics, male and female alike, died in ways that symbolized the extinction of their speech. At the stake, fire (a metaphor for speech) consumed the witch and her capacity to speak. Alternatively, fiery words were drenched permanently by drowning. Hanging simultaneously choked the ability to speak and the speaker.

Speakers as Hysterics

If speech did not assign a woman to the streets, it could consign her to the sickbed or the insane asylum; expressive, emotional female speech was seen as a symptom of a diseased uterus or a distressed mind.

Meddling in the male world was sometimes more than the mind of woman could bear. So, for example, Puritan leader Jonathon Winthrop recorded the case of a woman who had lost "her understanding and reason." She had given herself "wholly to reading and writing and had written many books." Her wits might have been spared had she "attended her household affairs and such things as belong to women and not gone out of her way and calling to meddle in such things as are proper for men whose minds are stronger."[59]

Hysteria, a disease peculiar to women, was first identified by Hippocrates, who drew its name from the Greek word for uterus. Because a woman was dominated by her reproductive role, any signs of mental distress could readily be attributed to a malfunctioning womb.

The symptoms of an hysteric were partially speech-based: She paid too little attention to detail, expressed too much emotion, and was flamboyant. These displays of feeling supposedly weakened physical endurance and endangered a woman's ability to bear strong children. To combat hysteria and other kindred nervous disorders, the University of Pennsylvania's Dr. S. Wier Mitchell pioneered the notorious "Rest Cure" in the nineteenth century. The cure, which consisted of mandated rest and reprogramming, required that patients place their bodies and minds in Mitchell's care. During their convalescence, he instructed them in ways of controling their urges to express their feelings to others.[60]

Charlotte Perkins Gilman, a nineteenth-century feminist, was among

Mitchell's patients. In her novel *The Yellow Wallpaper*, she indicts the psychiatrist for collusion with the protagonist's husband in the destructive "cure." The room to which Gilman was confined for Mitchell's treatment was the nursery, a location that symbolized the cure's attempt to reduce the patient to childlike submissiveness.[61] If she were to regain her health, Mitchell told his patient, she should dedicate herself to her home and child and "never touch pen, brush or pencil as long as you live."[62]

Mitchell's ministerings were mainstream. In 1853, in a book titled *On the Pathology and Treatment of Hysteria,*[63] Robert Carter advised that "If a[n hysterical] patient . . . interrupts the speaker, she must be told to keep silent and to listen; and must be told, moreover, not only in a voice that betrays no impatience and no anger, but in such a manner as to convey the speaker's full conviction that the command will be immediately obeyed." The "rest cure" was, in short, a form of sensory deprivation interrupted by aggressive attempts at behavior modification.

At one level, however, hysteria was a way of escaping the womb/brain double bind. The rest cure specified that the patient use neither brain nor uterus. Complete physical and mental rest, and with them a suspension of both the "duties" of wife and mother and of "intellect," were the prescribed regimen.

"No longer did she devote herself to the needs of others, acting as self-sacrificing wife, mother, or daughter," writes historian Carroll Smith-Rosenberg. "Through her hysteria she could and in fact did force others to assume those functions . . . Through her illness, the bedridden woman came to dominate the family to an extent that would have been considered inappropriate—indeed, shrewish—in a healthy woman . . . Consciously or unconsciously, they had thus opted out of their traditional role."[64]

The respite was temporary. Until and occasionally after the practice was outlawed by statute, women in the United States whose speech defied such cures risked being treated like Mrs. Packard, labeled crazy, and institutionalized. So too in Britain, "Victorian madwomen were not easily silenced," writes Elaine Showalter, "and one often has the impression that their talkativeness, violation of conventions of feminine speech, and insistence on self-expression was the kind of behavior that had led to their being labeled 'mad' to begin with . . . Mortimer Granville was concerned that female lunatics were always 'chattering about their grievances' or else involved in 'an excess of vehement declaration and quarreling.' He recommended that the women be set to work that would keep them too busy to talk."[65]

Early feminists recognized the purposes served by institutionalizing such women. "Could the dark secrets of insane asylums be brought to light," writes Elizabeth Cady Stanton in *Eighty Years and More* in 1898, "we should be shocked to know the great number of rebellious wives, sisters, and daughters who are thus sacrificed to false customs and barbarous laws made by men for women."[66]

The disproportionate use of lobotomy on women translates the same indictment into another form. A number of scholars have concluded that psychosurgery was not administered as much to cure an illness as to minimize behavior considered inappropriate. In a woman, "inappropriate behavior" included aggressive speech and action and overt sexuality.[67]

Social stereotypes aided Dr. Mitchell in his role as tongue-depressor. Aristotle sanctified the belief that women lack full deliberative capacity (*bouleutikon*).[68] Accordingly, women were "more void of shame and self-respect, more false of speech [and] more deceptive than men."[69] That view persisted. In his sixteenth-century *Arte of Rhetorique* (1560), Thomas Wilson illustrates a figure of speech he was describing by citing "a woman babbling, inconstant, and ready to believe all that is told her."[70]

The Manly and the Effeminate Speaker

When hormones did not swamp the mind, the mind sapped the uterus of needed energy. Since speech draws on intellect and cognition, prohibiting speech protected a woman's reproductive health. Underlying that assumption was a corollary, the idea that expending sexual energy robs both sexes of eloquence—the male drained by coitus, and female by conception, gestation, and birth. So for Roman men whose voices were shrill and "womanly," Quintillian recommended sexual abstinence.[71] For women, the alternatives cut closer to the bone: choose barren wombs or barren brains. A seventeenth-century eulogist for the childless but intellectual Duchess of Newcastle praised her as one of the exceptions to her frail sex "who have Fruitful Wombs but Barren Brains."[72] Maesia Sentia, a first-century Roman orator, was labeled an androgyne by one of her leading admirers, Valerius Maximus.[73] (Perhaps opponents of women's rights should have invited feminists to speak. Had public speaking proved the powerful contraceptive they supposed, feminist orators as a species would have gone extinct, victims of self-induced sterility.)

For millennia, theorists asserted or implied that nature had privileged the speech of men. Effectiveness and manliness were synonyms. As Margaret Fuller noted in 1843, a friend who used the phrase "a manly woman" meant it as a high compliment.[74] "Manly" speech was deified, "effeminate" or "womanly" speech devalued.

Once again the distinction between the manly and the effeminate was rooted in the conviction that men and their speech were inherently superior because discourse was governed by their minds. The speech of women was defective because it originated in their emotions. The widespread notion that the man is the head of the family, and the woman the heart, translated to the notion that man had the right and obligation to control female speech.

Driven by emotion, womanly speech was thought to be personal, excessive, disorganized, and unduly ornamental. Driven by reason, the manly style was regarded as factual, analytic, organized, and impersonal. Where womanly speech sowed disorder, manly speech planted order. Womanly speech corrupted an audience by inviting it to judge the case on spurious grounds; manly speech invited judicious judgment.

The view that public virtues are the by-products of the manly was underscored by Elizabeth I, who proclaimed that were she "turned out of realm" in her petticoat, she would be able to live anywhere in Christendom. She told the troops braced for the assault by the Spanish Armada that "I have but the body of a weak and feeble woman; but I have the heart of a King, and of a king of England, too."[75] Since the heart, not the head, was presumed to control female speech, Elizabeth's kingly heart certified that her claims were trustworthy, not effeminate. Only a sovereign uttering manly speech could claim credibly that should Britain be invaded: "I myself will take up arms—I myself will be your general, judge and rewarder of every one of your virtues in the field."

Although some men delivered "effeminate" discourse and some women mastered "manly" speech, theorists held that speakers generally stayed true to their gender. Those women willing to forego reproduction for the conception and delivery of ideas were presumed to aspire to be men. If a wife "still wants to appear educated and eloquent," noted Juvenal, "let her dress as a man, sacrifice to men's gods, and bathe in the men's baths."[76]

Under rare circumstances, a woman's "manly" speech won praise. But even then, the credit for such an accomplishment was given to the men who had taught them, which was the case with Laela and Hortensia, the daughters of famous Roman orators. Hortensia delivered a widely acclaimed speech in 42 B.C. in which she argued: "You have already deprived us [women] of our fathers, our sons, our husbands, and our brothers on the pretext that they wronged you, but if, in addition, you take away our property, you will reduce us to a condition unsuitable to our birth, our way of life, and our female nature."[77] Like Elizabeth, whose "heart" was that of a king, the eloquence of these famous daughters was attributed to the bloodline that enabled them to transcend their gender. And it should be noted that by speaking as a stand-in for an absent male, and doing so to "preserve her female nature," Hortensia clinched her chances for a favorable reception.

The state was civilized by manly speech, corrupted by effeminate speech. Not only could speaking cost a woman her ability to bear children, effeminate speech could cost the body politic its capacity to bear arms. Apparently believing that female speech would drain the nation of its testosterone, opponents of women's rights claimed that "the transfer of power from the military to the unmilitary sex involves national emasculation."[78] To save the country from the dustbins of history, women would have to return to their dustpans and aprons.

The distinctions between manly and effeminate speech is long-lived. The Romans distinguished between a manly style, which was revered, and the effeminate style, which was reviled. However, what constituted manly and effeminate speech has varied from decade to decade and century to century.

Some in Cicero's time thought his loose structure, high level of repetition, and general "timidity" effeminate, for example.[79]

By the late nineteenth century the scientific style was enshrined as manly. "The eloquence of Mr. Adams resembled his general character, and formed, indeed, a part of it," noted Daniel Webster. "It was bold, manly, and energetic."[80] Nineteenth-century textbooks lauded the "plain, manly, oratorical style."[81]

The notion erupts in unexpected ways today. In 1986, Judge Joseph J. Chernauskas ruled that Catherine Pollard, a 68 year old from Connecticut, could not be a female Boy Scout leader. As a role model, said the judge, the adult male "speaks with a male body motion . . . a strictly male bellowing of outrage, disgust, commendation, warning, encouragement, dissatisfaction—all a matter of speech and communication in its male role manifestations."[82]

By wrapping a condemnation of effeminate speech in the language in which eloquence was defined, theorists ensured that if a woman rose to speak she would embrace "manly" norms of discourse, thereby in some important sense counterfeiting her identity. From the late 1800s to the present, the women's movement debunked pejorative uses of "womanly." The movement's attacks drove the labels effeminate and manly from the vocabularies of theorists. But before the notion of a manly style lost currency, female speakers had absorbed its norms.[83]

The Bind Becomes a Blessing

By embracing such "manly" norms, women created an ironic situation for their descendants. The age of television, not even envisioned as the Grimke sisters took to the circuit in the nineteenth century, would invite the style once spurned as "womanly." In the television age, men would have to learn and women recapture the "womanly" style.

Since in the course of normal household events, speech was required from even the most docile woman, those who wielded the labels whore, heretic, hysteric, and witch defined a mode of appropriate female discourse. The silence/shame bind did not prohibit forms of communication available to women within the private sphere. In that environment, women cultivated the socially accepted forms of communication employed to nurture the young and preserve the family. Conciliation was the mother's role, storytelling her province, conversation with those invited to the home an acceptable intellectual outlet, letter-writing usually socially approved, occasionally mandated.

Whether the product of nature or nurture, choice or censorship, a distinct "feminine style" consistent with traditional notions of femininity emerged. "Structurally, 'feminine' rhetoric is inductive, even circuitous, moving from example to example, and is usually grounded in personal experience," noted rhetoric scholar Karlyn Kohrs Campbell. "Consistent with their allegedly poetic and emotional natures, women tend to adopt associative, dramatic, and narrative modes of development, as opposed to deductive forms of organization. The tone tends to be personal and somewhat tentative, rather than objective or authoritative."[84]

This form of personal, intimate style was further fostered by radio and then by television. As I argue in *Eloquence in an Electronic Age,* the intimate medium of television favors a conciliatory style over a combative one. Here the style once demeaned as effeminate is desirable.

Because girls identify positively with their mothers and then recreate the mothering role in their lives, some psychologists believe that their capacities for nurturance and empathy are more developed than boys'. Accordingly, they argue, it is natural for a woman to define herself through her social relationships.[85] By contrast, male exposure to the military and to competitive sports may engender such traditional male values as aggressiveness.[86] This demonstrates itself in language. Whether shaped by nature, culture, or some combination, men traditionally have been more comfortable than women in a combative "debate" style. Lecturing, arguing, pivoting on claims from reason or logic, and demanding or providing evidence are more typically male than female behaviors.[87] Consistent with these findings, women use less hostile verbs than men.[88] Men are more likely than women to engage in verbal dueling.[89] The messages of women are less verbally aggressive and tend also to be more pro-social—particularly in their stress on relationships rather than autonomous action.[90]

"When I came to be a law teacher," notes Carrie Menkel-Meadow, "I had to begin to explain this world with which I had some difficulties to other people. I had increasing difficulty describing the structure of what I saw in a way in which I felt comfortable. I increasingly wanted to bring people into it as I started to talk about what I did not like about it. So as a teacher, I was struck by the fact that I was speaking in two voices: the voice that I *had* to speak in, in order to be a teacher in a predominantly male institution, and the voice which attempted to tell my students what I really thought."[91]

Emotionalism Becomes Self-Disclosure

The intimate medium of television requires that those who speak comfortably through it project a sense of private self, unself-consciously self-disclose, and engage the audience in completing messages that exist as mere dots and lines on television's screen. The traditional male style is, in McLuhan's terms, too hot for the cool medium of television. Indeed, some theorists believe that where men see language as an instrument to accomplish goals, women regard it as a means of expressing internal states. In conversation, men are more prone to focus on facts and information, women on feelings. In group settings, men are more likely to concentrate on accomplishing the task, women on maintaining the harmony and well-being of the group.[92]

Once condemned as a liability, the ability to comfortably express feelings is an asset on television. Women are more likely than men to verbally indicate emotion.[93] This does not mean, however, that women respond to events more emotionally. Instead, it seems that women are more disposed to display their reactions in emotional terms.[94] Whether because of the ways they have been rewarded or how they are wired or both, women are advantaged in conveying emotion by their more expressive faces and body movements and their general skill in deciphering the nonverbal clues of others.[95] Consistent with their harmonizing tendencies, women look more at the person they are conversing with than men do.[96] Overall, females are more empathic than males;[97] they tend to both give and receive more emotional support than men.[98]

The inability to disclose some sense of private self in an intimate mass medium has proven a barrier for many men in politics today. (Ronald Reagan is a noteworthy exception to the rule, and Bill Clinton an enterprising aspirant.) In this venue, the "manly" style is a noose. Traditionally, scholars have assumed that the speech of men focuses on action and conquest of their environment where that of women is concerned with relationships and creation of cooperation and harmony. The self-disclosure of men and women is consistent with these differences.[99] Where men tend to disclose about *goals* related to instrumental needs,[100] women—serving expressive or affiliative needs—reveal *themselves*.

The impersonal nature of male speech is evident in male disposition toward using numbers to describe.[101] By contrast, for women "[t]he implied relationship between the self and what one reads and writes"—and, presumably, says—"is personal and intense." Consequently, some scholars argue that women's novels and critical essays are more often autobiographical. Indeed, some go so far as to claim that "[b]ecause of the continual crossing of self and other, women's writing may blur the public and private."[102]

Because the mass media are fixated on differences between the private and public self of public figures, a comfort with expressing instead of camouflaging self—or at the minimum an ability to feign disclosure—is useful for a politician. The utility benefits females. And because the broadcast media invite an intimate style, their conversational and narrational skills also advantage women.

Generations of women honed these skills within the circumscribed boundaries of the private sphere. Refining an intimate, conversational, narrative style while her husband, father, and brothers perfected the impersonal speech to the public audience, the "distaff side" engaged in intimate communication with friends, sisters, children.

Denied access to public forums, women developed facility in such private forms of communication as conversation, storytelling, and letter-writing. In many families, for example, the woman is the family's correspondent. By the seventeenth century, letter-writing had become a pastime of women of leisure. Consequently, "[m]any of the English women prose writers of the seventeenth century are known for their correspondence.

The first English woman whose letters were preserved and eventually published was Lady Brilliana Harley (1600–43)."[103] In a world in which the impersonal was publicly prized and identified as the norm in male communication, intimate self-disclosure and its expressive forms were devalued and therefore exercised by women without fear of censure. It is unsurprising then that "[t]he earliest extant autobiography in English was written by a woman, Margaret Kempe (c. 1373). Since she was illiterate, this work was dictated."[104]

Securing a space in which women could demonstrate their intellectual abilities—whether on the printed page, in the lecture hall, or in the classroom—was, however, a prerequisite of making the argument for equality. Only when such space was available could women undercut the self-fulfilling prophesy that condemned them for an inability to do what they were denied the wherewithal to accomplish.

The salons of seventeenth-century France and eighteenth-century Britain were a move in that direction. They "created spaces for the exchange of ideas" as well as an occasion to discuss "women's nature and their role in society. . . . In these spaces, where intellectual men and women treated each other as equals, the inclination was to answer the questions raised by the centuries-old dispute in favor of the equality of women, at least on the abstract, theoretical level."[105]

Women perfected the art of conversation in the salons of France and at literary breakfasts and evening conversational parties in Britain. Noting the skills of the eighteenth-century hostess Madame du Deffand, the English wit Horace Walpole wrote, "I have heard her dispute with all sorts of people, upon all sorts of subjects, and never knew her to be in the wrong. She humbles the learned, sets right their disciples, and finds conversation for everybody."[106] The influence of the salons is evident in Molière's caricature of *Femmes Savantes*.

Storytelling became another rich arena for women. The cliché "old-wives' tales" reminds us that, traditionally, women are a family's storytellers. So, for example, Goethe credits his father with engendering in him "the seriousness in life's pursuits," but praises his mother for transmitting "the enjoyment of life, and love of spinning fantasies."[107]

It was a short move from such intimate forms as "gossip," "storytelling," "conversation," and "letter-writing" to the sentimental novel, a literary form whose common style and focus on such daily occurances as love meant that it could be authored by those denied access to a classical education. It had a precursor in the poems of twelfth-century female troubadours, which employed the "straightforward speech of conversation."[108] Since educated men spurned the oral form of the ballad for the prestige of learned written poetry, women became the custodians of the ballad tradition. "[B]allads are old-wives' tales which were able to develop and change in authentically feminine ways mainly because men left them alone."[109]

Similarly, because the sentimental novel was considered "low" art, it was also a form presumed unworthy of serious men and appropriate for women. And its focus on family matters placed it within women's traditional purview.

When this narrative skill was seen as a liability, its mistresses were condemned as gossips. Sixteenth-century critic Steven Guazzo noted that although gossip was a vice common to many, "it is most familiar with certain women."[110] Such gossips retell the misfortunes of their neighbors in "speeches." "Have you not hearde the hard hap of my unfortunate neighbor," they ask. "[A]nd thereupon making the storie, they rehearse howe the husbande by means of his servant, took her tardie in her hastie business. Then they tell (the details of) the wall, and the way whereby her lover got downe: next, how cruelly her husband beate her, and her maid, and thinke not that they leave anything behind untolde, but rather will put too somewhat of their own devise."[111]

The indictment of the female gossip survived the centuries. "The second kind of female orator," writes Addison in the *Spectator,* "are those who deal in invectives, and who are commonly known by the name of the censorious. . . . With what a fluency of invention and copiousness of expression will they enlarge upon every little slip in the behavior of another! With how many different circumstances, and with what variety of phrases, will they tell over the same story!"[112]

Contemporary psychologists and anthropologists now recognize that most of the storytelling categorized under the pejorative "gossip" is in fact an essential glue in building a community and holding it together. The talent for capturing ideas and life-lessons in brief dramatic narratives is one that has been cultivated for generations by women—whether they are conveying neighborhood news, telling their children bedtime stories, or otherwise transmitting the manners and mores of the community to others. Women continue to exercise these skills. In both primitive and advanced cultures, women are the repositories of parable-like dramatic vignettes, concise stories that transmit the common wisdom from woman to woman and generation to generation.[113] Here again, television rewards these traditional feminine skills. Just as its bias favors self-disclosure and other aspects of women's conversational style, so it prizes narrative skills that once were pilloried.

The natural compatibility between the "womanly" style and television is not the only factor propelling candidates toward that style. Two bodies of evidence invite the "womanly" style as the natural marriage of political substance and expression. Both favor female candidates over males.

The gender gap revealed that female voters differ from their husbands, brothers, and fathers on humanitarian issues and matters of war and peace.[114] Moreover, as I note elsewhere, female candidates are more credible than males on those "human" issues that tie intuitively to a maternal role. These include nutrition for infants, food stamps, aid to the elderly, Social Security, and initiatives that would prevent offspring from dying in war.[115]

Style and substance coalesced. Not only were women more inclined to personal speech, but they were more inclined to favor issues that lent themselves to such speech. Not only were women ready to oppose military intervention, but they also were ill disposed to hostile verbs, agressive verbal behavior, and clear refutative postures. Not only did women favor a nurturant, incorporative style, but they also supported programs that nurtured.

At the same time, women's sense of political efficacy increased and with it their disposition to participate in politics. Although *The American Voter* found a gap between men and women in 1960, by 1976 women and men of the same age were reporting similar attitudes about their political power.[116]

Female candidates for local and statewide office responded to this convergence by returning to a more personal style. Bella Abzug gave way to Barbara Jordan; Barbara Jordan to Geraldine Ferraro, Nancy Landon Kassebaum to Carol Moseley Braun, Patty Murray, Dianne Feinstein, Barbara Boxer, Kay Bailey Hutchison, Ann Richards, and Christine Todd Whitman.

Since society approved their use of the "manly" style in public but the "womanly" style in private, many women entered the televised age proficient in both. Increasingly, female candidates felt comfortable blending the strengths of each style. Barbara Mikulski, elected to the Senate from Maryland in 1986, "fights" for humanitarian causes and comfortably combines data-giving and dramatizing.

And, as Ruth Bader Ginsburg did at her confirmation hearings, women in power are using their voice in the cause of women. "Indeed, in my lifetime, I expect to see three, four, perhaps even more women on the high court bench, women not shaped from the same mold, but of different complexions," Ginsburg told the Judiciary Committee. Nodding their agreement were Senators Carol Moseley Braun and Dianne Feinstein.

5

Double Bind Number Three: Sameness/Difference

IF everyone everywhere woke up tomorrow assuming that men and women ought to be treated in the same way in the same situations, a lot would change. Teachers would call on girls as often as boys. Women would earn the same salaries as men with comparable degrees. Ken would have joined Barbie in claiming that "math is tough"—or neither would have uttered the phrase.

Women would be licensed to leer at the anatomy of men or no longer be subject to similar scrutiny. A First Gentleman would be as likely to be found at the side of the president as a First Lady. And 60-year-old women would be cast in the role of lovers in Oscar-winning films as often as 60-year-old men. Fathers would be found in the laundry as frequently as mothers. Fathers and mothers would not assume that sons are born with a carburetor gene and daughters with a chromosomal disposition to bake cakes. And CBS News president Richard Salant would have expressed reservations about hiring an "intelligent" and "well-read" male reporter because he was too handsome just as he harbored reservations about employing Diane Sawyer because she was "too beautiful."[1]

There are, of course, physiological differences between men and women. And, as the continued presence of both on the planet attests, there are also areas of complementarity. But whether innate psychological differences exist is less certain.

In any event, asking whether men and women are the same or different obscures the fact that both sameness and difference are in the eye of the beholder. At the level of DNA strands, each of us is unique. At a more universal level, we are each human and as such fundamentally like other humans, whatever their gender.

Preoccupation with questions about essence obscures a more important question: Should men and women be treated as if they are the same? And, if so, when? The distinction is important because differential treatment can produce differences. The girl who is assumed to be poor in math, for example, is not given opportunities either to prove her ability or develop it.

Nor is the assumption of *sameness* penalty-free. That assumption was operative in a five-year study sponsored by the National Institutes of Health, which found that taking an aspirin every other day cut the incidence of heart attacks among a group of 22,071 people. All 22,071 were men—even though heart disease is also the number one killer among women. Specialists in women's health told a House of Representative's committee on June 18, 1990, that men and women are not biologically similar enough to justify the conclusion that what works for men will work for women.

In pragmatic fashion, women's rights advocates tailored their arguments to their audiences, featuring *difference* to win access to education and the ballot, and *similarity* in their efforts to secure equal opportunity in employment. But the devil could cite similarity and difference for his purpose as well. Similar treatment of husband and wife did not necessarily produce equal economic outcomes in divorce. And the "special protections" women obtained in the name of *difference* limited access to well-paid jobs. As Myra Bradwell found in 1873, the argument from difference could also be used to keep a woman in her assigned place, in her case in the home and out of the courts.

In the nineteenth century, practicing law required approval from the court. After studying with her attorney husband and passing the Illinois bar exam with high honors, in 1869 Myra Bradwell took the next step and petitioned for a license. Assuming a fundamental difference between men and women who had passed the bar exam, the Illinois Supreme Court said no.

Although Illinois lifted the barriers to women practicing law in 1872, Bradwell sought to bar such exclusions throughout the United States by taking her case to the Supreme Court. Its ruling was handed down in 1873. One Justice's concurring opinion in the ruling has become legendary.

"Man is, or should be, woman's protector and defender," wrote Justice Bradley in rejecting Bradwell's appeal. "The natural and proper timidity and delicacy which belongs to the female sex evidently unfits it for many of the occupations of civil life. The constitution of the family organization which is founded in the *divine ordinances* as well as in the nature of things indicates the domestic sphere as that which properly belongs to the domain and functions of womanhood. The harmony, not to say identity, of interests and views which belong, or should belong, to the family institution is repugnant to the idea of a woman adopting a distinct and independent career from that of her husband. . . . The paramount destiny and mission of woman are to fulfill the noble and benign offices of wife and mother. This is the law of the Creator. And the rules of civil society must be adapted to the general constitution of things and cannot be based upon exceptional cases."[2]

During her lifetime, Bradwell would see sisters-in-law practice before Justice Bradley's Supreme Court. Here the pioneer was Mrs. Belva Lockwood. Informed that she would not be permitted to practice before the High Court, she reacted without the "timidity and delicacy" Justice Bradley had divined in women. Instead she took her case to Congress. If the Creator cringed at this violation of "the general constitution of things," She didn't make her feelings apparent in the congressional cloakroom. Backed by Myra Bradwell's newspaper, the *Chicago Legal News,* Lockwood succeeded in pushing legislation that opened the U.S. Supreme Court to women. It became law in 1879.[3]

Yet the echoes of *Bradwell v. Illinois* persisted. In 1966, Mississippi's Supreme Court justified exclusion of women from juries. "The legislature has the right to exclude women so they may continue their service as mothers, wives, and homemakers, and also to protect them (in some areas they are still on a pedestal) from the filth, obscenity and noxious atmosphere that so often pervades a courtroom during a jury trial."[4] In that opinion, however, was the whisper of impending change. Women in "some areas"—presumably including Mississippi—"are *still* on a pedestal" said the opinion, which would be set aside by the U.S. Supreme Court. Where "pedestal" was the preferred location for proponents of protections based on difference, "level playing field" was the arena of choice for the advocates of equal treatment. One isolates, the other engages. For the pedestal polishers, separate and superior was a desirable vantage point.

In the years between Nixon's second term and Bush's first, society's view of a woman's place began to shift in favor of the playing field. In general election years from 1972 through 1988, the National Election Studies included a question that asked about women's roles. "Recently," said the question, "there has been a lot of talk about women's rights. Some people feel that women should have an equal role with men in running business, industry, and politics. Others feel that women's place is in the home. . . . Where would you place yourself on this [seven point] scale, or haven't you thought much about it?" In 1972, just under 50% of the sample accepted the strong egalitarian position on equal roles; in 1988, it approached 70%.[5]

Woman as Other

The prime problem posed for women by equality/difference resides in the question, different from or equal to whom? Poignantly raised during the abolition movement, "Ar'n't I a Woman?" survives as a famous refrain of the women's movement. It does so by confronting an issue women's rights advocates continue to struggle with: By what standard is an individual assessed? This chapter will argue that if the answer is "by one's similarity or difference" to some other supposedly "normal" group, the person in question has been handed a rigged coin. No matter how many times it is tossed, the results are the same. "Heads I win," for the person in power; "tails you lose," for the person who lacks it.

The woman who in slavery was known as Isabella probably did not actually ask, "Ar'n't I a woman?" in her famous 1851 speech at the Women's Rights Convention in Akron, Ohio.[6] The language is more likely that of historian and poet Frances D. Gage who presided at the convention and recalled the speech from memory twelve years after Sojourner Truth delivered it. As important for my purposes is that Gage assumed she could put words in Truth's mouth, a presumption that invites the question raised in the last chapter: In whose voice and language is history conveyed? The synoptic statement that we identify Truth with today was probably written by an educated, white middle-class woman.

Nor do we know with certainty that Truth delivered the lines that Gage records as "Nobody eber helps me into carriages, or ober mud-puddles, or give me any best place."[7] Accounts written at the time do agree that Truth did say words to the effect that "I have as much muscle as any man, and can do as much work as any man. I have plowed and reaped and husked and chopped and mowed, and can any man do more than that? I have heard much about the sexes being equal; I can carry as much as any man, and can eat as much too, if I can get it. I am as strong as any man that is now."[8]

Whether Truth spoke a version of it or not, the question "Ar'n't I a woman?" is particularly apt because Truth received neither the privileges of white women in carriages nor the pay of the men, whether black or white, with whom she labored. In both settings she was "other." As Nobel Prize-winning novelist Alice Walker wrote of one of her characters, "She had nothing to fall back on; not maleness, not whiteness, not ladyhood, not anything. And out of the profound desolation of her reality she may well have invented herself."[9]

"Ar'n't I a woman?" prefigures the contemporary argument of feminists of color who criticize their "pale" sisters for assuming that all women identify with the concerns of those who are white, middle class, and heterosexual.[10] In its argument for Truth's capacities as an individual rather than a member of a group, the question forecasts the notion that no woman (or man for that matter) is essentially and invariably the same as any other or as women or men as a group.

For much of history, most of those who wrote and spoke premised their arguments on a quite different set of beliefs. Just after the turn of the century, Charlotte Perkins Gilman indicted "the same tacit assumption: man being held the human type; woman a sort of accompaniment and subordinate assistant, merely essential to the making of people. She has held always the place of a preposition in relation to man . . . before him, behind him, beside him, a wholly relative existence—'Sydney's sister,' 'Pembroke's mother'—but never by any chance Sydney or Pembroke herself." The culture was, as Perkins Gilman indicated in the book's title "androcentric."[11]

In the discussions of equality or difference, men remained the standard. How could women be judged to be competent if they were individuals invariably seen as an instance of a type called "women" and presumed, by virtue of membership in that class, to be a defective version of men?

For centuries, defect was the dominant filter through which women were defined. For Freud, they were "castrated," for Aristotle and Aquinas, "misbegotten." Women were emotional, not rational, controlled by their uteruses instead of their brains.

As the debate progressed in the nineteenth century, women's rights advocates embraced the claim that protecting the home and cleaning up government required the influence of women in the public sphere. Once there, women either stressed their biological difference and claimed special treatment, or emphasized their fundamental similarity to men and argued for equal treatment.[12]

Whether women are in fact different or not is the subject of ongoing controversy. In the 1970s, feminist psychologists offered constructs such as "androgyny" in an effort to restructure what in the past were discussions of sex differences. As the placement of *andro* before *gyne* implies, androgyny historically envisioned the perfected *man*. Feminist critics worried that the concept would be taken to amalgamate highly valued "masculine" attributes with traditionally devalued "feminine" ones, without questioning the existence or relative value of either set. Others asked whether the ideal of androgyny would praise "feminine" traits in men while mandating "masculine" ones in women. And some wondered why the supposed "feminine" and "masculine" traits should be treated as co-equal and complementary. Why not instead assume that the "feminine" were superior?[13]

In the discussion of similarity and difference, Carol Gilligan's *In a Different Voice: Psychological Theory and Women's Development*,[14] published in 1982, was either a landmark or a landmine. Her argument that women speak in a different moral voice seemed to offer some scientific support for the claim to difference. As important, it took traditional liabilities and interpreted them as previously unrecognized assets. Where men use an ethic of justice, argued Gilligan, women employ an ethic of care.

Interlaced in the debate about the merits of her book[15] were political concerns. Would the book reinforce stereotypes used to deny women opportunity? "I am troubled by the possibility of women identifying with what is a positively valued feminine stereotype," notes Catharine MacKinnon. "It is 'the feminine.' It is actually called 'the feminine' in the middle chapter of the [Gilligan] book. Given existing male dominance, those values amount to a set-up to be shafted."[16]

Read against a century of conflict over equality and difference, those concerns made sense. Proponents of women's rights fell into two camps. Both argued that specifiable biological differences, most centering on a

woman's unique role as childbearer, did exist. One claimed that the resulting differences dictated that women be given protections denied men. The
other school held that women's reproductive capacities did not signal other
differences; instead the supposed differences were exclusionary, mere rhetorical constructs crafted by those intent on maintaining a "woman's
place." Therefore, in every area but reproduction, women and men should
be treated as equals.

Equality and Difference in the Suffrage Debate

Some of the earliest statements of women's equality in the United States
took rhetorical form in resolutions promulgated at a national meeting of
women's advocates at Seneca Falls in 1848.

"We hold these truths to be self-evident: that all men and women are
created equal . . ."[17] read "The Declaration of Sentiments" promulgated
there. In this important reconstruction of the Declaration of Independence, "equality" was a radical rejection of the idea that women were
defective, and an affirmation of fundamental sameness. Abolitionist and
suffrage advocate Wendell Phillips agreed. For him, the woman's rights
movement had a single objective: "to deal with the simple question which
we propose—how to make the statute-book look upon woman exactly as it
does upon man."[18]

The position drew support from the theorizing of Harriet Taylor Mill
and John Stuart Mill, who argued that "the principle which regulates the
existing social relations between the two sexes—the legal subordination of
one sex to the other . . . ought to be replaced by a principle of perfect
equality, admitting no power or privilege on the one side, nor disability on
the other."[19]

"[W]oman is man's equal," proclaimed the third resolution at Seneca
Falls. Indeed such equality "was intended to be so by the Creator."[20]
Resolution 10 affirmed that "equality of human rights results necessarily
from the fact of the identity of the race in capabilities and responsibilities."[21] But Resolution 5 chided "man" for "claiming for himself
intellectual superiority" while not according "to woman moral superiority."[22]

The principle of equal capabilities and responsibilities proved a hard sell.
Opponents of women's rights would transmute this assertion to say that
advocates of equality sought not the same treatment as men, but wanted
instead to *be* men. So, in a pattern that would often be repeated in American history, feminists seeking the vote attempted to achieve their goals by
stressing their difference from men. As guardians of the moral order and
mothers of the nation's citizens, argued the suffragists, women would
purify the public sphere with their ballots.

The shift recast antagonists as allies. In the reconstruction, suffrage was
simply the means for mothers and wives to bring the virtues of the home to
government. Having remounted the pedestal, the suffragist could now

counter the Mrs. Arthur M. Dodges of the world who contended that "If women keep their eyes upon their homes and those within them, the ballot box will take care of itself; if they keep their eyes upon the ballot box, the homes will not." And "suffragists are for the destruction of the home."[23]

By portraying themselves as stalwart defenders of home and husbands, suffragists made it more difficult to depict them as unsexed, unwomanly, frustrated spinsters, or their antithesis—libertines. "Housekeeping" took on new meaning. In the United States, Hull House's Jane Addams claimed that "[C]ity housekeeping has failed partly because women, the traditional housekeepers, have not been consulted as to its multiform activities."[24] "What will save the home?" asked a National American Woman's Suffrage Association poster. "Votes for Women."

Consistent with their claim to be different from men in beneficial ways, female leaders stressed their role as helpmeets and mothers. Florence Nightingale was "The Lady with the Lamp"; Eleanor Roosevelt, her husband's legs. *Scientific American* described Madame Curie as "unassuming, plainly but neatly dressed, womanly and motherly in appearance. . . . She remains just plain Madame Curie, working for the good of humanity and for the expansion of scientific knowledge."[25]

This domestic mandate resonated, in part, because it was the banner under which women first moved outside the home without serious controversy. Home and the public sphere had initially been bridged by women who engaged in benevolent work among the poor and disadvantaged. Their efforts forged what Linda Kerber terms "a synthesis" that made it possible for women to enter "politics without denying women's commitment to domesticity."[26] "Not only did they [benevolent women] sew and knit for the poor just as they did for their own families, they also behaved as good mothers were supposed to: rewarding virtue, attempting to cure bad habits, and concentrating special attention upon children and old women, precisely as they did in their homes."[27]

The argument that women would carry the virtues of home and hearth to the ballot box didn't mean, of course, that women, in general, would see themselves in that role. Or in any event that political scientists would. For example, in the first edition of their classic *The American Voter* published in 1960, Angus Campbell and his colleagues reported that "Mothers of young children . . . are consistently less likely to vote than are fathers of young children across all levels of education. . . . The presence of young children requiring constant attention serves as a barrier to the voting act."[28] What was their explanation for this difference in voting and perceived ability to affect the system? "[W]hat has been less adequately transmitted to the woman is a sense of some personal competence *vis à vis* the political world."[29] In this view women are ineffective citizens not through inherent flaw, but because others have failed to socialize them properly.

Additionally, the political scientists explained, "The man is expected to be dominant in action directed toward the world outside the family; the woman is to accept his leadership passively. She is not expected, therefore, to see herself as an effective agent in politics."[30]

Whether these interpretations reflect the ways in which women and men saw themselves in 1960, or how male political scientists thought they ought to see themselves, is another question. In either case or both, the emphasis was on difference—but not of the sort championed by the suffragists.

Getting access to the public sphere and then being treated as an equal and effective agent there were, of course, two different things. In political practice, women who had envisioned themselves cleaning up politics instead found themselves consigned to clerical work. In 1964, a position paper by women in the Student NonViolent Coordinating Committee (SNCC) protested incidents including:

> . . . Two organizers were working together to form a farmers league. Without asking any questions, the male organizer immediately assigned the clerical work to the female organizer although both had equal experience in organizing campaigns.
>
> Although there are women in Mississippi project who have been working as long as some of the men, the leadership in COFO is all men.
>
> . . . A veteran of two years' work for SNCC in two states spends her day typing and doing clerical work for other people in her project.[31]

Similarly, as late as 1972, a study of party leaders in California found that women leaders were functioning, whether by choice or not, in a role that paralleled the traditional role of wife and mother in the family. So, literally, the idea of women "cleaning up" government turned into an all-too-literal housekeeping role. "In general, [the woman] is relegated to, or relegates herself to, a supportive role of more or less selfless service to her family or party, while the male partner or co-partyist pursues a career in the outside world."[32] The stories women tell of their rise to power would suggest that they were more relegated against than relegating.

Meanwhile, the claim that they could clean up government put women not on a par with men but on a new and fragile pedestal. The assumption of female virtue, explicitly invoked to get the vote and ban the bottle, became justification for a double standard. A male candidate, consultants now conclude, can better weather charges of corruption than a female. And as Hartford, Connecticut, Mayor Carrie Saxon Perry notes about sexual conduct, "There's no such word as 'man-izer.' You couldn't get away with it. Never!"[33]

Although the bias is beginning to erode, women candidates are still perceived as less expert on those matters traditionally associated with the public sphere and more expert on those identified with the private. A 1991 survey found that "voters prefer women of both parties on abortion (a woman candidate adds 8 points to her party's performance on this issue), health care (7 points), education (8 points), children's issues (12 points), and meeting the needs of the middle class (5 points). Voters think women are weaker than men on foreign affairs (losing on average 3 points), on defense (Republican women, particularly, lose 7 points), and on crime (particularly Democratic women, who lose 6 points)."[34]

The double entendre in "take advantage of you" speaks volumes in a question asked of Geraldine Ferraro but not her Republican counterpart, George Bush, in 1984. "[D]o you think that in any way that the Soviets might be tempted to try to take advantage of you simply because you are a woman?" asked Marvin Kalb on *Meet the Press*.

Special Protections

Some advocates of women's rights tied the need for special protections to the claim to difference. So, for example, in the early twentieth century the Women's Trade Union League sought special protections for the physically weaker sex. The League argued that women who had strained their bodies and nervous systems produced feeble children.[35]

The Courts agreed. Upholding a Montana statute that exempted groups of one or two laundresses from a tax, in 1912 Justice Oliver Wendell Holmes wrote, "If Montana deems it advisable to put a lighter burden upon women than upon men with regard to an employment that our people commonly regard as more appropriate for the former, the Fourteenth Amendment does not interfere by creating a fictitious equality where there is a real difference."[36]

In another case, the Supreme Court accepted the notion of shorter hours for women on the assumption that a "woman's physical structure and the performance of maternal functions place her at a disadvantage in the struggle for subsistence . . . and, as healthy mothers are essential to vigorous offspring, the physical well-being of a woman becomes an object of public interest and care in order to preserve the strength and vigor of the race."[37] As infant mortality dropped and concerns about overpopulation rose, this sort of argument lost some of its power.

The sameness/difference quandary became the central struggle in the seventy-year-old battle over the ERA. The first efforts to pass an equal rights amendment in the 1920s were opposed by those supporting special legislative protection for women.

This division emerged from the victorious suffrage movement when the radical wing of the National American Woman's Suffrage Association (NAWSA) established the National Woman's Party. The NWP cast its lot with the case for equality. Suffragist Alice Paul set the party's task as passage of an amendment stating that "Men and woman shall have equal rights throughout the United States and every place subject to its jurisdiction."

The subsequent debate focused on opportunity in employment. Supporters of the amendment echoed the words of Wendell Phillips. "[E]very occupation open to men shall be open to women," argued the NWP, "and restrictions upon the hours, conditions and remuneration of labor shall apply alike to both sexes."[38]

On the assumption that equal rights and protective legislation for women were incompatible, an opposition movement emerged. Felix Frankfurter was among the legal scholars who championed special protection. In 1921 he wrote to a leader in the Women's Trade Union League saying that the ERA threatens "even the very life of these millions of wage-earning women."[39]

In 1923, all but three states needed for ratification had signed on. Yet the effort to ratify stalled. Fifty years later the same issue was at the heart of a new ERA debate. In 1971, the House passed a version that read "Equality of rights under the law shall not be denied or abridged by the United States or by any State on account of sex." In 1972, the Senate followed. Under ERA, employers could either give men the protections then accorded women or take them away from women. Proponents held that men would benefit. The amendment, they argued, would not eliminate protective legislation but extend it to both sexes.

Still, the fear that protections would be lost drove some anti-ERA sentiment. "Many antifeminist women were clearly self-reliant but suspicious of changes that they feared would make it difficult for them to fend for themselves," argue Mathews and De Hart. "As realists, they understood that their place in the work force differed from that of men; it was more difficult, less well rewarded, restricted to a special status, but shielded nonetheless from unfair male competition by protective legislation. Only assumptions about differences between men and women made those laws possible. . . . Even if many opponents of ERA had heard the argument that the amendment would extend protective legislation to men, they would not have believed it because of their own experience. 'Look, Honey,' one anti-ERA woman observed, 'these women don't need ERA. They need a union.'"[40]

Again, the amendment fell short of the number of states required. The differences on which anti-ERA forces pivoted in the 1970s were defined by the nine amendments offered by Senator Sam Ervin in 1972. The North Carolina Senator moved to exempt women from compulsory military service and combat, to preserve protective legislation, to exempt wives, mothers, and widows, to exempt laws affecting the responsibilities of fathers for child support, to exempt privacy from being affected, and to exempt sexual offenses. Each of Ervin's amendments was defeated.[41] Each would re-emerge in the rhetoric of the STOP ERA movement whose leaders, as a result of the Senate's action, argued that ratification would result in women being drafted and sent to the front lines, unisex restrooms, and no more alimony or child support.

So, for example, in Massachusetts, ERA opponents claimed that ERA would "'restructure society' through assault on 'the *immutable specialness* of the female role,' The result would be 'a unisex sameness.'"[42] In *The Power of the Positive Woman,* the leader of the STOP ERA movement argued that the women's liberation movement was set on "[e]limination of the role of 'mother.' Wives and mothers must be gotten out of the home at all costs to themselves, to their husbands, to their children, to marriage, and to society as a whole."[43]

The sameness/difference dichotomy even split those who advocated the ERA. Where suffragists had found a way to make difference and equality compatible—indeed mutually reinforcing—now ERA advocates fell into the sameness/difference dichotomy.

The Equal Pay Movement

The equal pay movement would face the equality versus difference argument anew. Employers used the difference of women's place in the family to justify her lower wages. "Employers assumed," writes labor historian Joanne Meyerowitz, "that all working women lived in families where working males provided them with partial support."[44] The argument from difference proved useful in the suffrage debate. But when applied to employment, it set a trap. Laws protecting women in the workplace opened the door to some jobs and barred access to others. Jobs opened to women carried less status and lower pay than those traditionally held by men. Fork-lift operators were paid more than secretaries.

In divorce cases too, the argument from difference set a trap. Some argued that the young child's primary caretaker should be given primary custody. Of course the primary caretaker was usually the mother. "Although gender-neutral in form," notes legal theorist Deborah L. Rhode, "these presumptions have not worked out that way in fact. And they come at a cost. As long as the vast majority of primary caretakers are women, privileging their custodial rights reinforced the assumption that parenting is largely the mother's responsibility. This assumption has circumscribed opportunities for both sexes. It has also helped to perpetuate burdens on mothers who obtain sole custody and has fostered the guilt and social ostracism that discourage women from declining that role."[45]

Granting that women were at their essence wives and mothers also made it more difficult to argue that household responsibilities should be shared. And the assumption that women have a superior aptitude for some things, men for others had far-reaching consequences in the marketplace. The bottom line was that the areas "natural" to women were not those traditionally associated with either high pay or leadership.

A lawyer in Philadelphia's Wolf-Black firm, Nancy Ezold specialized in defense law and commercial litigation. Nonetheless, she was assigned "low-prestige specialties . . . [and] denied an opportunity to prove her mettle in high-profile legal work."[46] When she was offered a partnership as a matrimonial specialist rather than as a litigator, she sued. Ezold's dilemma was akin to that of students who are told that they can't qualify for a job without experience but can't get experience because no one will give them the opportunity. The low-profile work she was assigned didn't provide the opportunity needed to establish her skills as a litigator. The skills she was permitted to refine qualified her for a more traditionally female role—matrimonial specialist. In terms of access and qualifications, "U.S. District Judge James McGirr Kelly ruled that Ezold was unlawfully held to a different, higher, standard than her male colleagues."[47] Ezold's employers appealed to the Third Circuit Court of Appeals, which reversed the ruling. Without comment, the Supreme Court in October 1993 refused to hear the case.

Just as difference was used to justify equal access to the ballot, so too was it invoked to justify equal pay. "Feared as a challenge to traditional gender roles, equal pay emerged as a way of enabling women to fulfill their family responsibilities," notes historian Alice Kessler-Harris.[48]

The Equal Pay Act has been on the books for more than a quarter century, but despite recent gains Department of Labor Statistics figures indicate that on average women in most jobs still earn less than men. Moreover, in general, the greater the percentage of women in a particular field, the less the jobs in that field pay. When women moved in large numbers into a once-male preserve—as clerical work originally was—the status of the jobs dropped. Under these circumstances, equal pay for male and female clerks will not raise women's salaries.

In House deliberations in 1963, the idea of comparable worth—a phrase that would have revolutionized women's standing—was dropped. Equal pay was not guaranteed for "work of comparable quantity and quality."[49] Segregating women and men in different job categories allowed continued salary discrepancies that discriminated against women. Men were paid more for doing "male jobs" than women were for jobs requiring comparable education and skills. In such a world, equal pay won't raise women's salaries. A study done under Title VII by the EEOC in 1970 "found that state laws protecting women from hazards in the workplace had actually 'protected' them from high paying jobs."[50]

The argument from difference had been costly. Only by assuring that comparable skills should be rewarded with comparable salaries are the salaries in the traditionally female fields likely to rise to match those in the traditionally male.

Questioning the Zero-Sum Game

From Eve to Hillary Clinton, debates over equality versus difference presupposed that empowering women disempowers men. Male-female relations, in this construction, are a zero-sum game. If I win, you lose, and vice versa.

The notion that wages given to women would be taken from men was a commonplace at the turn of the century. There is a set amount of money available for wages, went the argument. If women can be employed more cheaply than men, money will be moved from male heads of households supporting families to women. So, for example, Samuel Gompers told the American Federation of Labor that "In industries where the wives and children toil, the man is often idle because he has been supplanted, or because the aggregate wages of the family are no higher than the wages of the adult man—the husband and father of the family."[51] In other words, the family income would remain the same even if mother and father both toiled in industry. Thus, the husband and father lost standing, the family sacrificed a mother at home, but there was no net gain in earnings.

This assumption that women's gains are men's losses underlies much of the opposition to equality. Ruth Bader Ginsburg built her legal career on the argument that equality was not a zero-sum game. The results were rulings consistent with the premise of ERA proponents: Men would not lose under equal rights. Men as well as women were advantaged by equal treatment under law, she argued. The case of Paul Wengler is illustrative.

Paul's wife Ruth was killed in an accident while working for a drug company. Paul claimed workers'-compensation death benefits. Had it been he who had been killed, his wife would have been given the benefits. As a widower, he was denied compensation. He would have been eligible, said the law in question, only if he were mentally or physically incapacitated from earning a living or could prove that he had been dependent on his wife's wages. Had the situation been reversed, the benefits to his wife would have been automatic. Ginsburg argued for benefits for Paul on the grounds that failure to do so constituted unequal treatment that also demeaned the wife's work and devalued her contribution to the compensation system.

When Stephen Weisenfeld's wife died giving birth to their only child, he found that he could not collect Social Security benefits. She had been the wage earner in the family. Had he been the wage earner and died, she would have been given benefits. Ginsburg noted that "She paid the same social security taxes a male wage earner pays. Because Congress did not see her in the wage earner role, however, her employment did not secure for her family the protection Congress afforded the family of a male wage earner." Ginsburg won a unanimous court victory. "Since the Supreme Court's 1975 *Wiesenfeld* case," she notes, "the widowed mother's benefit has been effectively replaced by a widowed parent's benefit. Both men and women gained from the change; there was no loss to either sex."[52]

What Ginsburg did was undercut the anti-ERA, anti-equal pay argument that had held that if equality were assured, women would be deprived of protections. Instead, argued Ginsburg, men would gain without women losing.

By treating men as "other," Ginsburg's cases reversed the traditional assumption. The reversal casts the relationship between men and women in a new light. In this context, women do not hoard their privileges but reach out to extend them to disadvantaged men. The move invites reciprocity. At the same time it implies that women are disinclined to repay past injustices in kind. And it undercuts the assumption that claims for equality produce a zero-sum game.

The Equal Treatment Standard

In the past decade, two court cases evaluated the claim of large retail chains that women were underrepresented in better paying jobs by choice, not because of discrimination. The first, involving Sears, supported the idea that women behaved differently than men in the labor market. The second, concerning Lucky Stores Inc. and decided in August 1992, found that assertion implausible. The differences between the two cases are instructive.

EEOC v. Sears Roebuck & Co. is a lightening rod. Brought under Title VII of the Civil Rights Act of 1964, the Sears case set on the table the EEOC's claim that the retailer had discriminated against women, relegating them disproportionately to noncommission, and hence lower-paid, jobs. The "crux of the issue," argued the attorneys for Sears, was "[t]he reasonableness of the EEOC's *a priori* assumptions of male/female sameness with respect to preferences, interests, and qualifications."[53]

The EEOC case was built not on the grievances of specific women, however, but on statistical aggregates. Neither the EEOC nor Sears could legitimately speak of specific women, for there were none before the Court, or of women in general, since even if discrimination had occurred, it had not affected all of Sears's female applicants. As a result, the statements of proof are oddly impersonal. At their root was an epistemic question: How can you know why some but not others wound up in commission sales? Each side offered a smoking hypothesis.

The experts who testified for Sears situated their arguments in a claim to difference; those called by the EEOC argued from fundamental similarity. The statistics offered by the EEOC and undisputed by Sears showed a pattern of concentration of males in commission sales.

The case is not as simple as some accounts make it seem. Women were employed in commission sales, but not in the same proportion as men. Sears, in other words, had clearly not excluded all women. "The irony is, of course, that the statistical case required only a small percentage of women's behaviors to be explained," writes Joan W. Scott. "Yet the historical testimony [offered on behalf of the E.E.O.C.] argued categorically about 'women.'"[54] The EECO claimed that between 1973 and 1980, 40% of those applying to Sears were women qualified for commission sales work, but only 27% of the applicants got such jobs. Seventy-two percent of those in noncommission sales during that time were women. They received only 40% of the promotions into commission work.[55]

How, the court was asked, should we account for this disparity? Had women disproportionately chosen noncommission sales, as Sears argued, or, as the EEOC contended, had Sears' selection processes foreclosed an option for women who otherwise would have taken those jobs?

To generalize from historical examples to current worker attitudes, Sears offered polling data summarized by sociologist Irving Crespi. Using polls from the mid-1930s to 1983, Crespi concluded that "(1) men were more likely than women to be interested in working at night or on weekends; . . . (3) men were more likely than women to be interested in sales jobs involving a high degree of competition among salespersons; (4) men were more likely to be interested in jobs where there was a chance of making more money, even though it involved a risk of losing the job if they did not sell enough; and (5) men were more likely than women to be motivated by the pay of a job than by the nature of the job and whether they like it."[56]

By failing to discredit the reported polls, the EEOC left itself in a difficult position. The Court could presumably generalize from the defendant's survey results to the attitudes of the women and men in Sears' applicant pool. But was this characterization of men's and women's historical attitudes applicable to the employees who sold slacks and—in the view of the EEOC but not of Sears—aspired to sell carburetors?

Crespi's conclusions were compatible with those of feminist historian Rosalind Rosenberg, who disputed the EEOC assumption that "women and men have identical interests and aspirations regarding work." "Historically," she contended, "men and women have had different interests, goals and aspirations regarding work."[57] Among the sources Rosenberg drew on was Carol Gilligan's *A Different Voice* and Nancy Chodorow's *The Reproduction of Mothering*.[58] Her position, in other words, gathered support from, if indeed it was not shaped by, the work of the so-called relational feminists.

From a study of 12,000 couples done by Philip Blumstein and Pepper Schwartz[59] in the early 1980s, and from the work of Gilligan and Chodorow, Rosenberg concluded that "Women's participation in the labor force is affected by the values they have internalized. For example: (a) Women tend to be more relationship-centered and men tend to be more work-centered. Although both men and women find satisfaction and a sense of self-worth in their jobs, men are more likely than women to derive their self-image from their work. Most employed women continue to derive their self-image from their role as wife and mother. Women tend to be more interested than men in the cooperative, social aspects of the work situation."[60]

But Rosenberg's statement also drew evidence from the work of feminist labor historian Alice Kessler-Harris. And it was to Kessler-Harris that the EEOC turned for rebuttal. The premise underlying her testimony is that "choice can be understood only within the framework of available opportunity."[61] "The argument that women are only interested in certain kinds of work reflects women's perceptions of opportunities available to them which are themselves products of employers' assumptions and prejudices about women's roles," she maintained. "It constitutes evidence that discrimination in fact exists in the work force."[62]

Women and men are fundamentally similar, argued Kessler-Harris.

They both gravitate toward jobs that pay well. "History's evidence clearly indicates that substantial numbers of women have been available for jobs at good pay in whatever field those jobs are offered, and no matter what the hours. Failure to find women in so-called nontraditional jobs can thus *only* [emphasis mine] be interpreted as a consequence of employers' unexamined attitudes or preferences, which phenomenon is the essence of discrimination."[63]

District Judge John Nordberg concluded that the case for difference was more cogent than that for similarity. The witnesses for the EEOC, he wrote, "described the general history of women in the workforce, and contend essentially that there are no significant differences between the interests and career aspirations of men and women. They assert that women are influenced only by the opportunities presented to them, not by their preferences. They often focused on small segments of women, rather than the majority of women, in giving isolated examples of women who have seized opportunities for greater income in nontraditional jobs when they have arisen. . . . More convincing testimony in this area was offered by Sears expert Dr. Rosalind Rosenberg. Dr. Rosenberg testified that, although differences between men and women have diminished in the past two decades, these differences still exist and may account for different proportions of men and women in various jobs."[64]

The 1992 case involving Lucky Stores was more clear-cut. The women who filed suit had sought and been denied work in the better paying, male-dominated grocery and produce sections, and had been barred from training opportunities that led to management positions.

Like *EEOC v. Sears,* this was a class-action suit. But the actual stories of plaintiffs in the Lucky case made it easier to show direct discrimination. And, in the Lucky case, there was no plausible rationale, based in different patterns of male and female socialization, that would explain why women might prefer to work a cash register or in the bakery or deli instead of in the grocery or produce sections, or why they should prefer nonmanagement to management positions. It was easier, in other words, for Sears to explain why women might not gravitate toward the selling of tools and auto parts than for Lucky to explain why they would not want to work in grocery or produce.

The lawyer for the plaintiffs in the Lucky Stores case wove his way through the difference/equality thicket by demonstrating that unequal results had been caused by dissimilar treatment. The Sears case showed unequal results without marshaling the specific evidence needed to demonstrate that they had been produced by treating women and men differently.

The Lucky plaintiffs won. To settle, the chain agreed to pay almost $75 million in damages to those denied promotional opportunities and to put up $20 million for affirmative action programs.[65]

Nothing in the Sears case repudiates the equal treatment standard. In the eyes of the judge, the EEOC failed to establish that unequal treatment had occurred.

The Female Candidate: Difference Without Special Protections

In politics, unlike employment, the argument from difference has helped. Just as their opponents cast women as the same or different depending on which worked, so too women's advocates learned that they could adjust the frame to their advantage.

As laws mandating school attendance for children and the like bridged the divide between private and public, the experience traditionally associated with women became increasingly relevant to governance. The residue of the belief that women are different in ways tied to a woman's "traditional" role is seen in findings that female candidates are expected to deal more efficiently than men at tasks allied to women's traditional strengths—family matter, consumer protection, and education.[66]

In fact, regardless of party, women in Congress are more likely than their male counterparts to support family and medical leave and access to abortion. All the Democratic women in the House and 86% of the Democratic men supported the Family and Medical leave Act in 1993. On the Republican side, the bill garnered the support of 45% of the women and 21% of the men. It passed. On the vote to federally fund abortions for the poor, 86% of the Democratic and 50% of the Republican women sided with the bill; 53% and 6% of the men in the two parties did. That bill lost.

Another aspect of difference played into the hands of female candidates. In a 1993 survey, only 7% of those polled reported a great deal of confidence in the people running Congress.[67] With "gridlock" a byword in politics and the popularity of politicians approaching that of used car salesmen and advertisers, being "other" and "different" became an advantage for women. Their strength was further bolstered by their numbers. Since women constitute more than 50% of the electorate, opponents who played to traditional notions equating *difference* with *defect* risked a backlash. The four female senators elected in 1992 won with a majority of the votes of both women aged 18 to 29 and over 60.[68]

In the past, the argument from difference made it more difficult for women to secure leadership positions, particularly in areas not identified with "women's issues." But when those in power are condemned as "insiders" corrupted by that power and in bad need of term limits, being "different" is no longer a bane but a blessing. In 1992, for example, female challengers picked up an average of 5 points in the polls against their male opponents. The result was the election of Barbara Boxer, Dianne Feinstein, Patty Murray, and Carol Moseley Braun.

A national survey in September 1991 had forecast the advantage. It found that voters view women as "agents of change who will make government work."[69]

"The biggest difference that I can see in American electoral politics right now," noted political consultant Ann Lewis in the fall of 1993, "is that the issue of gender difference has gone from being a liability to an asset. It used to be that a woman candidate's first goal was getting up to even so that she could compete on a level playing field. You had to prove she was capable of doing the job. Now with public discontent about politics as usual, being in government is no longer as asset. There is unhappiness with conventional politics. People think it leaves them out. Women are still considered outsiders. Because they had to struggle to get where they are, voters think: 'women candidates are more likely to know what my life is like and less likely to forget about me.'"[70]

But the loss of Attorney General Mary Sue Terry in the 1993 Virginia gubernatorial race suggested that a woman identified as an incumbent carried traditional liabilities. In that contest, Republicans successfully tied the Democrat Terry to what voters perceived as the failed Democratic administration of incumbent Governor Doug Wilder and Senator Chuck Robb.

As candidates, notes Ed Goeas, the pollster in Christine Todd Whitman's successful 1993 race for the New Jersey governorship, women are perceived to be more honest and compassionate than male candidates. But they also have to deal with two weaknesses. They must prove that they are strong and smart enough. Goeas argues, and other Democratic and Republican consultants agree, that the question in gubernatorial races is, Are they *strong enough* to run a state? In races for Congress, the question is, Are they *smart enough* to legislate?

Although Whitman won, Goeas praised her opponent's consultants for running a "brilliant campaign." They "basically took a strength which was compassion and through playing class warfare [the Democrats stressed that Whitman comes from a wealthy family] raised questions about whether compassion was there. . . . On the strength side . . . everyone thought the issue was assault weapons. It wasn't. The issue was [whether] she was going to be strong enough to put the emphasis on crime that the voters wanted to see."[71]

Despite Terry's loss, consultants on both sides of the New Jersey and Virginia gubernatorial races of 1993 agreed that Republican Christine Todd Whitman in New Jersey and Democrat Mary Sue Terry in Virginia could presuppose their credibility from the start of their races.[72]

Overcoming the Difference/Equality Bind

In 1973, one hundred years after Myra Bradwell was told that the law court was not a suitable place for a woman, the Court held that sex classifications had been used throughout history "to put women not on a pedestal, but in a cage."[73] In that decision, Justice Brennan challenged the assumption that her biology dictated that a woman be treated differently. "[T]he sex characteristic frequently bears no relation to ability to perform or contribute to society. As a result, statutory distinctions between the sexes often have the effect of invidiously relegating the entire class of females to inferior legal status without regard to the actual capabilities of its individual members."[74]

While equal treatment became the watchword in most cases, in some important ways it was set aside in favor of other issues and principles assumed to be more fundamental. When reproductive discussions are the focus, privacy, not equality, is at issue and equality/difference questions do not apply. So, for example, *Roe v. Wade* was decided not on sex discrimination grounds, but on the premise that a woman's right to privacy supersedes state interests until the third trimester.

In other areas, equal treatment fell short as a means of protecting women's rights. Based on what would work for women in a given situation, various theorists began to offer alternative notions of equality. Recognizing that equal treatment of women in divorce cases ultimately disadvantaged women who stayed out of the labor force to raise children, some began to argue that when individuals *are not comparably situated,* equality of result, not equality of treatment, should be the test. If individuals are not "in fundamentally the same position . . . the result of applying rule-equality may be to further perpetuate result-inequality," argues Martha Albertson Fineman. "Rules that focus on result-equality, by contrast, are attempts to ensure that the *effects* of rules as they will be applied will place individuals in more or less equal positions. Such rules are constituted to take into account the different structural positions of women and men in our society and seek to achieve parity in position between individuals."[75]

So, for example, equal division of assets in a divorce settlement might be unfair and produce an unequal result for a spouse who has shared disproportionate responsibility for childrearing and in the process foregone economic and educational opportunities that might have eased that person's return to the labor market. Achieving equal results—two people comparably situated to survive economically—might dictate unequal distribution of resources to put one of the two back in the labor market.

Overall, however, the courts and the Congress have threaded their way through the complications of equality/difference by establishing equal treatment as a test. Accordingly, the law specifies that women cannot be discriminated against in employment, housing, federally funded education, credit, or at the ballot box. Women would also, as the court held in *White v. Crook* (1966), be guaranteed the right to serve on juries. And, as the Court concluded in 1994 in *J.E.B. v. Alabama T.B. ex. rel.,*[76] equal protection bars using peremptory challenges to exclude jurors on the basis of their sex.

Still, in subtle ways the assumption of defect persisted in the workplace; such biological differences as a woman's capacity to bear a child would be treated as if they were analogous to forms of temporary male disability. So, for example, the Pregnancy Discrimination Act of 1978 held that "women affected by pregnancy, childbirth, and related medical conditions shall be treated the same as other persons not so affected but similar in their ability or inability to work." Since men lack the ability to become pregnant, and men are the norm, pregnancy in a woman is analogized to a medical disability in a man to ensure equal treatment.

Discrimination is not an issue, however, if sex can be shown to be a bona-fide occupational qualification "reasonably necessary to the normal operation of that particular business or enterprise (703e of Title VII)," as it was in excluding a woman from the job of prison guard.[77]

What was sauce for the goose was sauce for the gander. Consistent with the cases carried into court by Ruth Bader Ginsburg, the same standards of equal treatment would be applied to men as well as women. That was the result when the Courts were asked to decide whether the friendly skies would become friendless if men were offering coffee or tea. In *Diaz v. Pan American World Airways*,[78] the airline argued that female flight attendants were better able to soothe the anxiety of fliers than male attendants. Speaking for the airlines, psychiatrist Dr. Eric Berne posited a double bind for flight stewards. "He explained that many male passengers would subconsciously resent a male flight attendant perceived as more masculine than they, but respond negatively to a male flight attendant perceived as less masculine, whereas male passengers would generally feel themselves more masculine and thus more at ease in the presence of a young female attendant."[79] By this logic, female travelers would presumably perfer male attendants, and men would prefer female pilots!

The Circuit Court would have none of it. It held that such stereotypic notions of sex roles did not make sex a bona-fide occupational qualification for airline attendants. Men could not be excluded from the job of flight attendant.

Nor could Southwest Airlines justify barring male flight attendants. Despite the fact that "[i]ts T.V. commercials feature attractive attendants in fitted outfits, catering to male passengers while an alluring feminine voice promises in-flight love," sex appeal wasn't relevant, said the Court. "Southwest is not a business where vicarious sex entertainment is the primary service provided. Accordingly, the ability of the airline to perform its primary business function, the transportation of passengers, would not be jeopardized by hiring males."[80]

In the domain of childrearing, women and men also would receive equal treatment. Consistent with that notion, Congress decided in 1933 that both mothers and fathers would be eligible for leave under the Family and Medical Leave Act.

But even in areas such as employment, in which sex discrimination was barred by law, womb/brain remnants persisted in the form of sexual harassment.

A test of the applicability of equal treatment occurred in late 1993 when the high court heard a Nashville worker's claims that her boss in a forklift firm had repeatedly subjected her to off-color jokes, sexual innuendo, and sexually demeaning remarks. Among other affronts, Teresa Harris was called "a dumb-ass woman" and subjected to the boss's suggestion that he and she negotiate her raise at the Holiday Inn. The company president, Charles Hardy, told her "You're a woman, what do you know?" After dropping objects and asking female employees to pick them up, he commented on their cleavages. He also asked Harris to retrieve a coin from inside his front pants pocket. Under such circumstances, her attorneys claimed, she had no choice but to quit. After doing so, she sued under Title VII of the Civil Rights Act of 1964.[81]

In December 1993, the Supreme Court decided *Harris v. Forklift Systems Inc.* 9–0 for the plaintiff. In the Court's opinion, Justice Sandra Day O'Connor spoke not about women or men but about employees: "A discriminatory abusive work environment, even one that does not seriously affect employees' psychological well-being, can and often will detract from employees' job performance, discourage employees from remaining on the job or keep them from advancing in their careers."[82] In a concurring opinion, Ruth Bader Ginsburg identified the crucial issue as "whether members of one sex are exposed to disadvantageous terms or conditions of employment to which members of the other sex are not exposed."[83]

The fact that women have come a long way was evident in the calm reaction of the business community. As one AT&T spokesperson put it, "This decision is only a blow to yahoos."[84]

6

Double Bind Number Four: Femininity/Competence

MORE than two decades ago, a ground-breaking study by psychologist Inge Broverman and her colleagues concluded that the characteristics psychologists defined as feminine were inconsistent with maturity and primarily negative. These characteristics included being easily influenced, emotional, and illogical. By contrast, the traits seen as masculine were those associated with mental health and psychological maturity: being direct and logical, able to easily make decisions.[1]

This is the crux of what I call the femininity/competence bind. After the double binds of womb/brain, silence/shame, and equality/difference are dispatched, we still confront a bind that expects a woman to be feminine, then offers her a concept of femininity that ensures that as a feminine creature she cannot be mature or decisive.

The assumption that women cannot be both feminine and competent, and its code phrase "tough or caring," are vestiges of the binds treated in the last three chapters. Those who exercised their brains and brawn in public were thought to be tough, active, analytic, decisive, competent, and masculine; those who exercised their uteruses with the attendant responsibilities in the private sphere were identified as nurturant, passive, warm, and feminine.

Such supposedly innate female characteristics function, in Lipman-Blumen's term, as "control myths" that forestall behavior in the self-interest of those who cleave to them. Women who primarily define themselves in these terms are less likely to crusade for rights in the public sphere than others.

The echoes of the speech/shame bind are evident here as well. Forms of speech that in 1897 signaled the shame of "sexual inversion" or lesbianism are among those today that hint that a woman is "too masculine" to be accepted in the professional world.

The American Bar Association's Commission on Women focused attention on the assumption that femininity and such traits as decisiveness and leadership are at odds. Attorneys told the group, which was chaired by Hillary Rodham Clinton before "Hillary" was a household word, that "women walk a fine line between being regarded as too feminine (and thus not tough, lawyer-like, or smart) or too tough (and thus unfeminine or not the kind of women male colleagues feel comfortable relating to)."[2]

The words commonly associated with the femininity/competence bind are "too" and "not . . . enough." The evaluated woman has deviated from the female norm of femininity while exceeding or falling short of the masculine norm of competence. She is too strident and abrasive or not aggressive or tough enough. Or, alternatively, she has succumbed to the disabling effects of the feminine stereotype of emotionalism.

"So a girl is damned if she does, damned if she doesn't," observes linguist Robin Lakoff. "If she refuses to talk like a lady, she is ridiculed and subjected to criticism as unfeminine; if she does learn, she is ridiculed as unable to think clearly, unable to take part in a serious discussion: in some sense, as less than fully human. These two choices which a woman has—to be less than a woman or less than a person—are highly painful."[3]

This double bind draws energy from our tendency to think in dichotomies characterized as masculine or feminine, and then set in a hierarchical relation to one another with the masculine thought superior and the feminine inferior.[4] Here the idea that women are defective persists.

Women's vocal pitch is even deemed deficient. The "high, female voices" of the early female West Point cadets, for example, were mocked by their male peers. "'Good morning, sir,' she would say, with the prescribed deferential salute. 'It was a good morning until you got here, bitch,' he would mimic, in falsetto. Squad leaders would order the women to 'lower' their voices, to stand in front of the mirror and practice shouting 'loud and mean.'" "I got my voice a little lower," a cadet told Betty Friedan. "Then I realized if my voice was loud enough to be heard, did I really have to sound like a man?"[5]

The deeper male voice, however, has long been assumed as the norm for exercise of leadership. Over time, such conventions as use of a male announcer (the voice-over) on television reinforce the connection between lower pitch and authority. As a result, female candidates are coached to increase their credibility by lowering their pitch. Among those following that advice was British Prime Minister Margaret Thatcher.

"If the norm is male," writes a contemporary theorist of gender, "women will always be the other, the deviant. Superior or inferior, she is not the same. She is caught in a catch-22. If she attacks the problem by trying to be male, she will be too aggressive. If she attacks the problem by trying to be female, she will be the ineffective other."[6]

In managerial positions, notes Rebecca Stafford, president of Monmouth College, "you always run into what we call the 'catch 22' situation in which if you're tough, you are considered to be nasty and mean and if you are not tough then you are too emotional."[7]

"Women are not called tough leaders, they are called bitches," Stafford adds. The words used to describe comparable male behavior are positive. "A woman is bitchy, and a man knows what he wants. A woman is aggressive and harsh, and a man is directed and goal-oriented," says Dr. Pam Douglas, a cardiologist at Harvard Medical School.[8] What is valued in a man is considered defective in a woman. In sociologist Robert K. Merton's formulation, "in-group virtues become out-group vices."[9]

As a result, Rosabeth Kanter argues, women are "often measured by two yardsticks: how *as women* they carried out the sales or management role; and how *as managers* they lived up to images of womanhood."[10] By this test a woman who does her job well in a nontraditional field may be evaluated negatively.

At the root of it, as Simone de Beauvoir recognized more than forty years ago, is the long-lived historical tendency to see woman "defined and differentiated with reference to man and not he with reference to her. . . . He is the Subject, he is the Absolute—she is the Other."[11] And because she is a woman, and hence "other"—not the norm—her competence will be questioned, not assumed. "[T]o be a woman, if not a defect, is at least a peculiarity," wrote de Beauvoir. "Woman must constantly win the confidence that is not at first accorded her."[12]

As problematic is the fact that this ongoing challenge to one's competence may produce counterproductive behavior. "To feel the weight of an unfavorable prejudice against one is only on very rare occasions a help in overcoming it," suggests de Beauvoir. "The initial inferiority complex ordinarily leads to a defense reaction in the form of an exaggerated affectation of authority." So, for example, "Man is accustomed to asserting himself; his clients believe in his competence; he can act naturally: he infallibly makes an impression. Woman does not inspire the same feeling of security; she affects a lofty air, she drops it, she makes too much of it."[13]

A Higher Competence Threshold

Stories told around the water-cooler as well as statistics confirm that a man's competence is more likely to be presupposed, a woman's questioned. So, for example, "63% of the woman lawyers surveyed in Massachusetts said court workers had asked them, 'Are you an attorney?' That was more than double the proportion of men who said they were asked the same question." And a survey of New York lawyers found that "female experts must give many more credentials than male experts."[14]

Social scientists have confirmed that, in general, women are assumed to be less competent than men.[15] Their findings are reflected in the stories told by women in professions as dissimilar as politics and beer brewing.

So, for instance, in Dianne Feinstein's 1990 gubernatorial campaign, her staff concluded that "the hardest task they faced during the campaign was the need to establish Feinstein's credibility as a leader over and over and over again, despite her credentials as mayor of San Francisco and her charisma. Yet the public had no trouble seeing her mild-mannered Republican opponent, Pete Wilson, as a leader, for as a white male he fit the public's most comfortable leader image."[16]

"I was in the process of proving every day that I was sane," prize-winning correspondent Eileen Shanahan recalls. "Sane and competent. A real professional. I was covering tax legislation. I was covering the budget. I was covering the Federal Reserve. I was covering all that stuff the male power structure respects. I think you have to prove it in an ongoing way. Don't put that in the past tense. I think you simply have to submit ongoing evidence that you are rational and professional. Never mind if you've got a twenty-, thirty-, or forty-year history. Are you sane today, or have you gone into a feminist tizzy? I really think that evidence has to be there. I don't see any signs that the need for that has abated."[17]

"I find that I think I have to be better prepared, that I have to have more knowledge," notes Sunny Jones, Western Government Affairs Manager for the Miller Brewing Company. "I have to have the arguments perhaps a little better rehearsed because I have watched these other guys and I think 'God, you are getting away with murder.'"[18]

Such testimony is supported by studies. In 1968 a classic experiment asked men and women to evaluate the intelligence, persuasiveness, and style of a set of essays. The subjects were told that the essays were written either by John T. McKay, J. T. McKay, or Joan McKay. Although they were identical, Joan's essays received consistently lower ratings from both men and women than John's. In 1985, the study was replicated, as were the results.[19]

Other analyses confirm that the same cues are evaluated differently in men and women.[20] Assertiveness is valued in men, but not in women.[21] Where attractive men in managerial positions are assumed to have achieved their status through native ability and hard work, attractive women are assumed to have slept their way to the top.[22] When women use qualifiers in their statements, their credibility suffers; not so for men.[23] The same cues (e.g., poor eye contact, vocalized pauses) are interpreted differently in men and women. Men with poor eye contact are seen as shy; women are deemed incompetent. Female professors have to pass a higher threshold test as well. "Male teachers who are responsive are 'really good guys'; women who are responsive are just being women."[24]

In the process of assessing competence, women are held to a different standard. Witnesses told the ABA's Commission on Women that "women must still work harder and be better than men in order to be recognized and succeed." Female candidates believe the same is true for them. "All too often there is a contradiction between the attributes voters expect in a candidate and what they want in a woman. Ambition is a plus in a man but a drawback in a woman. Men should be tough, but strength in a woman is threatening. Male candidates should be aggressive, women compliant and deferential."[25]

The femininity/competence bind is sustained by a long history of sexual stereotyping that both men and women bring to their sense of self and relationships with others. Note the adjectives that both men and women thought more desirable in one sex than the other in an influential study done in the mid-1970s:[26]

Masculine items	*Feminine items*
Acts as a leader	Affectionate
Aggressive	Cheerful
Ambitious	Childlike
Analytical	Compassionate
Assertive	Does not use harsh language
Athletic	Eager to soothe hurt feelings
Competitive	Feminine
Defends own beliefs	Flatterable
Dominant	Gentle
Forceful	Gullible
Has leadership abilities	Loves children
Independent	Loyal
Individualistic	Sensitive to needs of others
Makes decisions easily	Shy
Masculine	Soft-spoken
Self-reliant	Sympathetic
Self-sufficient	Tender
Strong personality	Understanding
Willing to take a stand	Warm
Willing to take risks	Yielding[27]

(Modified and reproduced by special permission of the Publisher, Consulting Psychologists Press, Inc., Palo Alto, CA 94303 from *Bem Sex-Role Inventory* by Sandra L. Bem. The Bem Sex-Role Inventory is distributed by Mind Garden, a service of Consulting Psychologists Press, Inc. Copyright 1974 by Consulting Psychologists Press, Inc. All rights reserved. Further reproduction is prohibited without the Publisher's written consent.)

Such ascriptions are the rhetorical remains of the womb/brain and silence/shame double binds. Although psychologists commonly label such "masculine" traits instrumental and "feminine" traits expressive, in fact the former are traditionally associated with the public sphere, the latter with the private, the former with the brain, the latter with mothering, the former with the productive, the latter with the reproductive. Both lists contain words that entail forms of expression: the first expresses the intellect, the second "the heart"; the first the "rational" faculties, the second the "intuitive" or "emotional."

Such gender stereotyping is reinforced by the widespread assumption that the so-called expressive and instrumental characteristics are bipolar opposites,[28] when instead they can and usually do characterize the same person.[29] But because the male is the norm, his competence is assumed and we rarely question his expressive capabilities. A woman's competence is not assumed; her deviation from the supposed masculine "norm" is more likely to be noticed and, once noticed, penalized.

Trying to satisfy this complex set of expectations is impossible. Women are penalized both for deviating from the masculine norm and for appearing to be masculine. When women try to establish their competence, they are scrutinized for evidence that they lack masculine (instrumental) characteristics as well as for signs that they no longer possess female (expressive) ones. They are taken to fail, in other words, both as a male and as a female. When women excel at female (expressive) behavior, the significance of their skills is undercut by the fact that these are devalued behaviors.

There is evidence that such categories are beginning to erode when men are condemned for being too aggressive or, alternatively, not sufficiently sensitive.

Language That Sustains the Bind

Our language reflects the fact that, historically, women engaged in public activity have been suspect. The word "misogyny" was coined to describe hostility toward women but also as a means of fighting back. Where the word "misogyny" has been around for a long time, "misandry" only appeared in the most recent supplement of the *Oxford English Dictionary*.[30]

Here is another residue of the silence/shame bind. As we have seen, the range of language available to condemn female speech and action is wider than that available to criticize a man. Before being acknowledged as competent, women are held under a more powerful microscope for longer periods of time and examined through different lenses. So, for example, two of the three adjectives used by a *Washington Post* reporter to describe newly elected Canadian Prime Minister Kim Campbell's personality are not likely to have been applied to Brian Mulroney. She is, said the reporter, "at times brittle, defensive and haughty.[31]

Whether vulnerable to its labels or not, women are vetted through this catalogue of condemnation. In the process, of course, its hold is reinforced on the reviewer, the person reviewed, and the gawking voyeurs. When the pejoratives are found inapplicable, the woman is praised, implicitly if not explicitly, for transcending one of the curses of her gender. Even friends and supporters do it. "There was no tinge of that dreaded word always laid on women: 'emotional,' said Lynn Hecht Schafran of the NOW Legal Defense Fund [about Ruth Bader Ginsburg]. 'There was nothing shrill.'"[32]

Although the label "sexist" constrains the words men other than Rush Limbaugh publicly apply to women, such language survives. Occasionally women use it against each other. To situate her argument that she is both competent and feminine, Georgette Mosbacher, cosmetics company CEO and spouse of a former Bush cabinet secretary, claims that she fits none of the stereotypes that constrain women in Washington.

Neither is she a wife who fades "into the woodwork" or one of "the few women power brokers, often from Congress, who were positive they had to be tough, acerbic, and combative to attract any attention. Known for their shrill one-liners, they helped to reinforce everyone else's conviction that if this was the alternative, it was far better to be a doormat."[33] One might wonder whether "shrill" and "acerbic" are not code words indicating that Mosbacher disapproves of the speakers' ideology. And why the false alternatives? Surely there is some terrain between and beyond being "shrill, tough, acerbic, and combative" and being a doormat. Are there actually any women in Washington who are either?

Similarly, gender-based inferences bubble through Bush press aide Kristin Clark Taylor's mind as she meets for the first time with Vickers Brian, a woman close to her in age, who is showing her the ropes in the White House. "'There are obviously certain people on the White House staff whose names you better commit to memory starting today, if you don't know them already,' she [Vickers Brian] warned. 'I'll do you a favor and go through the list and highlight the most important names. It really is in your best interest to remember those.'"

If spoken by a man, these words could be heard as patronizing. They probably wouldn't invite questions about the speaker's home and marriage. "I wondered," writes Clark Taylor, "if Vickers had a husband or a family. I wondered if they liked her much."

"'Craig Fuller is chief of staff,' she [Vickers Brian] said, drawing a thick yellow line through Craig's name and enunciating her words slowly, as if I might not comprehend otherwise. She was a tad condescending, and more abrasive than I was accustomed to." Are the words or tone abrasive or is "abrasive" a code that translates: This is a woman behaving in a way I find offensive?

"'Thanks, I know,' I answered back, sugar sweet. 'He's the one who brought me into the White House.' 'Don't get petty now, girl,' I thought to myself. 'No catfights allowed on the first day.'"[34] Women expect other women to be supportive, not distant. Because women are expected to be maternal, co-workers and subordinates assume that, unlike men, sisters will be compassionate and warm. So, for example, male and female lawyers both tend to define female judges as too tough, lacking compassion.[35]

Women are more accustomed to being patronized by men than by women. What is at issue here is appropriate communication between women of the same age and presumed status. If Clark Taylor objects, she risks being accused of engaging in petty female infighting—"a catfight"— reinforcing the stereotype that says that women in power don't work well with other women.

Women are also able to characterize themselves in ways that would be dismissed as sexist if employed by a man. "Now speaking here at M.I.T. confronted me with a dilemma," Camille Paglia told an audience. "I asked myself, should I try to act like a lady? I can do it. It's hard, it takes a lot out of me, I can do it for a few hours. But then I thought, *Naw*. These people, both my friends and my *enemies* who are here, aren't coming to see me act like a lady. So I thought I'd just be myself—which is, you know, abrasive, strident, and obnoxious. So then you all can go outside and say, 'What a bitch!'"[36]

What Camille Paglia, Georgette Mosbacher, and Kristin Clark Taylor have in common is not age, race, or profession, but a disposition to describe women in terms not usually used to describe men. Implied in these usages are dichotomies that crumble under examination. What differentiates Paglia from Mosbacher and Clark Taylor is that her statement is self-reflective and ironic, a send-up of the terms themselves.

Just as the words we use to describe men and women reveal the dissimilar categories through which we assess their behavior, so in the boardroom, at the bench, in the political arena, as well as the bedroom, women and men are judged by different standards. When the qualifications, predilections, or peculiarities of a man are examined, he is likely to be treated as an individual, not as a representative of the category "man"; a specific woman, by contrast, is likely to be scrutinized as an instance or exemplar of "women." But only when she is faltering. As example after example in this chapter will show, the successful woman—and by that I mean the woman who transcends the double bind—will be viewed as an exception. And when evaluating others, not oneself, "a woman's success is more likely to be explained by external factors like luck or ease of task, or by high effort, an internal but unstable factor, whereas a man's success is more likely to be attributed to high ability. The reverse is true for explanations of failure; men are said to fail because of hard luck, a hard task, or low effort, whereas women are said to fail because of low ability."[37]

Ways of Sustaining the Bind: the "Exceptional" Woman

If femininity and competence in the public sphere are mutually exclusive, one instance of a woman who is both should be sufficient to up-end the dichotomy. That, of course, is not the way things work. Accomplished women regularly challenged such assumptions. "I never saw such a woman [as George Eliot]," wrote historian John Fiske to his wife in 1873. "There is nothing a bit masculine about her; she is thoroughly feminine and looks and acts as if she were made for nothing but to mother babies. But she has a power of *stating* an argument equal to any man; equal to any man do I say? I have never seen any man, except Herbert Spencer, who could state a case equal to her."[38]

The response of those whose expectations were violated was surprise and relief. "It certainly must have been a relief for the women of the country to realize that one could be a woman and a lady and yet be thoroughly political," wrote Agnes Meyer to Eleanor Roosevelt.[39]

Yet the stereotype persisted, to be challenged again and again. "After the portrayal of her as a ruthless natural autocrat it was stunning to meet her; to see with what composure and courtesy she managed her entourage and her Cabinet colleagues," noted British Labor leader Michael Foot of Indira Ghandi.[40]

The reality of George Eliot, Indira Ghandi, and Eleanor Roosevelt will function as proof of female competence only if their capacities are seen as suggestive of what women can do. And, of course, they weren't. They were instead, to offer a cliché, taken as the exceptions that proved the rule.

So, for example, in 1897 physicist Max Planck recalled that his experience with the few women who had audited his course was "favorable." "On the other hand," he added, "I must keep to the fact that such cases must always be regarded as exceptions. . . . Generally, it cannot be emphasized enough that nature herself prescribed to the woman her function as mother and housewife, and that laws of nature cannot be ignored under any circumstances without grave danger—which in the case under discussion would especially manifest itself in the following generation."[41]

And in an echo of the womb/brain bind, the exceptions were praised for being more manly than men. "A story—which, as far as I know, is all it was—once went the rounds of Israel to the effect that Ben-Gurion described me as the 'only man' in his cabinet," wrote Golda Meir in her autobiography. "What amused me about it was that obviously he (or whoever invented the story) thought that this was the greatest possible compliment that could be paid to a woman. I very much doubt that any man would have been flattered if I had said about him that he was the only woman in the government!"[42]

Battling the Bind

The femininity/competence bind has been assaulted by advocates of women's rights in four mutually reinforcing ways. First, those vulnerable to it have minimized its power by acknowledging it, while those who have won the battle to establish competence are adding their voices to the growing chorus condemning it. Second, scholars have been working to undercut the assumption of defect that reinforces it. Third, the courts have held that use of it against women employees is a violation of the protections of Title VII of the Civil Rights Act. And finally, as women move into leadership positions, their varied styles are widening the lens through which other women are seen, at the same time that their demonstrable competence is shattering the bind. When their numbers are sufficiently high—a percentage psychologists place at between 15 and 25%—their performance will dispatch the bind.

The Thresholds of Competence

Although she had been "tough" enough to serve as a successful member of Congress as well as the vice-presidential nominee of her party in 1984, Geraldine Ferraro was crossing a new threshold for women. Unlike a member of Congress, a president in time of war becomes the commander in chief. "Are you tough enough?" was a question that plagued Ferraro in her 1984 vice-presidential bid.

But if one adopts rhetorical moves suggesting "strength," one risks being in the situation of then UN Ambassador Jeane Kirkpatrick, who heard her speeches dismissed as lectures, her rebuttals of charges against the United States as "confrontations." "Terms like 'tough' and 'confrontational' express . . . disapproval at the presence of a woman in arenas" that require assertiveness, concluded Kirkpatrick.[43] And, I would add, what is "analytic" in a man is "coldly calculating" in a woman.

The result supposedly penalizes women who are "too tough" as well as those who are "too traditionally female." The second burden prompted Republican Senator Nancy Landon Kassebaum to quip that someday she was "going to hit someone over the head for calling me diminutive and soft-spoken."[44]

Even though in this dilemma Kirkpatrick seems to have lighted on one horn and Kassebaum the other, each found a way to establish her credibility in the public domain before the femininity/competence bind could be set up. Kassebaum did it in the way women have traditionally used to gain political power: by invoking a family legacy—in her case as the daughter of a revered former governor and presidential aspirant. Kirkpatrick took an alternative route, gaining credentials from major universities and writing important books. Each entered the public domain with the protection of assumed competence.

Other women who exercise power evoke descriptions from observers revealing that they have, in fact, surmounted the femininity/competence bind. So, for example, Hillary Rodham Clinton's success before congressional committees elicited descriptions noting that she was both aggressive and disarming. Similarly, Roslyn Carter was tagged the "steel magnolia," and British Prime Minister Thatcher was dubbed "The Iron Maiden." For some, this phrase evokes images of a medieval torture device, but for many it signals the co-existence of toughness and femininity.

Once women pass over the competence hurdle, their positions enable them to confront the double bind head-on. When Governor Jim Florio attacked his Republican opponent Christine Todd Whitman's proposal to cut taxes by 30%, Whitman responded, "Maybe he thinks women still can't be trusted with money."[45]

What Senator Nancy Kassebaum, scholar-statesperson Jeane Kirkpatrick, and Governor Christine Todd Whitman have in common is public communication—meta-communication—about the femininity/competence bind. By identifying and indicting stereotypes, such rhetoric helps dismantle their power.[46]

Implicitly, these women are also challenging one of the assumptions that lies at the crux of the bind: that a person must fall into one or the other of two invariant categories—masculine or feminine. The bind ignores the possibility that there is a third category, labeled androgyny by some, that incorporates the characteristics of both. "If 'Good Managers' Are Masculine, What Are 'Bad Managers'?" asks a provocative article by Gary N. Powell and D. Anthony Butterfield published in 1984.[47] The answer is surprising. "In contrast to a stereotypical view of the good manager as masculine, bad managers were seen by business students as low in both masculinity and femininity, or in nonstereotypical terms."[48] Femininity, in other words, may not be the obverse of masculinity.

Such findings are consistent with the notion of Sandra Lipsitz Bem that

> it is possible, in principle, for an individual to be both masculine and feminine, both instrumental and expressive, both agentic and communal, depending upon the situational appropriateness of these various modalities; and even that it is possible for an individual to blend these complementary modalities in a single act, being able, for example, to fire an employee if the circumstances warrant it, but to do so with sensitivity for the human emotion that such an act inevitably produced.[49]

Increasingly, women in power are inviting audiences to dismiss the masculine/feminine, agentic/expressive dichotomy for the notion that female leaders can do as Bem envisioned and employ both traditionally masculine and feminine characteristics. At the 1992 Democratic convention, for example, California State Treasurer Kathleen Brown credited her experiences as a wife, mother, and daughter with teaching her the values of thrift, honesty, tolerance, and hard work. She then acknowledged "I, like all women, have been shaped by these very female experiences. But I have also been shaped by those conflicting messages. We've been told, 'Be strong but not strident. Get a job but take care of the kids. Be compassionate but shed no tears.' Yet our strength and compassion is the best part of what we women can bring to our party and our country. But don't get me wrong. Just because we are caring doesn't mean we can't make tough decisions. We do it every day. We're tough enough to run school boards and tough enough to run board rooms. We're tough enough to stand alongside men in combat, and we are tough enough to stand up to an all-male Senate Judiciary Committee."

The statement directly alluded to the testimony of Anita Hill. It also called to mind the slogan Dianne Feinstein translated into an upset primary victory over John Van de Kamp in California: Tough and Caring. Those—including many male reporters—who couldn't conceive of the terms as cohabitants transmuted it to "Tough BUT Caring." Yet the ad, showing Feinstein with children in one shot, with police in another, standing for abortion rights and for the death penalty, invited audiences to suspend their assumptions about women and judge this woman on her positions. This ad is credited by consultants on both sides of the contest with closing the deficit Feinstein was suffering in the polls.

Those arguing that men and women alike should be both instrumental and expressive can find cause for hope in a replication in the mid-1980s of Sandra Bem's original study. Once again respondents were given the opportunity to distinguish between traits desirable in men and women and those that actually characterize men and women.[50] In the comparative format, 63% of the male and female judges thought that being loyal was desirable in both men and women, 57% of the men and 63% of the women thought that being cheerful was desirable in both. And 47% of the men and 37% of the women believed that both men and women should be sensitive to the needs of others and warm. The percentages of men and women indicating that supposed masculine traits were equally desirable in men and women is as follows.

	Men	*Women*
Self-reliant	30	33
Defends own beliefs	50	37
Independent	27	30
Athletic	18	3
Assertive	40	30
Strong personality	43	47
Forceful	13	3
Analytical	44	44
Has leadership ability	17	20
Willing to take risks	10	25
Makes decisions easily	37	50
Self-sufficient	27	30
Dominant	10	10
Masculine	3	0
Willing to take a stand	27	33
Aggressive	20	7
Acts as a leader	10	13
Individualistic	40	50
Competitive	17	10
Ambitious	40	50[51]

On other fronts, scholars in a wide variety of disciplines are challenging the idea that masculinity is the norm.

Denying the Incorporative Male

The idea that man is the standard exists both explicitly and implicitly, from coverture—the nineteenth-century notion that men (fathers and husbands) legally represented women—to use of the word *man* to include women (*mankind*). But in contemporary textbooks, newspapers, books, magazines, and political speech the generic "he" has given way to "he or she" or "they," and "mankind" has been replaced with "humankind," "people," or "humanity."

Agitation for change was a key element in historian Mary Beard's important 1946 book *Women as Force in History.* "For hundreds of years the use of the word 'man' has troubled critical scholars, careful translators, and lawyers," she wrote. "Difficulties occur whenever and wherever it is important for truth-seeking purposes to know what is being talked about and the context gives no intimation whether 'man' means just a human being irrespective of sex or means a masculine being and none other."[52]

The change finally occurred in the wake of social scientific evidence, produced in the main by women, that the social consequences of excluding women were detrimental. Women, for example, were less likely than men to see themselves described in want ads specifying a search for the supposedly generic "he." Female students were less apt to view themselves as scientists and engineers when textbooks showed only male exemplars.

Using *man* as a generic can have other pernicious results. Until recently, not even female scientists objected to the omission of women from the large-scale physiological studies funded by the National Institutes of Health—a flaw not remedied until the final years of the Bush administration under NIH head Bernadine Healy. "[N]either I nor so many others had noticed it before," observed Carol Gilligan in 1985. "It is such an enormous design flaw. How can organizations such as the National Institute of Mental Health and the National Institute of Child Health that pride themselves on reviewing research, have funded, year after year and for incredible amounts of money, studies that had all-male samples and never have seen this as a design problem?"[53]

In spring 1992, during the final year of the Reagan-Bush administrations, NIH made an announcement out of keeping with a backlash hypothesis. The Women's Health Initiative, the largest research study ever proposed by the Institutes, would devote $625 million to a fourteen-year study of more than 160,000 women between the ages of 50 and 79. The study would probe the causes of cancer, heart disease, and osteoporosis and the effects of estrogen replacement therapy, calcium, vitamin D therapy, and diet. Two years later the Centers for Disease Control and Prevention announced plans to open a new office, The Office of Women's Health to oversee projects on spouse battering, breast cancer screening, and research into female-controlled contraception.

The subtle but pervasive notion that women are biologically and socially defective contributes to the hold of the femininity/competence bind. Under pressure from scholars, assumptions of defect are giving way to assumptions of constructive difference. Recent thinking in biology, linguistics, and film studies is illustrative.

Biology

Although the womb/brain bind has been consigned to the history books, the assumption that women are driven by biology and are biologically defective persists in subtle ways.

"While man has been categorized in terms of a generally limitless potential, for rational thought, creativity, and so on, woman has been viewed as functionally determined by her reproductive role, and her actual and potential abilities perceived as stunted."[54]

Ideas about feminine passivity and defect are deeply embedded in the rhetoric of scientific literature, as Ruth Herschberger pointed out in the 1940s. "The patriarchal biologist employs *erection* in regard to male organs and *congestion* for female. . . . erection is too aggressive-sounding for women. Congestion, being associated with the rushing of blood to areas that have been infected or injured, appears to scientists to be a more adequate characteristic of female response. . . . While robbing us of some of our illusions about father science, the discussion [of the fallacies in this construction] may have a salutary effect upon poets, who have expressed fears that the language was losing its flavor and its myth-creating qualities."[55]

Where the sperm starred in patriarchal biology, the ovum played the lead in Herschberger's reconstruction. "The male sperm is produced in superfluously great numbers," she writes, "since the survival of any one sperm or its contact with an egg is so hazardous, and indeed improbable. The egg being more resilient, and endowed with solidity, toughness, and endurance, can be produced singly and yet effect reproduction."[56]

Although Herschberger's rendering broke the ground, it had little immediate effect on the ways in which the texts described the female role in reproduction. The idea of defect was pervasive. When the stomach lining is shed, for example, as it is routinely in both men and women, medical texts describe it as regeneration and renewal; when the lining of the uterus is shed, as it is in menstruating women, medical texts describe it as degeneration, sloughing, disintegration.[57] "One can choose to look at what happens to the lining of stomachs and uteruses negatively as breakdown and decay needing repair or positively as continual production and replenishment," one critic wrote in 1987. "Of these two sides of the same coin, stomachs, which women *and* men have, fall on the positive side; uteruses, which only women have, fall on the negative."[58]

Centuries of assumptions were finally sloughed off in September 1993 with publication in *The Quarterly Review of Biology* of a revised view of the process of menstruation. The monograph, authored by MacArthur award-winner Margie Profet of the University of California, argues that "menstruation evolved as a mechanism for protecting a female's uterus and Fallopian tubes against harmful microbes delivered by incoming sperm."[59] In Profet's view, the uterus functions proactively and positively, discarding the outer lining of the uterus and any pathogens lingering there and bathing "the area in blood, which carries immune cells to destroy the microbes."

Among other things, the theory explains why women using IUDs have heavy menstrual periods. And the theory suggests that physicians should not respond to inexplicable uterine bleeding by trying to suppress it, but should interpret it as a possible sign of infection.

Profet's theory throws over the assumption of defect by transforming the supposedly passive uterus into a vigilant agent. In the absence of an approach that saw menstruation as an adaptive physiological phenomenon, "functional mechanisms such as menstruation, are sometimes even viewed as defects," wrote Profet. "The seventeenth-century anatomist Regnier de Graaf (for whom the Graafian follicle of the ovary was named), hypothesized that 'The menstrual blood escapes by the feeblest parts of the body, in the same way that wine or beer undergoing fermentation escapes by the defective parts of the barrel.' Although researchers and menstruating women in modern societies may find this particular hypothesis absurd, they have not abandoned its underlying theme: that menstruation is a process of an inefficient body that repeatedly wastes, rather than recycles, its nutritious blood-filled uterine lining."[60]

When this publication was greeted throughout the community of biologists with astonished words of praise, it reaffirmed the important role that women have and will continue to play in the reinterpretation of women's biology and aptitudes. Interestingly, Profet had foresworn the PhD, a credential long denied women, on the grounds that it dampens creativity.

Linguistics

In 1922, the Danish linguist Otto Jespersen claimed that women have less extensive vocabularies than men, often use hyperbole and strong adverbs indiscriminately and without regard to the actual meaning of words, and "break off without finishing their sentences, because they start talking without having thought out what they are going to say."[61] Male speech, for Jespersen, was the norm; female speech, deviant. His claims were based on stereotypes (although not gargantuanly unusual ones), not social science.

Half a century later, linguist Robin Lakoff took a different tack, arguing that women used what she termed "powerless" speech but attributed it to women's lack of power. In one of the first book-length treatments of women's discourse, Lakoff claimed that women typically use more hesitations and pauses than men, employ more "empty" adjectives such as "charming," use more tag questions—ending sentences with phrases such as "isn't it" that supposedly seek approval, and employ more hedging forms such as "maybe" and "perhaps."[62]

Lakoff's women are victims who are taught a language that "submerges" their "personal identity" by denying them the means of expressing themselves strongly and also "encouraging expressions that suggest triviality in subject matter and uncertainty about it."[63] Women are the agents of their own undoing. "[W]omen prejudice the case against themselves by their use of language."[64] The cover of her book shows the picture of a woman's face with a bandage over her mouth.

Feminist scholars responded that women do not differ from men in some of the ways identified by Lakoff, that some differences are assets, not deficits, and that Lakoff did women an injustice by treating male speech as the norm and female speech as deviant. Scholars, they argued, ought not perpetuate the assumption of women as defectively other.

Others questioned whether women do in fact use more tag questions and suggested that they are not so much an expression of powerlessness as instances of cooperative, facilitative conversation.[65] Similarly, Jennifer Coates argued that women use hedges such as "perhaps" cooperatively, not powerlessly.[66] Where Lakoff saw weakness and a deficit, her critics perceived constructive forms of cooperation.

When women use language differently, the difference, in other words, is not a deficit. "She takes male language as the norm and measures women against it," wrote Dale Spender, "and one outcome of this procedure is to classify any difference on the part of women as 'deviation.' Given these practices, it is unlikely that Lakoff could have arrived at positive findings for women."[67]

Film Studies

In the 1970s feminist film critics began to argue that popular culture has divined and positioned women as "lack" and "other." The camera invited men and women to see from the man's point of view. "Mainstream cinema's contradictory/complementary representations of women as either idealized objects of desire or as threatening forces to be 'tamed' are not attempts to establish female subjectivity," argued E. Diedre Pribram, "but rather reflect the search for male self-definition."[68]

This scholarly focus on woman as victim was not long-lived. "[B]y the mid-1970s, feminist historians and literary analysts had turned away from a preoccupation with explaining women's victimization and towards a 'recuperative' project which soon found evidence of significant female power evident in records of daily life and in literary texts by both male and female writers."[69]

A parallel shift occurred in the popular debate in the early 1990s with the publication of Naomi Wolf's *Fire with Fire*.[70] Where Wolf's first book *The Beauty Myth*,[71] published in 1991, was subtitled *How Images of Beauty Are*

Used Against Women, her second, published in 1993, carried the banner *The New Female Power and How It Will Change the 21st Century.* Passive voice had become active; those against whom power had been used were now its custodians. *Fire with Fire* even contains a chapter titled "Victim Feminism Versus Power Feminism." In it Wolf castigates those who create "an identity" from victimization—which, as I read it, is what she had done in her earlier work with such claims as "[t]he contemporary ravages of the beauty backlash are destroying women physically and depleting us psychologically."[72]

Also having second thoughts are those who pioneered the argument that battered women who kill their husbands are victims. Law professor Elizabeth Schneider, who helped craft that legal defense, notes that it is "a two-edged sword. Many battered women lose custody of their children because judges see them as helpless, paralyzed victims who can't manage daily life." And if *victim* is one pole defining women, then the scale by which they are measured creates a double bind for battered women. "[I]f a woman seems too capable, too much in charge of her life to fit the victim image, she may not be believed," adds Schneider.[73]

The problem with seeing women as victims is that the label invites resignation and passivity, not redress. At the same time, the term hints that its wearer lacks the capacity to exercise the power women as a group have gained in centuries of transcending double binds.

Title VII

By including protection from sex-based discrimination in employment in Title VII of the Civil Rights Act of 1964, Congress handed women and men the legal remedies to break the femininity/competence bind. This protection was won by accident—when an opponent of the act inserted it in an unsuccessful effort to defeat the overall bill.

The definition of sex-based discrimination was expanded and refined during the supposed "backlash" decade of the 1980s. In 1980, the Equal Employment Opportunity Commission (EEOC) issued guidelines defining sexual harassment as a violation of Title VII. In 1986 the Supreme Court, in *Meritor Savings Bank v. Vinson,*[74] identified two categories of harassment. Outlawed were "quid pro quo" harassment, which occurs when a supervisor requires sexual submission as a condition of receiving employment advantages, and "hostile environment" harassment, which bans the creation of a hostile work environment on sex-based grounds.

In 1989, the femininity/competence double bind came before the Supreme Court for review in the Title VII case of Ann Hopkins against Price Waterhouse. The case demonstrates the ongoing hold of the bind, the impact of the scholarly attacks on it, and the power Title VII of the Civil Rights Act gave the courts to overthrow it.

For five years, Ann Hopkins had been working in the Office of Government Services at Price Waterhouse in Washington, D.C. In 1982, she had every reason to believe she would be given a partnership in the accounting firm. None of the others considered for partnership in 1983 had brought in more business or billed more hours. The contract awards Hopkins secured with the Department of State and the Farmers Home Administration were worth between $34 and $44 million.

The firm's senior partner, Joseph Connor, described the State Department contract, a computerized system capable of managing State department transactions worldwide, as a "leading credential" that made it possible for Price Waterhouse to win other business from the federal sector. One official with whom she dealt at the State Department described her as "extremely competent, intelligent . . . strong and forthright, very productive, energetic and creative." Another praised her "decisiveness . . . and intellectual clarity." In other words, both the bottom line and customer satisfaction said that Ann Hopkins was a star.

Her productivity was not the only thing that set Hopkins apart from others at Price Waterhouse. In 1982, seven of Price Waterhouse's 662 partners were women. Eighty-eight employees would be nominated for partnership that year and forty-seven selected. Hopkins was the only woman in the group.

She didn't make the cut. Instead of admitting her to partnership, the firm postponed consideration. For practical purposes, the nominee with the most billings had been turned down. Her efforts to find out why revealed that she had been ensnared in the femininity/competence double bind. The firm's chair advised her to "take charge less often." Her strongest supporter urged her "to walk more femininely, wear make-up, have her hair styled, and wear jewelry."

When the court required Price Waterhouse to disclose the notes of its reviewing partners, it found an ongoing pattern of stereotyping. Earlier female candidates had been characterized as "curt, brusque, and abrasive" and "Ma Barker." One partner was on record commenting that he would never vote for a female partner.

Hopkins herself elicited such remarks as:

> [S]he is consistently annoying and irritating—believes she knows more than anyone about anything, is not afraid to let the world know it. Suggest a course at charm school before she is considered for admission . . . .

> [S]he may have overcompensated for being a woman . . . .

> She uses "foul language."
> [I]f you get around the personality thing she's at the top of the list or way above average."

> She "had matured from a tough-talking, somewhat masculine hard-nosed mgr. to an authoritative, formidable, but much more appealing lady partner candidate."

> She had "a lot of talent" but required "social grace."

One clear hint that stereotyping was at work was contained in the statements of partners who admitted they had limited contact with Hopkins, but nonetheless offered categorical assertions about her personality and conduct. She is *"universally* disliked" (emphasis added) by the staff, said one. She is *"consistently* annoying and irritating," observed another.

The firm's senior partner urged Hopkins to "undertake a Quality Control Review, which would allow her to work with more partners, demonstrate her skills, and allay concerns about her ability to deal with staff." Although Hopkins followed his advice, she was not subsequently proposed for a partnership. Reading the writing on the ledger, she resigned in January 1984 to set up her own firm. And she took one more action. Under Title VII of the Civil Rights Act of 1964, she sued.[75]

Her attorneys argued that the promotion was denied not on merit but because she violated stereotypical female behavior. The U.S. Court of Appeals found such "convergent indicators" of stereotyping as "the extremely small number of female partners at the firm; the absence of any other female candidates among the 88 nominated along with Hopkins; the exaggerated and extremely intense negative reactions of Hopkins' critics to behavior that supporters perceived as positive; the ambiguous criteria the firm used to evaluate a candidate's personal qualities; the absence of complaints from Hopkins' clients; and the positive assessments of Hopkins in areas where performance could be measured objectively."

Backing Hopkins was The American Psychological Association, which weighed in with a summary of the scholarly literature. Drawn into its brief was the work of those scholars who had followed the pioneering path set by Thompson [Woolley] at the turn of the century. "Norms specify behaviors that are thought to be not only characteristic of each sex, but also desirable and encouraged," argued an APA brief. "For example, females who display 'womanly' traits and males who display 'manly' traits are more favorably evaluated and judged more psychologically healthy than those who do not. Conversely, those who engage in what is perceived to be cross-sex behavior can be the victim of social sanctions."

Speaking to criticisms made of Hopkins, the brief added, "A woman who speaks aggressively, strides across the room, and wears no-nonsense clothes is perceived to be insufficiently feminine. Such discrepant individuals are psychologically 'fenced off' from the rest of the group into a subcategory, often one that is negatively evaluated. In the present case, Ms. Hopkins' supporters described her behavior as outspoken, independent, self-confident, assertive, and courageous. Her detractors interpreted the same behavior as overbearing, arrogant, self-centered and abrasive."[76]

The result, according to the APA brief, was a "double bind situation." If women "are viewed 'as women' they are frequently denied access to high power positions because their presumed attributes cause them to appear incapable or their performance is ascribed to something other than competence. This is particularly the case if coworkers convey, even in subtle ways, their lack of support for a female leader. If, however, they are perceived as engaging in 'masculine' behaviors deemed essential for the job, they are considered to be abrasive, or maladjusted. In many cases, then, the achievement oriented woman is caught_whatever her behavior, it bodes ill for her career."[77]

At the district court, one of the dissenting judges betrayed the powerful hold of the very assumptions the APA was unmasking. Judge Williams questioned the credibility of an expert witness—Dr. Susan Fiske, a social psychologist and Associate Professor of Psychology at Carnegie-Mellon University—who argued that the comments of the partners were stereotypic. "To an expert of Dr. [Susan] Fiske's qualifications, it seems plain that no woman could *be* overbearing, arrogant or abrasive," wrote the judge. "[A]ny observations to that effect would necessarily be discounted as the product of stereotyping. If analysis like this is to prevail in federal courts, no employer can base any adverse action as to a woman on such attributions." What is the likelihood, one might ask Judge Williams, that the triad—overbearing, arrogant, or abrasive—would have been applied to a man?

In its efforts to discredit Professor Fiske, the attorneys for Price Waterhouse applied the corollary of the double bind their clients had inflicted on Ms. Hopkins. Where Hopkins was criticized because she was competent but too "masculine," Fiske's competence was challenged on the grounds that her conclusions were arrived at "intuitively." Women, goes the stereotype, are intuitive rather than rational, with reason valorized and intuition dismissed as a way of knowing.

The Supreme Court chided Price Waterhouse's attorneys for their attempt to dismiss Professor's Fiske's conclusions as "intuitively divined" and as "intuitive hunches." "Nor are we disposed," they wrote of the lower court's ruling, "to adopt the dissent's dismissive attitude toward Dr. Fiske's field of study and toward her own professional integrity."

The Supreme Court agreed that stereotypes had operated in the decision to deny Hopkins a partnership at Price Waterhouse. "[I]t takes no special training to discern sex-stereotyping in a description of an aggressive female employee as requiring 'a course at charm school'" . . . and "if an employee's flawed interpersonal skills can be corrected by a soft-hued suit or a new shade of lipstick, perhaps it is the employee's sex and not her interpersonal skills that has drawn criticism," noted the high court.[78]

Indeed, the Court specifically described Hopkins's situation as a bind. "An employer who objects to aggressiveness in women but whose positions require this trait places women in an intolerable and impermissible Catch-22: out of a job if they behave aggressively and out of a job if they don't." "Title VII," announced the Justices, "lifts women out of this bind."

To determine whether Hopkins would have been promoted in the absence of the stereotyping, the High Court remanded the case to the lower court. There the judge ordered Price Waterhouse to admit Ann Hopkins to partnership, beginning July 1, 1990, and to give her the back pay she had lost.[79]

Under mandate from the Federal Advisory Committee Act and the Civil Rights Act of 1991, the Department of Labor's Glass Ceiling Commission continues to monitor the progress of women. The Commission minimizes use of the bind by holding out the possibility that such use will carry legal penalty. This should produce scrutiny to ensure that hiring and promotion are based on merit, not stereotyping, for, as an amicus brief filed by the American Psychological Association argued in the Hopkins case, "the opinion of a third party to the evaluation process, particularly that of a superior or colleague, can exert a major influence on the decisional process and the decision itself."[80]

Making the Exception the Rule

It will probably take a while before an Ann Hopkins is CEO of a major accounting firm. In Fortune 1000 companies, 16.9% of the total number in management are women and 6% minorities; at the level of vice president and above, 6.6% are women and 2.6% minorities.[81] An exploration of SEC reports from 806 public Fortune 500 industrial and Fortune 500 service companies from July 1991 to December 1992 found that while 55% had a woman on their board of directors, only 16% had more than one. Women and minorities held "864 (about 9%) of the 9707 total board seats at the companies studied," with white women accounting for 5.5% of the seats and white men 88.6%[82]

Demographically, however, men are now "other." In 1992, the nation's 131 million women constituted 51% of the total population. Sheer numbers help to guarantee a large entry pool into education, business, and government.

In 1991 "almost equal proportions of the adult men and women 25 years and older had attained a high school diploma. . . . In 1991, almost equal shares of women and men . . . had finished between one to three years of college." In the white adult population, however, only 19% of the women had completed four or more years of college, compared to 24% of the men. If recent patterns of graduation hold, however, the percentage of women will soon match that of men with college degrees, for in that 1991 survey "24% of the 25–34 year olds had finished four or more years of college, regardless of gender." (Women of other races did not fare so well: only 12% of African-American women and 9% of the Hispanic women had completed four or more years.) The income gap separating college-educated men and women has also closed a bit. In 1986, women with bachelor's degrees "made about 35% less than their male counterparts; in 1991, they made 31% less."[83]

In several traditional male bastions the numbers of women in the entry pool is high enough to warrant cautious optimism about the future. Although only 8% of the partners in the country's 250 largest firms are women, 42% of those now in ABA-accredited law schools are women.[84] In 1990, women earned 14% of the bachelor's degrees awarded in engineering, 30% of the computer science degrees, and 48% of the degrees in biological science. Since 1986, more than half of the Master's degrees have been awarded to women. Just over 3000 more men than women earned PhDs in 1991,[85] and 47% of those seeking MBAs were women.[86]

According to psychologists, women are at highest risk of stereotypic appraisal when they form less than 15 to 25% of a management level. When women move in large numbers into upper management, as in many professions they are now poised to, the evaluative norms will change. In theory at least, women will then be seen not as women managers, but simply as managers. Meanwhile, in settings in which a minority makes up less than 15% of the work force, psychologists say that the members of that group have the psychological equivalent of "solo status." Indeed, one study has shown that female applicants to a given field are disadvantaged when they form less than 25% of the applicant pool.[87]

As the American Psychological Association argued in its amicus brief for Ann Hopkins, "Singular or rare individuals attract more attention, are evaluated more extremely, and are more likely to be perceived as enacting stereotyped roles, and are believed to have a greater, sometimes more disruptive, impact on the group."[88] When men are in "solo" positions, they are disadvantaged as well.[89] As important, women with "solo" status are less happy in their jobs than women whose workplace includes women in senior positions.[90] So the women waiting in the wings in such firms may decide to wait elsewhere.

One might surmise that the socialization patterns will change with the advent of large numbers of women-owned businesses. The Women Business Owners Foundation reports that by the end of 1993, "there will be more Americans employed in women-owned businesses than [in] the Fortune 500 companies."[91] A Small Business Administration Study shows that women are starting small businesses one and a half times faster than men. By the turn of the century, women will own 40% of all U.S. small businesses.[92]

This trend, together with increased numbers of women in the pipeline for corporate executive positions, prophesies eventual change—however slow in coming. As women ponder their future, two other social trends should increase their comfort in executive positions and accelerate their acceptance by others. The first is the increasing female presence on the faculties of colleges and universities through which virtually all of society's future leaders move. The second is a dramatic increase in the number of women in powerful, visible, and once "male-only" positions in government and the media.

We are more likely today than at any time in the past to turn on the TV or open the newspaper to read something by or about a powerful, visible woman. Political reporters Robin Toner and Maureen Dowd regularly claim front-page space in the *New York Times*, as do Dana Priest and Ann Devroy in the *Washington Post*. Eleanor Clift is both a *Newsweek* correspondent and part of the *McLaughlin Group*.

The 1992 campaign forecast a change in who speaks for whom. Torrie Clarke and Mary Matalin took Bush's case to the public; Mandy Grunwald and Dee Dee Myers advanced Clinton's. In the White House, Myers continued her work as Clinton's press secretary. And of Clinton's first 500 appointments to high-level posts, women made up nearly 40%.[93]

In politics as well as business, women are holding posts once stereotyped as male. Lynn Martin followed the lead of FDR's Francis Perkins as Bush's Secretary of Labor. Clinton's pathbreakers include Attorney General Janet Reno, Energy Secretary Hazel O'Leary, Health and Human Services Secretary Donna Shalala, Office of Management and Budget Director Alice Rivlin, and EPA head Carol Browner. Joycelyn Elders succeeded Antonia Novello as Surgeon General. The UN representative Madeleine Albright isn't a "first"; her predecessors included Jeane Kirkpatrick. And in the Clinton inner circle is the architect of the administration's most important domestic initiative, national health reform—Hillary Rodham Clinton.

Where, until 1971, only three women had served as a Justice's law clerk in the Supreme Court,[94] in 1993, there were two women on the court itself.

These appointments convey many messages. One is that women are not cut from identical bolts of cloth. The appointment of Ruth Bader Ginsburg to the Supreme Court gives those seeking female role models a second there, different in style, temperament, and ideology from the first. Elected officials also display such differences. C-SPAN brings its viewers moments on the House and Senate floor with women as dissimilar in background and bearing as Barbara Mikulski and Nancy Landon Kassebaum—one who voted against confirming Clarence Thomas, one who voted for.

Women also hold top on-camera positions in network news. More women than men anchor CNN's 24-hour *Headline News*. Candy Crowley reports on politics for CNN. On CNBC two women co-host *Equal Time*. NBC's top political correspondents are Lisa Myers and Andrea Mitchell. Connie Chung is now Dan Rather's co-anchor at CBS. Each of the three network morning news shows is co-hosted by a woman and a man. This move at the national level actually follows a local trend—most local broadcast news shows have male and female co-anchors.

Former *Face the Nation* host Lesley Stahl has a regular slot on a program with top network ratings—*60 Minutes*. Cokie Roberts is a regular on *This Week with David Brinkley*. Linda Werthheimer hosts NPR's *All Things Considered*. And as the lead wire-service reporter covering the White House, Helen Thomas effectively opens and closes the president's news conferences. A 1992 survey concluded that women hold 19.4% of the top editorial and managerial positions in the nation's newsrooms[95]—a percentage that now moves toward 25%.

The extent to which the femininity/competence bind has been overcome by these reporters was evident in the 1992 presidential campaign, when Independent candidate Ross Perot responded to pointed questioning by

In summer 1993, Connie Chung was named co-anchor of CBS Evening News. Chung and her co-anchor Dan Rather responded with poses suggesting that they had instead just announced their engagement. *TV Guide* captioned this photo "The Honeymooners . . . Rather and Chung at the press conference announcing TV's new pecking order." (Mitsu Yasukawa, New York, © 1993 NEWSDAY.)

female reporters by remarking that they were trying to prove their "masculinity." Translation: Get back in your bind. They refused to back down. Instead of being hurt by the charge of lack of femininity, they responded with competence. Indeed, the best broadcast reports on Perot in the 1992 campaign were Lisa Myers's reports on NBC and Lesley Stahl's *60 Minutes* examination of his claim that George Bush planned to sabotage his daughter's wedding.

While such visible changes are heartening, women nevertheless remain underrepresented in the newsroom, a problem not readily remedied in a time characterized by the demise of some newspapers and the downsizing of staff at others. "One thing that has not changed much in American journalism, to our surprise," write a team of scholars profiling the newsroom in the 1990s, "is the percentage of women working for all different news media combined. In spite of rapidly increasing enrollments of women in the U.S. journalism schools during the 1980s, and the emphasis on hiring women since the late 70s, the overall percentage of women has remained virtually unchanged."[96] A Freedom Forum survey found that 34% of all journalists are women, a stable figure in the past decade.[97]

As women and men are socialized through higher education, they are seeing women behind the podium. In 1991, Eve's offspring made up 32% of full-time faculty, with 15% at the full professor rank, 28% in associate professorships, and 40% assistant professors. But in academe as in business, salary disparities persist. In 1991 male full professors averaged $59,180 per year, female $52,250. The student in the classrooms, however, sees not a salary but a competent woman knowledgeable in her field and in control of part of the student's academic destiny.[98]

Increasingly, women are heading colleges and universities. "In 1992, 348 women were chief executive officers of [private and public] colleges and universities, a 22% increase from 286 in 1984. . . . In 1975, only 16 public institutions were headed by women, by 1992, 164 institutions were."[99] As I write this, four of the University of Pennsylvania's twelve deans are women, as are two out of three of its top administrative officers. Two of the four deanships, Nursing and Education, have traditionally been open to women—indeed in Nursing a male dean is the exception—but the women deans in Fine Arts and Architecture, Arts and Sciences, and Communication are less usual. And the University of Pennsylvania became both the first Ivy League school to be headed by a woman interim president—Claire Fagin, in 1993—and by a female president, when Judith Rodin was inaugurated in 1994. Rodin is the fourth woman to head a major research university, following Hanna Gray and Donna Shalala, who headed the University of Chicago and the University of Wisconsin, respectively, and Nannerl Keohane, who was inaugurated as the eighth president of Duke University in October 1993.

And for those keeping score, there is progress as well. In the past decade the number of female conductors of American orchestras has risen from a dozen to fifty-seven. When a European conductor told Marin Alsop, conductor of Denver's Colorado Symphony, that "women conductors should stick with chamber orchestras because they can't conduct weighty music like Bruckner," she responded, "Why not? Our batons weigh the same."[100]

In 1992, the number of women in the U.S. Senate tripled to six; the House added twenty-four female representatives. Where in 1973 sixteen women were serving in the House, holding 3.7% of the seats, in 1993 there are forty-seven, holding 10.8%.[101] In 1993 1517 women are state legislators.[102] Eight states have females serving as attorneys general; thirteen have female secretaries of state.[103] Three states are headed by women governors; 175 women hold mayorships in cities with over 30,000 in population.[104]

The public is apparently becoming accustomed to representation by women in government. "The reason there aren't more women in public office is not that women don't win, but that not enough women have been candidates in general elections," a comprehensive study of office-seekers conducted by the National Women's Political Caucus concluded in 1994. "When women run, women win as often as men."[105]

A central message of the Hill-Thomas hearings was conveyed not by words but by pictures: Hour after hour viewers saw the questions being asked by an all-male, all-white Judiciary Committee. Women were literally out of the picture, and voters took note. The 1992 elections changed that. On the committee in 1993 were Senators Carol Moseley Braun and Dianne Feinstein.

Similarly, in Texas, Rose Spector crafted a television spot pointing out that just as there was something wrong with the all-male Judiciary committee, there was something missing in the photo of the Texas Court. It too was all male. Although she was outspent by almost two to one, Spector changed that picture by becoming the first woman elected to the Texas Supreme Court.

Where observers in the past could marvel at a single George Eliot or Eleanor Roosevelt, they will now find if not hundreds then at least scores in their livingrooms and dens, courtesy of television. Every woman I know confesses that she is a bit startled to hear a woman's voice and see a woman's face associated with high governmental office. But what today seems startling is tomorrow's status quo. There are people, for example, who think that from the beginning of time Los Angeles sprouted palm trees, soup came in cans, and *Saturday Night Live* appeared at 11:30 P.M. EST on NBC.

The visual intimacy of television lures us into believing that we know the people—and hence the women—we invite regularly into our homes. We know them as different personalities, with different competencies and styles. Stereotypes cannot survive such an onslaught of disconfirming evidence. Arguing that George Eliot was an exception was possible because most people didn't know many great authors, male or female. We are getting to know tough and caring female senators and congresspersons, cabinet secretaries and first spouses, anchors and reporters, and justices of the Supreme Court. And they no longer enter our world, in Ruth Bader Ginsburg's phrase, as "tokens of one."

7

Double Bind Number Five:
Aging/Invisibility

"**D**O any of you stand a chance of making an unsolicited appearance in Senator Packwood's little memoir?" a reporter asked NPR's Cokie Roberts, Nina Totenberg, and Linda Wertheimer. The question referred to charges of sexual harassment levied against the Senator and possibly recorded in his diaries. Two out of three in turn denied the possibility because they were too old. They responded:

NT: Not unless he's been fantasizing.
LW: This is part of being older. I mean, I would say that in the last ten or so years any member of Congress that came after me would be somebody with an asterisk by his name that said "senile," "demented."
CR: The Members of the House—they're just babies! They're *soooo* much younger than we are. They call us "ma'am." They see us as dowagers.
NT: Speak for yourself, Cokie!

In January 1994 when this interview was printed, Roberts and Wertheimer were 50, Totenberg, 49. Packwood was 61. Yet what is probably midlife for the stars of NPR would in fact have been old age a century ago. The nineteenth-century portrait "A Study in Black and Gray," also known as Whistler's Mother, showed a woman in her forties.

For reasons too complex to explore here, ours is a society pervasively influenced by stereotypes about aging. They are as old as Western culture. Aristotle, for example, characterized old men as "cynical," "distrustful," "cowardly," "fearful," and "shameless." "Their fits of anger," he noted, "are sudden but feeble. Their sensual passions have either altogether gone or have lost their vigor."[1] Denied both citizenship and formal access to the public sphere, older women went uncharacterized. Women were spared as well by Samuel Johnson, who observed that "there is a wicked inclination to suppose an old man decayed in his intellect." That "wicked" inclination drew support from age-biased IQ tests that defined and measured abilities important during youth.[2]

Nestled among our stereotypes about aging are a rash of double standards. "If old people show the same desires, the same feelings and the same requirements as the young," remarked Simone de Beauvoir, "the world looks upon them with disgust; in them love and jealousy seem revolting or absurd, sexuality repulsive."[3]

If judging older people differently creates a double standard, adding gender to the equation multiplies the injustice. Fifty-year-old women, for example, are not seen to be as attractive as their younger sisters, and older men are thought more attractive than older women,[4] a view more often held by men than women.[5] Indeed, "youthful" is a synonym for "attractive" in an older person.

Consistent with the notion that women are supposed to reach middle and old age sooner than men,[6] the "best age" to marry, have children, and shoulder major responsibilities is assumed to be younger for women than men.[7] But, as we have seen, women must overcome the forms of sex discrimination that make early career advancement more difficult, and often pay a professional price for having children during their twenties and thirties. By the time they have surmounted these hurdles, they are faced with the penalizing stereotypes of female aging.

A second bind makes it difficult for women to recognize role models of their future selves. The dearth of older women in advertising, prime-time television, and films makes locating such models less likely. At the same time, the disposition of successful older women to downplay their age makes it more difficult to perceive them as older.

Until the mid-1970s, when the aging rights movement demonstrated its clout with the passage of a bill protecting workers from age discrimination until the age of 70, most age stereotypes were negative. At their core was the belief that age inevitably signals physical and mental decline. One survey concluded that the public views "older" people as uglier and more boring, talkative, prejudiced, and forgetful than "younger" ones.[8] Others found that we assume that older individuals are unattractive, poor company, ill, tired, and inactive.[9]

Undergirding the stereotypes is the assumption that those over a certain age form a homogeneous group of like-minded, comparably capable or incapable individuals. Instead, "the most powerful fact on which all gerontologists agree is that the elderly are heterogeneous on nearly all the attributes ever measured."[10]

Challenging the negative stereotypes were aging rights organizations including the Gray Panthers, The Older Women's League (OWL), and the American Association for Retired Persons (AARP). These groups drew support from scholars of the human life cycle and a growing number of gerontologists. Magnifying the political clout of the senior groups is the greater disposition of older persons to vote and, with the successes of Social Security, Medicare, and Medicaid, the decreasing likelihood that they will find themselves in poverty.

As "senior power" was successfully marshaled in defense of cost-of-living adjustments in Social Security, the stereotypic notion that older Americans are healthy, wealthy, and stealthy took hold. Some believe that it has supplanted the notions predicated on passivity and weakness. In their place, goes this hypothesis, is resentment of older Americans as a self-protective group eager to feather its nest, even if the cost is the economic and educational well-being of younger generations.[11]

Just how formidable seniors are is evident in a 1988 survey by Louis Harris and Associates that found that members of Congress, the Senate, and federal officials ranked the elderly above doctors, hospitals, employers, health insurers, and any other group in influencing health-care policy.[12] The countermovement, organized in 1985 under the title Americans for Generational Equity (AGE), is set on protecting the interests of younger and future generations.

In their advocacy of such causes as banning mandatory retirement, aging rights activists occasionally trafficked in stereotypes of their own, including the assumption that older Americans are invariably healthy, quick-witted, competent, and active. Some theorists label such views "positive age-ism."[13] The older person is just like the rest of the population, goes this position, which is called into question by the movement's own championing of home health care, adult care centers, Meals on Wheels, nursing home reform, and research into the causes and consequences of Alzheimer's Disease.

Indeed, the National Health Interview Survey concludes that those over 65 have "twice as much disability and four times the limitation in activity" of younger adults. "They account for 30% of all hospital discharges, 20% of all doctor visits, and one-third of the country's personal health care expenditures."[14]

The truth resides somewhere in between the unremittingly positive and negative, with no generalizations holding for most. People age differently; the process is a function of heredity and health care, among other factors. These differences prompted Bernice Neugarten to identify active older persons as "young old' and those who are more frail as "old old."[15] On the assumption that some common problems begin to emerge at age 85, The National Institute on Aging now identifies that group as "the oldest old."

Whether the stereotypes of aging are positive or negative, women are disadvantaged. Positive images of power and privilege may overshadow the fact that more than 15% of the elderly "still live in poverty and among women, minorities, and the oldest old these figures are considerably higher."[16] Because women outlive men, the oldest old are predominantly women.

Women also are more disadvantaged by negative social perceptions of their aging.[17] This notion defies the physical evidence. Since women on average outlive men by seven years, should they not be assumed to have a longer rather than a shorter "middle-age" than men? Shouldn't women outnumber men in "older roles" cast for prime-time television shows and in films? Neither is the case.

The dramatic difference was highlighted when word filtered through the press that a 59-year-old British woman had given birth to twins on Christmas Day 1993. 'Medical and public opinion is divided here," wrote a British-based AP reporter, "though mostly disapproving of the wealthy career woman, who recently married a 45-year-old economist."[18] The mention of her career hints, of course, that she chose not to have children earlier. Here no doubt was the stereotypic selfish sequencer.

The inclusion of career is interesting for two reasons. First, she has "recently" married. Second, the child was conceived with donated eggs fertilized by her husband's sperm and implanted. Since she was post-menopausal at the point of conception, we do not know whether earlier in her life she was able to conceive.

"Irresponsible," said some commentators. "Women do not have the right to have a child. The child has a right to a suitable home," opined British Health Secretary Virginia Bottomley.[19] Such remarks reeked of a double standard.

She might not live to see her children reach adulthood, others said. Yet when Senator Strom Thurmond fathered four children, after his marriage at age 66, congratulations were heard all around. Since a 59-year-old woman can expect on average to live another 24 years, the mother-to-be was likely to see her children through college. Moreover, because longevity is positively correlated with standard of living, her comfortable financial circumstances predicted that she would probably be around longer than that.

Others implied that the woman's actions were pathological. "The problem is that nobody seems to know what is pathological behavior anymore. When people get fixated on wanting a child, nobody says, 'Get your head read.' These women are going to have a terrible time when they finally meet the Grim Reaper. They're really out of touch with reality, and most women are not," observed feminist author Germaine Greer.[20]

The concerns seem misplaced. The woman had delivered healthy twins and the new mother was doing well. And since she was married to a 45 year old, he presumably could continue parenting if something happened to her before the children were grown. But assume that women are responsibile for childrearing, that older men marry younger women and not vice versa, and that women slip inevitably into terminal decline after menopause, and the objections make sense.

We seem not to have thrown off the nineteenth-century notion that "Everything that is peculiar to her [a woman] springs from her sexual organization."[21] By conceiving after menopause, the complaint seemed to be that the new mother had violated the natural law of her reproductive organs. The reason her age is more noteworthy than that of her husband is that men don't mark their age by some recognizable biological event such as menopause. Since it is only within the last century that most women lived beyond their late forties, our focus on the climacteric is comparatively recent.

If some in the nineteenth century held that women who ignored their husbands and children, practiced birth control, supported women's suffrage, or sought out higher education would have a difficult, disease-ridden menopause,[22] in the twentieth century, women were offered the corollary claim that "[I]n the well-balanced, educated, contented woman who finds her family, sexual and professional life fulfilling, *there may be no symptoms* [of menopause] *whatsoever.*"[23]

Dr. Robert Wilson, the apostle of estrogen, was more specific when he suggested that symptom-free menopause was the outward sign of an inward disposition toward devotion to husband. And in a tangle of binds, those whose unwomanly past produces a "negative" menopause become, in Dr. Wilson's phrase, "dull-brained but sharp-tongued." "It has been my observation, based on more than forty years of gynecological practice," wrote Dr. Wilson in his best-selling *Feminine Forever,* "that women who at some time were genuinely and profoundly in love with their husbands do not develop such extremes of menopausal negativism [as becoming 'a dull-minded but sharp tongued caricature of her former self']. This lucky minority—about 20 percent of the patients I have encountered—apparently feel that they have already fulfilled their womanly destinies. One of these women once told me, 'If I die tomorrow, I have had all a woman can ever hope for—true love.'"[24]

Consider the dilemma in which such rhetoric placed women. In order to confirm that she is well-balanced and contented, a devoted and loving wife, a sensible woman who is experiencing symptoms might well conclude that it's not to her advantage to reveal them—silence out of fear of shame. If she suffers symptoms, she should blame her behavior, not her biology. And between hot flashes, the medical texts would have her conjure up a symptom-free sister rhapsodizing about having attained "all a woman can ever hope for—true love."

Femininity/competence for the menopausal woman came into the national spotlight with the publication of Wilson's book in 1965. It was not a working uterus that made one feminine, implied the author of *Feminine Forever,* but the estrogen that the reproductive organs supplied before menopause. Wilson promised that women who used estrogen therapy could be recognized by their appearance, including their "supple breast contours, taut, smooth skin on face and neck, firm muscle tone, and that particular vigor and grace typical of a healthy female. At fifty, such women still look attractive in tennis shorts or sleeveless dresses."[25] When the capacity to bear children is gone at menopause, Wilson suggested, as had medical men before him, women were no longer women. "Some women, when they realize that they are no longer women, subside into a stupor of indifference. Even so, they are relatively lucky. The most heartbreaking cases, I feel, are those sensitive women who witness their own decline with agonizing self-awareness."[26] But the *Forever* Wilson promised was shorter-lived than it was for those who foreswore his treatment. When the rates of uterine cancer rose with Wilson's therapy, *Feminine Forever* lost favor. The estrogen replacement therapy now on the market combines estrogen with progesterone to mute the cancer-causing effect, and is marketed not as a fountain of youth but a protection against osteoporosis and heart disease.

In production as well as reproduction, the double standard holds. The author of a major study of flight attendants observed that "Even women in their thirties were occasionally called 'granny' or subjected to within-earshot remarks such as 'Isn't she about ready for retirement?' As one woman in her mid-thirties noted: 'There is definitely a difference, oh yes. The men take it for granted that they can work until sixty or sixty-five. The women work like dogs to prove they can still do the job. And then they have to fight the granny remarks.'"[27] Male flight attendants haven't been around long enough to plausibly posit older ones. When the thirty- or forty-something-year-old male pilot saunters down the aisle to the rest-room, he is, however, unlikely to be called "gramps"—more likely he will be seen as a glamorous, respected figure. And we are a long way from the day when fliers will not respond with surprise to a pilot whose first name is Ann, not Andrew.

Powerful men are sexy, sexy women are powerful, and these proposi-tions are not at all the same. If middle age can confer power and increased sex appeal on men, in women the reverse is held to be the case. Those dismissive remarks about middle-aged flight attendants are based in deeply held assumptions about the relation between appearance, femininity, and womanhood. When college students, for example, were asked to evaluate photos of men and women of various ages, they concluded that the women's "femininity" dropped as they aged but the "masculinity" of the men went unchanged.[28]

These reactions are amply precedented. In the first century, the satirist Juvenal wrote, "Just let a couple of wrinkles appear, let her skin become dry and start to sag, let her teeth turn black and her eyes go puffy—'Pack Your bags!' a freedman will cry, 'and be on your way.'"[29]

"Since witches are usually old women of melancholic nature and small brains, (women who get easily depressed and have little trust in God), there is no doubt that the devil easily affects and deceives their minds by illusions and apparitions that so bewilder them that they confess to actions that they are very far from having committed," observed the sixteenth-century Belgian physician Johann Wier in *De Praestigiis Daemonum*.[30]

In his 1509 *The Praise of Folly*, Erasmus wrote, "Now it is even more amusing to see these old women, so ancient that they might as well be dead. . . . They are as hot as bitches in heat, or (as the Greeks say) they rut like goats. They pay a good price for the services of some handsome young Adonis. They never cease smearing their faces with makeup. . . . They drink a lot."[31]

"When menstruation is about to cease, the period is called critical, 'the change, or turn of life, the climacteric period'; and many important changes take place in the constitution at this epoch," observed nineteenth-century physician Dr. Michael Ryan in *A Manual of Midwifery* [1841]. "The cheeks and neck wither, the eyes recede in their sockets, and the countenance often becomes yellow, leaden-coloured, or florid, and the women become corpulent, and lose the mild peculiarities of their sex."[32]

"Buy old masters," Lord Beaverbrook was supposed to have observed in the 1960s. "They fetch a better price than old mistresses."[33]

Age stereotyping of women now seeps into the popular culture through the press, advertising, film, and television. In the process, it sabotages our sense that growing older is a natural part of human existence, and under-cuts the likelihood that old age will be perceived and hence have the possibility of being a time of continued development.

In advertising, young women advertise products to older ones. Since youth is what is being sold, the strategy is all but inevitable. Christian Dior Perfumes offers a product called Capture Lift, a skin-treatment designed, say its producers, for menopausal and post-menopausal women. But, un-less it has been surgically induced by hysterectomy, the model touting the treatment is trunkloads of Tampax from menopause.

"She is not in the age group that would buy the product," concedes Caroline Geerlings, VP of marketing, in an interview with *Working Woman* magazine. Does the "beauty"-product industry depict women within sight of menopause, which on average occurs at 50? The interview concluded that Lancombe's 41-year-old Isabella Rossellini "stands as the lone representative of the huge 40-plus market among major cosmetics companies." Why? "Geerlings feels that women aren't receptive yet to older models, though that will come 'down the road.' Andrea Robinson, president of marketing for Alexandra de Markoff, hopes to reach that point soon. She's repositioned the brand—the company's new tag line is "For Women, Not Girls"—and plans to run ads in '94 featuring *now-39-year-old* Patti Hansen (emphasis mine)."[34]

Fear of aging is one of the coins of the realm for advertisers who see older Americans as a $500 billion market.[35] Since such products are mar-keting perceived outward signs of youth, showcasing gray-haired women with crepe-like necks plus smile- and eye-lines would undercut the prod-uct's viability. Implicit in the unwrinkled faces of the models is the idea Clairol made explicit in the 1970s: You're not getting older, you're getting better.

We are, of course, all getting older, and since one is not the obverse of the other, we may be getting better as well. Yet, as men age in commercials, they become more distinguished; as women age, they disappear. And it is as true now as it was in 1978, when I coordinated two hearings on Age Stereotyping and Television for the House Committee on Aging, that when older women do appear in ads they are more likely than younger women or men of any age to be touting products signaling bodily decline. So, for example, two of the most frequently seen older women on television are testifying to the utility of denture creme (Martha Raye) and a diaper-like product to cope with adult incontinence (June Allyson). Glue yourself to the television for days and you will not find a single ad claiming to revivify the face of an older man. Yet "wrinkle cremes" for women abound. Embracing the tie between youth, a wrinkle-free face, and attractiveness, women are more likely than men to have facelifts.

In general, older people are underrepresented in product ads. One study conducted in 1990 found that only 11.8% of those appearing in network commercials were over 50.[36] Over 70% of the older characters identified in the study were males. And older people seem to be enjoying an active life in only three of the 54 commercials in which they appeared.[37]

One place older women usually show up is in medical journal ads for anti-depressants and other psychotropic drugs. By reflecting mistaken medical assumptions, such pharmaceutical advertising perpetuates stereotypes, and when such ads portray psychological abnormality as a norm among middle-aged women, they invite doctors to misdiagnose them.[38] Scholars have found that even when other indicators, such as measures of "nonpsychiatric psychological disburbances," show similarity among the males and females in a study, "physicians classified significantly more females than males as disturbed."[39] Men who visit a doctor for depression are less likely to leave with a prescription than women.[40]

In television programming, women fare no better than they do in televised advertising. For more than two decades, George Gerbner has asked about the roles and rules that persist in daytime and prime-time television. In summer 1993, his conclusions differed little from those published a decade before: "Despite changes in styles, stars, and formats, prime-time network dramatic television presents a remarkably stable cast. As we have noted two-thirds of the cast are men."

If television portrayed the world as it is, however, older women would be cast more often and more prominently than older men. In the United States there are 100 women for every 83 men among the population aged 65 to 69. Among 85 year olds the ratio is even starker—100 women to 40 men.[41]

In his study, Gerbner found that women play one out of three roles in prime-time television, one out of four in children's programs, and one out of five of those who make news. They fall short of majority even in daytime serials. Those women who do appear tend to be concentrated in the younger age groups. As they age, their proportion of roles as "settled adults" declines significantly more than men's. Moreover, they age *faster* than men, and as they age they are more likely to be portrayed as evil or unsuccessful.

"Seniors of both genders are greatly underrepresented and seem to be vanishing on the screen instead of increasing as they are in real life," Gerbner points out. "As characters age they lose importance, value, and effectiveness. Visibly old people are almost invisible on television. Mature women seem to be especially hard to cast—and hard to take. They are disproportionately underrepresented, undervalued, and undersexed."[42]

When older characters do star in such prime-time shows as *The Golden Girls, Murder, She Wrote, Jake and the Fatman,* and *Matlock,* their sexuality is not as central an element in the plot as it is when the leads are younger. The actors playing the parts are usually in their late sixties, the "young old." Still, sexuality "is an important absence in the lives of most elderly television characters, and especially in the lives of elderly women."[43] When older characters are cast in the lead, '[t]wo universes . . . seem to be depicted— a female universe of older women where there are no men, and a male universe of older men where there are women but no older women."[44] And in the sex-obsessed soap opera, foreplay begins to fall off in one's forties. Only 3.5% of those having sex in the soaps are over 40.[45]

In films, older men are also more likely to star than older women. The Screen Actors Guild indicates that about seven of ten feature film roles go to men, with few good roles available for women over forty, while the forties, fifties, and sixties are a male actor's highest earning years.[46] Where men typically win their best-actor awards in their mid-forties, the best actress award is more likely to go to women in their thirties. And we can munch popcorn in a darkened theater and watch *Grumpy Old Men* but not *Grumpy Old Women*—although we are on occasion treated to *psychotic* old women. One is hard pressed to find a male lead in a film comparable to *What Ever Happened to Baby Jane?*

Notable exceptions to this pattern include *Driving Miss Daisy* (1989) and *Fried Green Tomatoes* (1991). The plot of the first unfolds when Jessica Tandy as Miss Daisy drives her car off the road, prompting her son to engage a driver for her. In both films the older lead character spends time in a nursing home, although only 20% of the actual population over age 85 is in an institution.[47]

Films also offer us December–June romances with men more often than women in the role of December. So, for example, Clint Eastwood (63) is romantically paired with 35-year-old Renee Russo in *In the Line of Fire.* And Sally Field, who in 1988 played Tom Hank's lover in *Punchline,* was cast as his mother in *Forrest Gump* in 1994.

Such disparities prompted actress Joanne Woodward, who recently overcame the stereotypes by starring as the romantic lead opposite James Garner in a TV movie, to observe of her husband, "While Paul [Newman] is supposedly becoming more and more attractive, I'm becoming an old wreck. . . . There's no female Robert Redford or Paul Newman."[48]

Similarly, 40-year-old actress Kim Bassinger observes that "men can just age and turn into Clint Eastwood and Paul Newman, and women just get over the hill."[49] Indeed, in *Indecent Proposal,* Redford, age 56, is paired with Demi Moore, age 31.

In the nation's newsrooms, the numbers of both male and female older reporters have actually declined in recent years. A recent profile concluded that "Those 55–64 years old have continued to decline since 1971, suggesting relatively fewer 'elders' in American journalism now as compared with the early 1970s. Whether that will change much in the next decade as many of those in the large 35–54 group exceed 55 depends on how many stay in journalism."[50]

"[A]s you get older, you're not going to have the same value to people in broadcasting," notes CBS News' Ruth Ashton Taylor. "I haven't had."

"Let's face it. If I were a man, they wouldn't have suggested they could make a great deal with me in terms of a semi-retirement, which is what I did at the retirement age. We've got Mike Wallace, we've got Morley Safer. We've got a lot of people out there, men all over the place, Hugh Downs, a lot of people who are older men. You analyze the way they look. They have wrinkles and they're plainly out there. I looked at Dan Rather last night and I thought, for one thing, somebody is going to chew somebody out because the lighting was terrible. His circles showed. That wouldn't be tolerated in a woman, that the circles show. You're going to have to either back off, get out, do something. A woman is much more vulnerable to criticism on the basis of how she looks."[51]

As 52-year-old ABC correspondent Carole Simpson noted at the 1993 convention of the Radio-Television News Directors Association, "You don't see any women 55, 60, except for Barbara Walters. But you do see a lot of men hanging in there. David Brinkley's 72, Hugh Downs is 75, but there aren't any 70-year-old women on television."[52] "I have never sat around watching news hearing men comment that Dan Rather is looking old," observes *Newsweek's* Eleanor Clift. "I have heard that about women. All of us of a certain age are waiting to see how the generation after Barbara Walters will fare."[53] "Do I expect to be in broadcasting as long as Mike Wallace or David Brinkley?" laughs NBC's Lisa Myers. "That's like asking whether I expect pigs to fly."[54] But, notes Lesley Stahl, in an opinion seconded by Lisa Myers, there aren't very many older men on TV now; Brinkley and Wallace are notable exceptions.

Role models aren't all that women lose to the combination of sexism and ageism. When breast cancer is the subject, the refusal to show older women can be deadly. The acknowledgment by Betty Ford and Nancy Reagan that they had breast cancer injected some much-needed realism into the public consciousness about the disease—in contrast to the images in many public health campaigns. "The biggest risk factor for breast cancer, more than family history or any of that, is age," observes cancer expert Dr. Susan Love. But stories about breast cancer, both on television and in print, usually feature "a twenty-five-year-old body. All the public service announcements about breast self-exam and about mammography show twenty-five year olds. . . . The trouble is breast cancer doesn't happen [often] in twenty-five year olds. It happens in sixty-five and seventy year olds, but we never see sixty-five-year-old bodies in these articles."

The result? "[W]e have twenty-five year olds coming into my office begging for mammograms, panicking, sure they are going to get breast cancer [when] they are not really at risk and mammography is actually dangerous for them. You have the sixty-five-year-old women who are not perceiving themselves at risk at all who are not getting mammograms when they are most useful. . . . Our ageism in the media has really affected very much the message that gets out about breast cancer. The chance of getting breast cancer at age thirty is somewhere around one in five thousand per year. By forty it is one in a thousand per year, fifty it is one in five hundred, by sixty it is one in two hundred and fifty, by seventy it is one in one hundred twenty-five. If you live long enough, it is one in eight, but it definitely goes up. The longer you live, the higher your risk."

Yet when reporters approach Love they ask, "'Can you find us a young woman with a couple of small kids who is just diagnosed because that is going to look more tragic and more exciting on the TV screen?' And I keep saying, 'No, I want to find a sixty-five year old woman' because that is just as important and just as tragic. We have to get away from the ageism in this disease."[55]

Other more subtle problems arise from ageism. When women internalize negative attitudes toward aging, the result is a self-fulfilling prophesy. At the age of 41, Simone de Beauvoir wrote in *The Second Sex*, "Whereas man grows old gradually, woman is suddenly deprived of her femininity; she is still relatively young when she loses the erotic attractiveness and the fertility which, in the view of society and in her own, provide the justification of her existence and her opportunity for happiness. With no future, she still has about one half of her adult life to live."[56] In her early fifties, she recalled "One day I said to myself: 'I'm forty!' By the time I recovered from the shock of that discovery I had reached fifty. The stupor that seized me then has not left me yet."[57]

The stigma of age is so strong that those over 65 in one study who identified themselves as "middle-aged" rather than "elderly" or "old" were more likely to be alive a decade later.[58] Perhaps a self-fulfilling prophesy *is* at work. That is the judgment offered by Betty Friedan, who writes in *The Fountain of Age:* "In fact, gerontologists have found some evidence that people who don't think of themselves as 'old'—and, certainly now most people don't at sixty-five, and many don't at eighty—do better in age."[59] Alternatively, those saying they are "old" may have so conflated ill health with aging that one functions as a synonym of the other.

In either case, the finding might cause one to respond with alarm to Germaine Greer's stark division of the world into old and young in her book *The Change*. "At fifty-three I am not to say that I am old. I am only as old as I feel, people tell me, and allowing yourself to feel old is wrong. Phooey, say I, for although in my self I *feel* neither young nor old I *know* that I am old. If you are older than most of the people on earth, it seems more than a little silly to persist in claiming to be young."[60] Contrast that view with Gail Sheehy's in *The Silent Passage:* "Today fifty is the apex of the female life cycle. . . . If forty-five is the old age of youth, fifty is the youth of a woman's second adulthood. In fact, we can anticipate living the same number of years after menopause as we have already lived as reproductive women."[61]

Part of what drives the notion that women are "old" at a younger age than men is the conflation of a woman's productive and reproductive lives. Lurking around the fringes of the consciousness of some are residues of the view of the nineteenth-century physician who noted that it was "as if the Almighty, in creating the female sex, had taken the uterus and built up a woman around it."[62]

Alternative ways of seeing the climacteric in women are evident in such titles as *The Change: Women, Aging and the Menopause,*[63] *The Silent Passage,*[64] and *Menopause: A Midlife Passage,*[65] The cover of *Menopause: A Midlife Passage* features a group of women standing in front of a high-voltage power substation holding a sign that reads, "They are not Hot Flashes. They are POWER SURGES."

Menopause has not always been regarded as an invariable signal of decline. Menopausal Comanche women, for example, were welcomed into spiritual domains earlier denied them. "They could handle sacred objects, obtain power through dreams and practice as shamans, all things forbidden to women of bearing age."[66]

Juvenal aside, "Both Greeks and Romans accorded honor to post-menopausal women, released from the potentially polluting menstrual situation," notes historian Lois Banner. "But a countervailing tradition also existed. For these people believed, as would Europeans in later ages, that menstruation was therapeutically beneficial for women, since it withdrew potentially damaging humors from the body. With the end to menstruation, the body had no way of dispelling these humors; they remained in place, capable of wreaking havoc, both mental and physical."[67]

So tightly associated were menopause and depression that, a mere generation ago, "nervous breakdown" or "involutional melancholia" was expected at that time of life. More recent research suggests that post-menopausal depression is more likely in those who have experienced bouts of depression before menopause and in those surgically propelled into "The Change" by a hysterectomy.[68]

When women in the United States accept and gynecologists perform needless hysterectomies, they collude to create a twentieth-century version of the womb/brain bind. Remove the uterus and induce mental problems. But here an abrupt change in the body's chemistry is the culprit. Because a hysterectomy precipitates a sudden and dramatic change in a woman's hormonal state with a resulting change in her sense of well-being, those who perform medically unnecessary hysterectomies are increasing the likelihood that their patients will suffer depression. The more gradual hormonal transitions of natural menopause produce mood-swings but seldom depression or, interestingly, even regret. In fact, one U.S. study concluded that among 2500 women aged 45 to 55 years surveyed over a five-year period, three-quarters were "either relieved or neutral" about menopause; only 3% regretted reaching it.[69]

The number of women snapped into menopause by medically suspect hysterectomies is startling. "[M]ore than one-third of all women in the United States today have had a hysterectomy by the age of 60. It is estimated that 33 to 72 percent of hysterectomies are not medically necessary. Only 8 to 12 percent are performed to treat cancer or other life-threatening conditions."[70] Women in the United States "are up to six times as likely to have a hysterectomy as are women from other medically sophisticated countries, an unjustifiable disparity," writes one expert.[71] Although the ovaries continue to produce important hormones well past menopause, the idea seems to persist that when the uterus and ovaries are no longer capable of reproduction, they are not only functionless but menacing.

The myths of menopause are one barrier to productive aging. Another is the generalized fear that associates aging with death and projects the trajectory from here to there as one of decline. Survey researchers have found that most people view the "elderly" as a group as more enfeebled than their own parents or grandparents. On the assumption that older people are less competent, those who are younger make fewer demands on them. Their elders, fearful that they are slipping into this perceived decline, exaggerate small changes in their own abilities, producing an effect on themselves "more debilitating than the change itself." "Any such fears the elderly have about themselves," wrote two scholars who studied the problem, "are exacerbated by the dearth of appropriate role models who could serve as counter-examples of what life after 70 might be like."[72]

And here the absence of large numbers of identifiably older characters in films and television is particularly problematic. The difficulty? Since product commercials identify gray hair and wrinkles in women as the outward signs of age, those who are, for whatever reason, comparatively wrinkle-free and bearing some hair color other than gray or blue aren't identified as older. "Angela Lansbury isn't old," a friend of mine remarked, speaking of the star of *Murder, She Wrote,* one of the few popular television shows featuring an older woman. "Her hair isn't even gray. Helen Hayes was old."

Fearful that age carries disproportionate penalties, women are more reluctant than men to reveal how old they are. Dr. Debby Then is a scholar at UCLA's Center for the Study of Women. An article in the *New York Times* quotes her speaking about the double standards confronting women as they age. The article, titled "Age, Beauty and Truth," includes this passage:

> Dr. Then first realized the negative impact of age when she turned 30. "This man I was seeing said, 'Now you're 30, and soon you'll be 35 and 40,'" she remembers. "People categorize you accordingly, and that's unfortunate."
> And how old is Dr. Then?
> "I'm in my 30's," she said.
> Isn't that the sort of answer that perpetuates the idea that there is something shameful about getting older?
> "Right," she said. "And I choose to accommodate the system."[73]

Judith P. Vladeck, senior partner in the law firm Vladeck, Waldman, Elias and Engelhard, specializes in sex-discrimination cases. In December 1993, she won a $16-million settlement for a group of clients. A full-page profile in the *New York Times* is headed "Workplace Discrimination? Don't Try it Around Her." Early in the piece, the reporter notes that she is 70 and encloses—in parentheses—"(I really wish you didn't have to mention my age.)"[74]

"I've had a couple of people writing profiles ask how old I am," says a thirty-something female journalist. "When age discrimination against women ends, I'll start telling."

In a world dripping with denial about aging, singer Patti La Belle provides a welcome counterpoint. In response to a question about age on *Good Morning America,* she says, "This is what fifty is!" "Well, it just gets better" responds the show's host, 51-year-old Charles Gibson. "It does," reports La Belle.[75]

Although, in the abstract, we know that many women in Congress and those on the Supreme Court and in the Clinton cabinet are post-menopausal, our own deeply held age stereotypes deny them the ability to function as models of successful aging. Nor does their rhetoric encourage it. Where they are ready, even eager, to take their positions as evidence that women can make it, it is more difficult to locate statements indicating that they have overcome an additional hurdle—that of age.

The aging/invisibility bind whispers that because such women are competent, successful, and attractive, they can't be older. And since they can't be older, they can't function as models of our future selves. By these standards, Janet Reno, 56, Hazel O'Leary, 56, Barbara Mikulski, 58, Nancy Landon Kassebaum, 61, Ann Richards, 61, Ruth Bader Ginsburg, 61, and Sandra Day O'Connor, 64, cannot be viewed as models of successful aging. Nor does Reno's repeated self-description as "an awkward old maid" do much to convey her satisfaction with life beyond menopause.

One who devoted her post-retirement years to fighting such age stereotypes was Maggie Kuhn, who founded the Gray Panthers in 1970 after she was forced into mandatory retirement. "I am 72 years old," she told a congressional committee in 1977. "The Gray Panthers remind people of all ages that all of us are getting old. . . . Old age can be the flowering of life, and we, who have survived in this age of change, have a particular responsibility to those who come after us. We have nothing to lose."[76]

Finding role models of our future selves challenges the aging/invisibility bind. Temporal power can come to those who pass the 50-year mark. Golda Meir, Indira Ghandi, and Margaret Thatcher headed their countries in their post-menopausal years. Supreme Court Justices Sandra Day O'Connor and Ruth Bader Ginsburg and Cabinet Secretaries Hazel O'Leary and Janet Reno are past what once was called "the dangerous age," as are former Texas Governor Ann Richards and over half of the newly elected women in Congress.

As the number of women in positions of visible leadership rises, the categories into which older adults are parsed should change. In the early 1980s, in the absence of other information, younger people cast older adults in one of three roles: grandmother [but not grandfather], elder statesman [but not stateswoman], and the uni-sex senior citizen.[77] Underlying such categories is the wider range of positive roles popular culture holds out for men. In days of yore, in the public sphere, the older man might be a wizard, the older woman a witch; the older man a sage, the older woman, a nag; the older man, a prophet, the older woman, a hag. The family provided an alternative: If she were not the wicked stepmother to Cinderella, she could become grandmother, even if that meant being eaten by the wolf in "Little Red Riding Hood."

Increasingly, female politicians are capitalizing on our positive associations with the role of grandmother. Her granddaughter Lily was a prominent feature in the speeches of former Texas Governor Ann Richards. Her twin grandchildren played a similar role in the rhetoric of California's gubernatorial aspirant Kathleen Brown. In the final days of her first senatorial campaign, Dianne Feinstein, baby bottle in hand, appeared cradling her infant granddaughter in a televised political ad. And so pervasive was press description of Golda Meir as grandmotherly that style manuals in newsrooms specifically singled out that usage for excision.

The grandmother is, of course, an extension of the wife and mother role traditionally held to be acceptable for women. But by acting in unexpected ways, "grandmothers" are startling audiences into rethinking the stereotypes and moving beyond the categories available to classify women. So, for example, Dawn Clark Netsch shot a game of eight-ball in an ad that propelled her toward her party's nomination for the Illinois governorship. "Pool sharks are not supposed to look like Dawn Clark Netsch, the bookish, silver-haired, 67-year-old woman who won the Illinois Democratic primary," noted the *New York Times*.[78] The ad "helped alter her image from that of a stuffy schoolmarm to a down-to-earth contender."

To some extent, the range of roles available to both men and women is culture-bound. Where the role of grandmother is universal, that of headmistress and nanny are more typically British. The frame into which Mrs. Thatcher was cast also had a number of peculiarly British sexual nuances. "In popular understanding of her, Mrs. Thatcher's dominance became her most obvious characteristic," writes Webster. "She dominated her colleagues, the Conservative party, House of Commons debates, interviews, national life and, increasingly, as she was represented in the media, the international stage."

"Mrs. Thatcher's dominance was sometimes incorporated into a female image, as nanny, matron and headmistress."[79] "There were a number of variations on this theme, as men played child to Mrs. Thatcher's nanny, schoolboy to her headmistress, masochist to her sadist, conjuring up a picture of a particular upper-class male sexuality with resonances of punitive nannies, public school floggings, homosexuality, and overwhelming guilt and taste for punishment."[80]

George Bush was never cast as a headmaster or Mitterand as grand-father, of course; nanny and grandmother don't have male correlatives (except perhaps for the term "avuncular," which is ordinarily reserved for retired elder statesmen). When Bush and Mitterand succeeded, they were identified as statesmen.

Elderly stateswomen have, in fact, been around for a long time. Early leaders of the women's movement Elizabeth Cady Stanton, Susan B. Anthony, Margaret Sanger, and Frances Willard wielded influence into their sixties. Sarah Grimke remained active until her late seventies. Mary Baker Eddy, a founder of the Christian Science Church, was still heading her church in her seventies. "At the age of seventy-nine," writes her biographer, "Sarah (Grimke) trudged up and down the countryside, circulating and selling 150 copies of John Stuart Mill's *Subjection of Women.*"[81]

Elizabeth Cady Stanton published *The Woman's Bible* two weeks after celebrating her eightieth birthday. In 1900, at the age of 80, Susan B. Anthony "presided at three public sessions daily and at all the executive and business meetings, went to Baltimore and held a one-day's conference and made a big speech, addressed a parlor meeting, attended several dinners and receptions, participated in her own great birthday festivities, afternoon and evening, and remained for nearly a week of Executive Committee meetings after the convention had closed."[82]

In fact, the wily author of *Solitude of Self* turned the womb/brain bind into a useful counterargument for older women. Stanton described menopause "as an empowered force. Her 'vital forces,' formerly contained in her reproductive organs, wrote Stanton, were now 'flowing' to her brain, prompting her to leave her family for many months a year to pursue a career as a lecturer."[83]

Nor was Margaret Thatcher's capacity to govern during the presumed menopausal years a subject of serious controversy. "In 1943 a whole Parliamentary debate assumed that the menopause disqualified women between the ages of forty-five and fifty from war work," notes a British author, "and Dr. Russell Thomas talked of the 'tremendous lack of control of the arterial system' associated with the menopause, the 'great internal stress,' the head-aches and the palpitations. When Mrs. Thatcher stood for the leadership of the Conservative party some thirty years later, at the age of forty-nine, the menopause was scarcely mentioned, at least in public. When it was discussed, it was as a positive advantage, which meant that the reproductive system was using up less energy, so that a woman of Mrs. Thatcher's age could manage with only a few hours of sleep."[84]

Indeed, some argue that the physiological alterations of midlife provide women with the hormone needed to lead at the point at which men are suffering its loss. According to anthropologist Helen E. Fisher, "as menopause proceeds, levels of the female hormone estrogen steadily decline—unmasking natural levels of testosterone. Testosterone is one of nature's potent cocktails. The predominantly male hormone is linked with assertiveness in animals and people. Scientists have established that men (and male monkeys) of high rank often have elevated testosterone levels. In many societies, middle-aged women (with relatively high levels of testosterone) gain influence in political, religious, economic and social life. As Margaret Mead once said, 'There is no greater power in the world than the zest of a post-menopausal woman.' So it is likely that boomer women will attain authoritative positions as well."[85]

Underplayed by those who see a drop in estrogen as a cause, of course, are the years of hard work that positioned these women for the leadership roles they now hold. Still, with an empowering biological explanation in one hand and a list of older role models in the other, the middle-aged woman should be able to say, There with the grace of God go I.

Highlighting older role models are two major publications, *Modern Maturity,* the country's largest circulation publication, sponsored by AARP, and *Mirabella,* a fashion magazine for those over forty. Advertising in these magazines as well as in other media is increasingly being driven by the fast-approaching moment when baby boomers reach sixty. They will move into that decade under the protection of the Age Discrimination in Employment Act, which outlaws age discrimination in the workplace. Mandatory retirement has effectively been banned. And Social Security, Medicare, and Medicaid have broken the link between aging and poverty. At the same time, because women of that generation had the opportunity to choose when or whether they would have children, the likelihood that individuals will define themselves only or primarily as mothers is lessened.

Unsurprisingly, one result of such advances is a changed perception in well-being. The National Center for Health Statistics now notes no significant difference in the reported mental health of women in their twenties and those in their forties, fifties, or sixties.[86]

Assisting in this process, if they will break the taboo and speak their age, are visible national news reporters in or approaching their sixth decade. Among them are the beneficiaries of a 1971 ruling by the Federal Communications Commission that held that women should be given equal opportunity in broadcast hiring. They include: Carole Simpson, 53, Lesley Stahl, 50, Cokie Roberts and Linda Wertheimer, 50, Nina Totenberg, 49, and Diane Sawyer, 48. Where mandatory retirement claimed Eric Sevareid at 65, it no longer threatens productive older workers—including newscasters.

Leaders of the women's movement now in their fifties and sixties are making the case that one doesn't have to be a journalist or an activist to age well. "I started my quest for the fountain of age by simply looking for people who seemed to be 'vitally aging' as compared to the image of deterioration and decline that seemed to be the norm. . . . Even though our society's dread of age, its dreary or blanked-out image of aging, seemed to deny their very existence, even with so many elements of society seeming to conspire to prevent them from continuing to use their human abilities after sixty-five, I found that *they were everywhere,*" writes Betty Friedan.[87]

"Looking in the mirror," observes Gloria Steinem, "I see the lines between nose and mouth that now remain, even without a smile, and I am reminded of a chipmunk storing nuts for the winter. . . . But I have a new role model for this adventurous new country I'm now entering. She is a very old, smiling, wrinkled, rosy, beautiful woman, standing in the morning light of a park in Beijing. Her snow-white hair is just visible under a jaunty lavender babushka. Jan Phillips, who took her photograph, says she was belting out a Chinese opera to the sky, stopped for a moment to smile at the camera, and then went on singing. Now, she smiles at me every morning from my mantel. I love this woman. I like to think that walking on the path ahead of me, she looks a lot like my future self."[88]

The health of our future selves is likely to be protected by an emerging cadre of female medical professionals. One-fifth of those in the American College of Obstetricians and Gynecologists are now women. Half the medical residents in that area were women in 1991 and women now make up more than a third of entering medical school classes.[89] And more than a third of psychiatry residents are female.[90] All of this is important because female physicians appear to take care of women's unique needs better than men. So, for example, "[w]omen are more likely to undergo screening with Pap smears and mammograms if they see female rather than male physicians, particularly if the physician is an internist or family practitioner."[91] Meanwhile, female medical scholars will predictably follow the path set by their foremothers and bring the force of the scientific method to bear on medical issues of special concern to women, backed by the Office of Research on Women's Health, set up by the NIH in 1990, and empowered by Congressional passage of the Women's Health Equity Act in 1991, which guarantees funding for study of women's health issues.

At the same time, the demographic facts of life will increasingly focus medical practitioners on questions of concern to older Americans. This is, after all, the decade in which female boomers will enter menopause. One effect is already apparent. What only a decade ago was a taboo topic— menopause itself—is now an inescapable part of women's magazines, and has been the subject of a number of well-received popular books.

Aging, with all its complexities, could even come out of the denture-cream-ads and into prime time. Older Americans watch more television on average than any other age group. And as the baby boomers reach retirement, they will become an audience whose economic clout will require that its programming demands be honored. Will that generation escape enough of the positive and negative age stereotypes to embrace and reward realistic portrayals of aging, in general, and of the aging female, in particular? The signs are hopeful. "Beauty isn't about looking young," declares a new promotion by Estee Lauder's Clinique. These ads no longer use "young" as a positive adjective and "old" as a pejorative. The caption in a Revlon ad featuring 50-year-old super-model Lauren Hutton says, "This is our prime time. Let's make the most of it." And a Nike ad positions an active 50-year-old woman, lines and all, at "the age of elegance."

8

Newsbinds

DUBBED the "czarina," Helen Kushnick reigned over Hollywood as Jay Leno's manager and executive producer of his late-night show. Then the woman widely considered the most important female in the entertainment industry was deposed, dethroned by a double standard holding that locking *Tonight Show* guests out of the competition was ruthless—even though ruthlessness was business as usual for others booking shows. A male competitor practicing the Kushnick technique had a firm hand; Kushnick had a strong arm. So the woman who guided Leno's career for eighteen years and had positioned him to assume the *Tonight Show* mantle was fired by NBC.

What lesson does the saga of Helen Kushnick offer us, other than the difference between wearing the pants and the pantyhouse? That even very successful women don't get the same breaks as men? It tells us that the problem is not only in the pundits and the media but in the powerbrokers who first whispered to the media, then to NBC, and finally to Leno—who reportedly thought of her as a mother—that Kushnick was a ball buster. Of course no one in polite company used that language. The code was "ruthless," "aggressive," "vindictive," "abrasive." She was even likened to George Bush's short-lived enforcer, John Sununu. Kushnick wrote her own epitaph, "I've been serving you steak dinners for almost eighteen years," she reportedly told Leno. "I just haven't bothered showing you how I slaughtered the cow."

Helen Kushnick is a case study. Those who wanted to destroy her trafficked hoary stereotypes into print and onto entertainment television. And the press bought them. Her treatment was not aberrational. Reporters are as disposed as the rest of us to more quickly condemn women than men, to find fault on wider grounds, and to walk away unaware that anything

164

untoward has happened. Although the consciousness raising of the 1960s changed press language, and stylebooks said thou shalt not do to a woman what thou wouldst not do to a man, the binds that tie persist in subtle and pervasive press practices. They feature maternity but not paternity, characterize female but not male speech, and scrutinize female sexuality under a lens worthy of Galileo. As important, the nation's news outlets quote women less often in youth and middle age than their brothers, and then forget to mention it when they die. The press, in other words, perpetuates the binds that tie.

The patterns are intriguing. First, Helen Kushnick was snagged by femininity/competence, and then, as she struggled to regain her footing, her credentials as daughter and mother were paraded in public. The womb/brain trap had been sprung, insinuating its way into the public sphere by way of an interview with her stepdaughters, who whined that Helen had been a wicked stepmother. She had alienated her own father and failed theirs as well. "The stepdaughters of Helen Kushnick, dumped as Jay Leno's TV producer, have no supportive words for the comedian's manager. Jamie and Beth Kushnick say their stepmom tried to keep them from seeing their father on his deathbed. 'She terrorized us most of our lives,' said Beth. Added Jamie: 'Her (recent) behavior is consistent. She alienated her own father.'"[1]

Although you wouldn't know it from reading the accounts of Kushnick's downfall, in the 1970s newspapers responded to the demands of the women's movement by adding sections on "sex" to their stylebooks. "A basic rule," notes the guidesheet for the *Philadelphia Inquirer,* "is that women and men should not be treated differently in the news columns unless their sex is relevant to the news. A typical violation of this principle occurred in stories that reported on Golda Meir in her role as Israeli prime minister and gratuitously described her as a grandmother. Physical descriptions of a woman in news stories are permissible only if a man, under similar circumstances, would be described comparably."

When editorial concern is focused on treatment of appearance and marital and parental status, other signs of the competence/femininity bind may go unnoticed. Kushnick, for example, is double-bound by the adjectives that describe her behavior. If the public accounts indicated that she had crossed a line previously unbroken by her peers, it might be easier to explain labels such as ruthless, aggressive, vindictive, and strong-armed as something other than exhibiting the press's discomfort with "tough" behavior in a woman as opposed to a man.

The stylebooks mirror the women's movement's efforts to see women identified not as adjuncts to men but as individuals in their own right. Under pressure from the movement, the married *Mrs.* was no longer invariably demarcated from the single *Miss* with all men tagged *Mr. Ms.* entered the stylesheets.

But that didn't mean that distinctions vanished. In 1977 the UPI *Stylebook* instructed reporters not to "use Mr. in any reference unless it is combined with Mrs.: *Mr. and Mrs. John Smith.*" An honorific would continue to be applied to women out of tandem. A married woman known by her maiden name would be identified as Miss (e.g., Miss Jane Fonda) on second reference with first reference simply listing the woman's full name. A woman known by her "married name" would be designated Mrs. (e.g., Mrs. Carla Hills). Women who had never been married rated a Miss unless they "preferred" Ms.[2] The AP guidelines said much the same.[3]

By contrast, the style manual currently in use at the *Philadelphia Inquirer* makes no distinction in its treatment of men and women. "In general, do not use the courtesy titles *Miss, Mr., Mrs.,* or *Ms.,*" it says. "Men and women are identified on first reference by their full names, and by last name thereafter."

Although the advent of *Ms.* sidetracked one means of making marital status more visible for women than men, others persisted. Among them is a journalistic fascination with the meaning of women's names. So, for example, reporters noted that Geraldine Ferraro did not use the name of her husband (Zaccaro) in public life but had voted under it (presumably as an extension of her private life). As we saw in Chapter 2, news accounts probed when, where, and why the woman most often identified as Hillary tagged herself "Rodham," "Clinton," or "Rodham Clinton." The *Wall Street Journal*/NBC News poll even asked in January 1993 whether the first lady should be known as "Hillary Clinton or Hillary Rodham Clinton"; 62% chose the former, 6% the latter, 28% said it didn't matter, and 4% weren't sure.[4]

These notices were nondenominational. When Tucker appeared in Marilyn Quayle's name after her husband left the vice presidency, it too attracted comment. With the silliness escalating, columnist Ellen Goodman intervened to ask, "Does anyone dare suggest men with names like William Kennedy Smith drop 'Kennedy' and only use William Smith? And furthermore, have you ever heard men like William Kennedy Smith referred to in the media as just William Smith? . . . Why does society feel a woman's name choice is open to debate? And why does society often reject and ignore her wishes once she makes that choice?"[5] After all, one might add, where was the debate over the Fitzgerald in John Fitzgerald Kennedy?

I do not mean to suggest that reporters and editors are doing much more than reflecting widely held and hence largely unquestioned assumptions about what is and is not interesting. But to the extent that language shapes as it reinforces, widespread journalistic practices can accelerate or retard forces at work elsewhere in society. By replacing *Miss* or *Mrs.* with *Ms.,* reporters eliminated a subtle distinction that made marital information seem more relevant for women than men. But by zeroing in on the names a woman but not a man chooses to use, they refocus on difference. Simultaneously, they also implicitly deny a woman the power to self-identify.

If one picture is worth a thousand Rodhams, the fact that women are more likely than men to appear in news photos with their spouses or with groups of others, including children, is of note as well. Although the treatment is not universal, such patterns in coverage persist, illuminating the residues of the double binds on which this book has focused.

Inherent in each is the notion that men are the norm and women, other. By treating women as exceptions, reporters reinforce this assumption. So, for example, the recurrent headlines identifying specific women as "woman" might prompt one to think that somewhere out there is a very busy female. Perhaps designating 1992 "The Year of the Woman" was journalism's way of celebrating her, but some in power saw it as another sign of exceptionalism. "The press called it 'The Year of the Woman,'" recalled Senator Barbara Mikulski. "I personally didn't like the title. Like we only get one year per century. Like the year of the dog. The year of the zebra."[6] The persistent place of "woman" in headlines also suggests that gender is a relevant category of evaluation for women, but not for men—a problem that made Ann Hopkins's life more difficult at Price Waterhouse.

This tendency also transcends geographic borders. "Woman Set to Succeed Mulroney: Defense Chief to be First Female Canadian Premier," read a front-page headline in the June 14, 1993, *Washington Post*. Nowhere is Kim Campbell's name mentioned. And on the jump-page the pattern is repeated. That headline reads "Woman to be Canadian Prime Minister." Even when "woman" isn't part of the headline, it may contain words that signal judgments about the appropriateness of the candidate's embodiment of her gender. A *New York Times* headline identifies Texas senator Kay Bailey Hutchison, for example, as a "Demure Survivor."[7] Similarly, gender is treated as important for a female politician's supporters. Women seeking office are more often described as part of a group and the gender of the supporters of female candidates is more often highlighted. So, for example, "Boxer and Feinstein were among a flock of women candidates this year," according to the *San Francisco Chronicle*.[8]

"Polls show Feinstein with equal support among male and female voters," notes the *Washington Post* without identifying her opponent's supporters by gender.[9] "It was a big night for Bay Area women as Barbara Boxer and Dianne Feinstein claimed historic victories in the state's two U.S. Senate races," observed the *San Francisco Chronicle*. Was it not also a big night for Bay area men who favored Democrats?[10]

"More women than men were undecided on the eve of the election, indicating a strong, last-minute surge for Murray," reported the *Seattle Times* in an article titled "The Big Winners: Women. Murray's Win: Clincher Was Her Gender." Yet the article itself reports that "Nearly half of all voters who made up their minds late were Chandler supporters."[11]

At the same time, and on a more subtle level, studies by the groups *Women, Men and Media* and *Women Are Good News*[12] consistently show that women are quoted as sources and appear on interview shows much less frequently than men. In January 1993, for example, the front pages of twenty newspapers cited women as sources only 15% of the time, even though more than a third of the stories carried a woman's byline.[13] Even when women do gain national visibility, they remain invisible to some. In

March 1994, when a Harris Poll asked viewers which network anchor they preferred, 30% said ABC's Peter Jennings, 29% chose NBC's Tom Brokaw, and 24% said they liked CBS's Dan Rather. Ignored by the pollsters and as a result unranked was Dan Rather's co-anchor, Connie Chung.[14]

But the by-product of such anonymity may be immortality, for women are also less likely to find themselves written up on the obituary page. Indeed, if the month of July 1993 is typical, women are more likely to die in the Washington, D.C., area than in Philadelphia. But the climate in Philadelphia is not as conducive to eternal life for women as it is in New York City. Where the *New York Times* printed 5.6 more obituaries of men than women, the ratio at the *Philadelphia Inquirer* was 3.1 to 1, and at the *Washington Post* 1.8 to 1.[15]

Before planning to migrate northward, women of dubious health might want to know that the *Washington Post* prints any obituary submitted if the deceased has lived in the area for a considerable period of time, while the *New York Times* denies a final farewell to all but the "recognizably note-worthy."

Womb/Brain

No matter how you dress it, maternity remains more relevant than pater-nity to those who put together the news pages. Indeed, a visitor unfamiliar with our reproductive mechanics might conclude that children are con-ceived, carried, and raised by women alone.

On June 5, 1993, for example, the front page of the *New York Times* showed a photo of law professor Lani Guinier at the podium, speaking at a press conference after President Bill Clinton withdrew his nomination of her as head of the civil rights division of the Justice Department. The photo also features her husband, Nolan Bowie, and their son Nicholas. Bowie is watching his wife. Nicholas is hugging his father. "With her," says the caption, "were her husband, Nolan Bowie, and her 5-year-old son, Nicholas." Why is the child referred to as hers and not theirs?

The photo, moreover, carries the additional caption "Abandoned but Defiant." Presumably "abandoned" is meant to characterize Professor Guinier. Would it have been used to describe a male nominee? Or is the combination of these cues meant to elicit a gut-level response to an aban-doned mother and child?

As Janet Reno found when she was nominated for the post of attorney general, being unmarried is as noteworthy as marriage and motherhood. Since the presence of children certifies that a woman is sufficiently warm to pass the femininity test, familial affections must be certified in other ways for her. "Ms. Reno has never married or had children," noted the profile in the *New York Times,* "a fact that half-jokingly has been cited as exempting her from the illegal nanny problem that sank Zoe Baird and Judge Kimba M. Wood. She remains close, *however,* to her two brothers, Robert, a reporter for *Newsday,* and Mark, a sea captain; her sister, Maggy, a county commissioner in Florida, *and various nieces and nephews* (emphasis added)."[16]

Like iron filings to a magnet, descriptions of appearance and notation of age seem inexorably drawn toward such identifications of marital status. So, for instance, *New York Times* columnist William Safire described Polish head of state Hanna Suchocka by saying, "This handsome, matronly, unmarried lawyer of 47 has the unmistakable presence of a head of state."[17]

All the Women Are White, All the Blacks Are Men, But Some of Us Are Brave is the title of a women's studies text.[18] Add "No Men Have Children," and that title could serve as the header in a newspaper stylebook. The reason becomes apparent in the *New York Times* on June 6, 1993. That Sunday's paper included two articles noting the appointment of Episcopal bishops. Both are "Special to the New York Times" and neither carries a byline. The headlines are similar. "Connecticut: Black Is Voted Episcopal Bishop"[19] and "Woman Is Voted Vermont's Episcopal Bishop."[20] A headline that read "Man Is Voted Episcopal Bishop" would certainly have provoked double-takes. White men are the norm in the news, the assumed back- and foreground in which women, African Americans, Hispanics, and Asian Americans operate. This idea has become a habit of mind. Why do nearly all of us take it for granted that the black appointee is male and the woman white? Both inferences are in fact correct. In each case we are demonstrating an ingrained awareness of the norms that frame our perceptions, norms reinforced by the language of the press.

Both *New York Times* articles report the age of the new bishop. Suffragan Bishop Clarence N. Coleridge is 62; the Rev. McLeod is 54. But where his age is disclosed in a final biographical paragraph, the article on her seems to treat age as a factor in the selection decision. "Bishop-elect McLeod, 54 years old, was elected on the third round of balloting." Both articles indicate where the bishop-elect was born, schooled, and is currently employed. But while we are well informed that "Mrs. McLeod . . . is married to an Episcopal priest and has five grown children," we put down the article on Coleridge without knowing anything at all about his marital or parental status.

The headlines invite the conclusion that what is important about these "exceptions" to the presumed norm is gender or race, not accomplishment or name. The texts then invite us to conclude that marital status and childbearing are important facets of a woman's life, but not a man's. The fact that her children are grown, moreover, certified the female Bishopelect as a sequencer and hence unimplicated in the womb/brain bind.

Are such headlines, references, photos, and articles atypical? To determine that, we analyzed all the print coverage found in the *New York Times, Washington Post,* and major state newspapers covering the federal elections featuring women in 1990, 1992, and 1993.[21]

We found that whether they were candidates or married to candidates, women in political articles are identified in spousal or parental roles more often than men. Whether in the political arena by choice or by marriage, men are more likely to be identified as holders of a profession.

Reporting in the *New York Times* on the 1992 contests is illustrative.[22] There the senators-elect are described this way:

Ms. [Barbara] Boxer, a 51-year-old Brooklyn transplant. . . .

. . . 59-year-old Ms. [Dianne] Feinstein . . . faces another race in 1994 . . . no doubt she will rely, as she has in the past, on the assets of her husband, Richard C. Blum, an investment banker.

The 42-year-old Ms. [Patty] Murray lives in Seattle suburb of Shoreline, where in addition to her two children, age 12 and 15, she cares for her aging parents.

Ms. [Carol Moseley] Braun, who is divorced and the mother of a 15-year-old son. . . .

Ben Nighthorse Campbell . . . at age 59, he is striking for his appearance, with steel-gray pony tail and string ties. . . .

Dirk Kempthorne, the 41-year-old mayor of Boise. . . .

Judd Gregg, 45, is Governor of New Hampshire and previously served in the House.

Lauch Faircloth, 64, a businessman and farmer. . . .

A six-term Representative and 50-year-old Democrat, Byron L. Dorgan. . . .

Robert Bennett, a 59-year-old Republican businessman. . . .

Russell Feingold, 39, has served in the Wisconsin state Senate for 10 years, where he specialized in judicial affairs.

Where occupation is a defining feature of the male candidates, it is a secondary characteristic of the females. Not a single male candidate is identified as married, divorced, or single. By contrast, we learn that Moseley Braun is divorced and Feinstein married. Two women but none of the men are identified as parents. Like the women, the male Native American is characterized differently, here by appearance.

The *New York Times* is not exceptional. The *Washington Post* offers its own very different frames of reference for the candidates for the Texas governorship in 1990: "A 57-year-old white-haired grandmother, Ms. Richards" and then "Mr. Williams, a West Texas oil man."[23]

Even in an article addressing Canadian prime minister Kim Campbell's charge that the media employ a double standard, the *Vancouver Sun* displays one. "Three Campbell biographies are in the works. She has faced questions about her past, present and possible future romantic links, and has had to explain why she has no kids," notes an article titled "Campbell sees double standard in media spin on her campaign." The reporter then adds, helpfully, "She tried, but no luck."[24]

Spouses of the candidates fare no better. When referring to a female spouse, the press is less likely to include her name, and, if it is given, it is usually only her first name and not her profession. So, for example, the *Los Angeles Times* mentions "Feinstein and Richard C. Blum, her investment banker husband" but "Seymour and his wife, Judy." In the same story, "[Barbara] Boxer's returns show that she and her husband, Oakland attorney Stewart Boxer. . . ."[25] The *Dallas Morning News* identifies "Mr. Krueger and his wife, Kathleen," but specifies "the bond business of Ms. Hutchison's husband, Ray Hutchison."[26]

The single exception occurs in the *Philadelphia Inquirer,* which identifies Senator Arlen Specter's wife as "City Councilwoman Joan Specter."[27] Here Specter may be the beneficiary of the paper's style guide; it notes explicitly, "Occasionally both husband and wife have titles that may be applied to distinguish them: *Senator Specter and Councilwoman Specter* are an example. But do not apply such a title to only one spouse, as in *Sen. Specter* and *Specter, Arlen Specter* and *Councilwoman Specter,* or *Sen. Specter* and *Joan Specter.*"

Speech/Shame

Even when a supporter comes forward to unmask the machinations of the womb/brain bind, subtle fallout from others may creep in. So, for example, in a *New York Times* column about Senate hearings on Zoe Baird's nomination to the post of attorney general, Anthony Lewis asked his readers to

> Think about the moment in the hearings when Senator Joseph Biden, chairman of the Judiciary Committee, asked Ms. Baird to state how many hours she was away from her child: when she left home in the morning and returned at night. Would he have asked that of any male nominee, for any job?
> Or consider Senator Dianne Feinstein's hectoring of Ms. Baird about not resting on her laurels as a magna cum laude graduate but knowing the streets. Would Senator Feinstein have said that to Edward Levi, a reserved intellectual who was the one outstanding Attorney General in the last 20 years?[28]

If Senator Feinstein had been [insert the male Senator of your choice], would Mr. Lewis have described his statement as "hectoring"? Why say that Biden "asked" but Feinstein "hector[ed]?" And why does a column about stereotyping women stereotypically characterize a female's speech?

Women and men do it. Conservatives and liberals do it. Columnists and reporters do it. Even academics do it. They do it in such public places as the op-ed page of the *New York Times.* What they do is subtly question the legitimacy of women's speech.

No review of a male nominee for an Oscar contained a line even remotely resembling the one penned by the *Washington Post's* Tony Kornheiser about the *The Piano's* star, Holly Hunter. "'The Piano' is a double win, as you get to see Holly Hunter naked, plus you don't have to hear her hideous squawk." In the movie, Ms. Hunter, who won the Oscar for best actress, plays a mute woman. Instead of a double win, I see Kornheiser's comment as a reflection of a double bind. The actress's voice and body are subjected by the critic to appraisal that implicitly disparages her talent and detracts from serious treatment of her accomplishment.[29]

Other instances of stereotyping are less blatant but no less pervasive. An article in *The New York Times Magazine* introduces us to two leaders in the fight against AIDS. Both are medical professionals. Their ages are comparable. Neither is widely known outside the medical community. Imagine reading this: "Dr. Andrea S. Fauci speaks with a hint of a Brooklyn accent, which is out of sync with the elegance of her appearance—well tailored, tidy, trim. At 52, she is scientist-cum-celebrity, ridiculed by Larry Kramer in the play 'The Destiny of Me,' lionized by George Bush in the 1988 Presidential debate as a hero." "At 50, he [Christopher M. Gebbie] has the air of the head doctor. There is something at once no-nonsense and fussy about him—his erect posture, his precise and proper answers, his tendency to correct an interviewer's questions. He is what he is—a doctor who worked his way up through the public-health bureaucracy, a divorced father of three children, who has willed himself to a better lot in life."[30]

I have, of course, taken the liberty of changing the gender of the medical maestros. The tidy, well-tailored, trim Andrea Fauci is actually Anthony. The no-nonsense, fussy Christopher Gebbie is instead Kristine. She is a nurse, not a doctor, and Clinton's new AIDS czar. The first clue to my mischief is the fact that the marital and parental status of the supposed man but not the supposed woman are specified. The second clue comes in the verbs, "lionized" and "ridiculed"—a bit beyond the stereotypical pale if tagged to a woman. By contrast, "no nonsense," "fussy," "erect posture," "precise and proper answers," and "tendency to correct questions" whisper generations of stereotyped schoolmarms and nannies—but not doctors.

In subtle and some more obvious ways, press coverage reinforces our notions of "appropriate" role. In that process, the speech of a woman is both more likely to be described and characterized, as if describing it opens a window on identity not provided by male discourse, or as if its novelty is worthy of special comment. So Ann Richards was "tart-tongued"[31] and Hillary Clinton's "cookies and tea" comments are labeled "outspoken."

Role, soul, and speech are interwoven in an account of the retirement of Pittsburgh mayor Sophie Masloff. "Voters have treated her like the Jewish grandmother that she is," wrote a reporter for the Associated Press, "sometimes with deference, sometimes with disdain. Now she's ending a 50-year political career today after opting against a re-election try. With a bouffant hairdo and screechy, spine-chilling voice, Masloff—'call me Sophie'—is no Rudolph Giuliani, the polished new mayor of New York."[32]

In characterizing women's speech, "outspoken" and "soft-spoken" are near the top of the grab-bag, but both suggest behavior outside the norm. Neither is necessarily a pejorative. "Ms. Reno," says the *New York Times*, "has a reputation as a fair but very demanding, outspoken, and at 6 feet 1½ inches tall, a physically imposing boss."[33]

Many of the old pejoratives persist, however, and their resonance with scholars and politicians is a reminder that the press more often reflects than creates social assumptions. "He [Boxer's opponent Bruce Herchensohn] comes off as reasonable and rational," said Sherry Bebitch Jeffe, a political analyst at the Claremont Graduate School. "He reminds me of Ronald Reagan. You might hate what he stands for, but he's a decent human being. He's very media friendly, while Boxer comes off overly shrill."[34]

And when Boxer's colleagues criticize her anonymously, they too focus on speech. "Yet some of Boxer's colleagues and congressional staffers speak of her privately in less flattering terms—strident, abrupt, annoying and shrill are among the terms they use," reported the *Los Angeles Times*. The Senatorial aspirant responded by meta-communicating. Such terminology, Boxer pointed out, often is reserved for women legislators. "If a man gets up and says: 'The spending amounts for military parts is an outrage,' it's called leadership."[35]

"Truly, her verbosity is her Achilles Heel. Make that mouth," wrote the *Montreal Gazette* of Prime Minister Kim Campbell. "Campbell never could learn to control it. . . . And Canadians expect to see some courtesy extended among opponents. Pierre Trudeau got away with ridiculing his enemies ("Zap, you're frozen"), but that was Pierre Trudeau. Campbell has complained that she is the victim of a double standard because she is a woman. She says male politicians are obliged [permitted] more of an aggressive stature than females who play the game."[36]

Equality/Difference

A woman in the news is subjected to unique biographical dissection. In 1993 a former cheerleader was elected to the Senate and a former baton twirler nominated to the Supreme Court. The public learned of these milestones in the careers of Senator-elect Kay Bailey Hutchison and Supreme Court nominee Ruth Bader Ginsburg because two influential newspapers chose to feature them. "At last, Justice for the Baton Twirler: Feminism came late to Ginsburg. Behind her firm belief is a careful approach," reads the headline in the "Review and Opinion" section of the Sunday July 4, 1993, *Philadelphia Inquirer*.[37]

The front page of the *Washington Post*'s Style section[38] abutted a photo from Hutchison's days as a University of Texas cheerleader with a recent photo of her holding a book. The collage was captioned "Kay Bailey Hutchison, a former University of Texas cheerleader, is the first woman from her state to enter the Senate."

The caption is not an aberration. "It's been called her cheerleader smile," opens the article by staff writer Sue Anne Pressley. "She flashes it—*there*—at the television cameras. But Kay Bailey Hutchison—former cheerleader at Las Marque (Tex.) High School, former cheerleader for the University of Texas Longhorns—declines to recite any of her best routines. 'I'd rather not,' she says."

Lost in the lead paragraph is Hutchison's service as Texas's state treasurer, a biographical tidbit reserved for a sentence *five paragraphs later* that

A collage, created by intercutting these two photos, appeared in the *Washington Post* on June 14, 1993. It was captioned "Kay Bailey Hutchison, a former University of Texas cheerleader, is the first woman from her state to enter the Senate." (Ray Lustig of the *Washington Post*.)

(Texas Student Publications, Inc. Records, The Center for American History, The University of Texas at Austin [CN 07729].)

features the fact that in high school she was prom queen. There, being prom queen is implicitly equated with running a company and superintending state revenues. "The former senior prom queen, former Houston television reporter, former state legislator, former peanut brittle manufacturer, former state treasurer will be sworn in today as the first female senator from the state of Texas."[39]

In subtle and obvious ways, the language of politics hints that women are out of their sphere. "California's Warm Brown," says a headline in the *Washington Post* about gubernatorial candidate Kathleen Brown. "Her

Dad was Governor, Her Brother was Governor. But Is Sister Kathleen Too Nice for the Job?"[40]

The press, the pundits, and the politicians analogize the campaign not to such sports as skating, tennis, and gymnastics, in which men and women compete on an equal footing, but to contact male-dominated sports such as football and boxing. As the tension in a campaign mounts, war metaphors add an additional point of reference to domains until recently denied women.

So, for example, the headlines in the Kansas Senatorial race report that "O'Dell Fires a Barrage at Dole,"[41] "Dole, O'Dell Spar over Senator's Status as 'Insider.'"[42] Sparring had gone on in Texas as well. "Richards Spars with Williams over Tax Issue, Race Generates Even More Heat in the Final Stretch," wrote the *Houston Post*.[43] "In Missouri, a scrappy campaign by St. Louis city Councilwoman Geri Rothman-Serot has forced first-term Sen. Christopher S. Bond to work hard for his campaign funds— but has yet to bring her within striking distance."[44]

In Illinois, the *Chicago Tribune* reported that "Debate has familiar ring, Williamson chides Braun, but neither can draw blood."[45] In Canada, "Kim Campbell brought her 'gloves-off' Tory leadership campaign to London [Ontario] Thursday night but she didn't lay a finger on her closest competitor, Jean Charest."[46] In California, "In a debate which more closely resembled verbal mud wrestling, Democrat Dianne Feinstein and Republican John Seymour hurled bitingly personal charges at each other."[47] Meanwhile, in Pennsylvania, "If Lynn Yeakel were a football team, yesterday would qualify as one heck of a pre-game pump-up."[48] And in Seattle, "Murray, in in her own campaign, sometimes doesn't seem ready for such rough-and-tumble politics."[49]

It is silly to expect a female rather than a male to land a knockout punch or draw the first blood, even if the male's yearly athletic workout consists of signing his tax return and the woman is an accomplished cross-country skier. But by describing the political world in terms we comfortably associate with even the most unathletic of men, women are subtly defined as creatures alien to that habitat.

There may be a second, more insidious effect. By evoking politics as sports that women can't "win," the metaphors may reinforce the assumption that electoral victory is less plausible for a woman.[50] The 1991 Greenberg-Lake survey found, for example, that "Voters, especially men, continue to assume women candidates cannot win, even as they are voting for them. Overwhelmingly, voters choose the male candidate of either party as the 'more winnable' *even when they are voting for the woman*" [emphasis added].[51] "The final barrier then for women," conclude the researchers, "is still looking like a winner in the male world of politics." The perception that a candidate can win helps attract the money and press attention that make winning possible.

Being an outsider, of course, has its advantages. The downside for the female candidate is having to establish that she is "tough" enough "to land a knockout punch" when no one in the audience can call to mind an incident in which a woman has ever achieved that supposed feat. At the same time, if she is "tough" enough, she risks evoking the competence/femininity bind.

But perils exist for men here as well. Playground rules set for boys are clear. Don't beat up smaller kids and don't hit girls. In politics the result plays out as a double bind for the male candidate confronting a female opponent. "Gender aside, analysts from both parties say Mr. Krueger must come out swinging if he hopes to be the victor in the June 5 runoff. But political experts caution that trying to bloody a female opponent could provoke a gender backlash. 'I think you can say almost anything to a man and not be hurt by it,' said former Texas Attorney General Jim Mattox, who lost to Ms. Richards in the Democratic primary two years ago. 'But I think you can cause overwhelming support to go in the other direction if you say something that is offensive to a woman'"[52]

The former mayor of Anaheim said he blames his loss to Marian Bergeson in the 1990 race for the Republican nomination for lieutenant governor on his reluctance to attack a female opponent. "The honest-to-God's truth is, I have a tough time picking on any woman," Seymour told the *Los Angeles Times* after his defeat. "That's the way my mom raised me. It's very difficult to come out swinging."[53]

Within the framework of metaphorical combat, the adjective of preference to describe female candidates is "feisty." Barbara Mikulski is "a feisty liberal/pragmatist"[54] and a "feisty Senator";[55] Lynn Yeakel is "feisty" and dynamic.[56] Where Senator Mikulski is characterized as "feisty," her opponent is identified in op-eds as "a very articulate conservative"[57] and "thoughtful."[58] Carol Moseley Braun was described as "apologetic yet feisty."[59] But where the tag once hurt, in 1992 some thought it helped. In a year in which voters were searching for outsiders who would fight the system, "feisty"—with its terrier-like connotations—was heard by some as an acceptable way of suggesting that a woman was "tough" and at the same time appropriately feminine. Pollster Celinda Lake observed, "Voters like feisty."[60]

Femininity/Competence

When political winners and losers are under consideration, a consistent pattern emerges. Instead of crediting a female candidate with competence, reporters and her opponents assume that she wins through the actions or negligence of others. By contrast, men win because they are skillful and competent.

Arlen Specter's victory was attributed both to the ineptitude of his opponent *and to his ability to capitalize on her mistakes*. It is not her mistakes per se that account for his victory, but action he took to translate them into votes. "Sen. Arlen Specter (R. Pa.), who once trailed Democrat Lynn

Yeakel by a substantial margin, capitalized on Yeakel's political blunders and outspent her by almost 2 to 1 to claim a narrow victory," noted the *Washington Post*.[61] "Yeakel's campaign was inexpert and often fumbling, while Specter ran a virtually textbook campaign," said the *Philadelphia Inquirer*.[62]

No such skills are accorded victorious women. "Consultant John Weaver, former executive director of the Texas Republican Party, said, 'We lost our election tonight. Ann Richards didn't win it.' . . . And true to form, it was something Williams did in the closing weeks that seemed to finally turn the tide in Richards' favor," adds the *Houston Post*.[63]

"In a year when voters put a premium on change and diversity, Braun was the right candidate at the right time," said the *Chicago Tribune*.[64] "But if she does win, she'll be the luckiest candidate I've ever seen," opined *Chicago Tribune* columnist Mike Royko.[65]

This brand of double standard is nonpartisan. "1993 may also be remembered as the year that the Republicans did not so much win the senate seat as the Democrats lost it," said the *Houston Chronicle* of the Hutchison-Krueger race in Texas.[66] "Kay Bailey Hutchison probably still cannot believe her good fortune. Few politicians are lucky enough to face opponents as inept as Bob Krueger proved to be," observed *Dallas Morning News* columnist Phil Seib.[67]

The same pattern occurs in coverage of women in sports. In basketball and tennis matches "[w]omen were also more likely to be framed as failures due to some combination of nervousness, lack of confidence, lack of being 'comfortable,' lack of aggression, and lack of stamina. Men were far less often framed as failures—men appeared to miss shots and lose matches not so much because of their own individual shortcomings (nervousness, losing control, etc.) but because of the power, strength, and intelligence of their (male) *opponents*."[68]

Where issues of competence take a back seat, issues of femininity are at the controls. Hillary Clinton's dilemma in the 1992 campaign was not anomalous. Nor is it of recent origin. On May 29, 1954, a front page *New York Times* obituary for star reporter Anne McCormick included the line, "In spite of all her genius for seeing, understanding and reporting, she was also a deeply feminine person and could not help being so and would not have wished not to be so."

An op-ed in the *Toronto Star*, describing the situation facing Campbell but not Charest, seems to sum up the double binds facing women:

We want politicians who will listen to us, but still act on their principles.

We want leaders who will speak for us, but put forward ideas of their own.

We want a prime minister who will give us a role in national decision-making, but demonstrate vision and courage.

Maybe we are asking for the impossible. Perhaps the very phrase "consensual leader" is an oxymoron.

How can a single individual be both accommodating and authoritative, both flexible and decisive?

This conundrum is at the heart of Kim Campbell's troubled leadership campaign.[69]

Even though most newspapers now operate under style manuals designed to vanquish gender stereotyping, it persists in the bits and bytes that form news. The prospect of change is evident, however, in the dramatic difference in newsroom response to two cases fourteen years apart. The first involved the forced resignation of *New York Times* reporter Laura Foreman, in 1977, for supposedly sleeping with and accepting gifts from a source; the second with a 1991 *New York Times* article on Patricia Bowman who accused William Kennedy Smith of raping her at the compound of Smith's famous uncle.

In the first case, the exposure of Foreman's sexual past went uncriticized in the newsroom of the *Philadelphia Inquirer* which ran the story about its former reporter. In the second, outraged staffers, female and male, elicited a public statement of regrets from the powers that be at the *New York Times* for an article probing Bowman's sexual history and character.

William Kennedy Smith's accuser, identified by name as Patricia Bowman, was profiled in an article in the *New York Times* April 17, 1991. The piece chronicled not only Bowman's sex life but her mother's as well. Bowman's mother, it reported, had had an affair with a man who eventually became the accuser's stepfather.

Despite a *Times* policy against quotation of negative material by anonymous sources, the article included them in its unflattering profile. Patricia Bowman "had a poor academic record at Tallmadge High School, said a school official who spoke on the condition he not be identified." She had, said another, "a little wild streak." The piece went on to say that she was an unwed mother, who barhopped and had done poorly ("an Ohio high school student with below-average grades") in school. Her child was not the offspring of a long-lived relationship, reported the article. "[I]n 1989 she had a brief affair with Johnny Butler, the son of a once-prosperous family here that owned a lumber company. The company has since declared bankruptcy. Mr. Butler was the father of her child, friends say. It is unclear why the couple did not marry."

One evening, noted the article, a chef kept a restaurant kitchen open late to prepare a meal for her. One might read the article to say that he expected sex to be the reward for the food. "One evening a few weeks before the incident at the Kennedy estate, Mr. Pontirole [the chef] recalled, she came in at midnight after the kitchen was closed and she complained about being hungry. So Mr. Pontirole went back to his pots and sauces and fixed his specialty, rigatone a la vodka, for her. Later he escorted her to a nearby bar, E. R. Bradley's, but was disappointed when she fell into conversation with several other men, according to Mr. Pontirole."

In short, if the woman had indeed been raped, she had asked for it. More likely it wasn't rape at all, but consensual sex involving a woman who clearly wasn't a virgin. Having said "yes" before, it was implausible that she had said "no" that night. Or, a compatible hypothesis, the so-called rape was an attempt to do what her mother had apparently done—use sex to snare a wealthy man.

In the binary world of uterus and brain, Bowman lacked intellectual wherewithal (poor grades) but the implication is that this underactive brain had as its corollary an overly active sex life. She must have been a bad mother: what else was she doing barhopping (public behavior), and picking up men, when she should have been home caring for her child? In the frame of speech/shame, she was Eve using speech to beguile men into giving her a meal, even though without seeds and a core, and then denying them paradise. Her back and neck injured in a car accident, she was vulnerable and fragile but at the same time wily and manipulative.

The article incarnated the double standard. Why an article on her and not on him? On her sex life and not his? On her grades and not his? On her high school conduct and not his? Why a probe of her mother's sex life but not her father's? Her mother's but not his?

(Not until almost a month later, on May 11, did a profile of Smith run.[70] And then it did not present him as it did her. The entire piece is contextualized by the notion inherent in the apology the *Times* would offer for its treatment of Ms. Bowman. "No examination of Mr. Smith's life and background could possibly resolve the disputed accounts of what happened that night in Palm Beach—his word against hers, with a verdict now up to a court." Where anonymous sources reported unconfirmed "wild" behavior about Bowman, the profile of Smith notes "rumors and gossip column items that ask whether the Palm Beach incident was part of a pattern of aggressiveness toward women in private" but dismisses them by saying "But all efforts by reporters for the *New York Times* to trace such rumors to their sources ended in vague and unverifiable accusation, hearsay or inconclusive silence." Where her academic record is probed, only the schools he attended—not his grades or academic conduct—are recorded.)

The response from the *Times's* reporters to the article on Patricia Bowman was clear. One hundred employees of the newspaper signed a petition expressing "outrage about the profile." Rebecca Sinkler, the editor of the Sunday *Times Book Review,* was among those who carried the protest to a meeting of editors.[71]

Two days later the editors responsible for the article met in New York with a large gathering of staffers concerned that the article had demeaned women, violated the *Times's* journalistic standards, and damaged the paper's reputation. Reporters in the Washington bureau listened on a speaker phone. "The anger and tension were palpable even to those of us in Washington," recalls a reporter there.

On Sunday April 21, *New York Times* columnist Anna Quindlen indicted the *Times* article. "If we had any doubt," she wrote, "whether there is still a stigma attached to rape, it is gone for good. Any woman reading the *Times*

profile now knows that to accuse a well-connected man of rape will invite a
thorough reading not only of her own past but of her mother's, and that
she had better be ready to see not only her name but her drinking habits in
print."

On Friday, April 26, the *Times* ran an editor's note of apology and an
article examining the controversy. The note confesses that "[M]any readers
inferred" that publication of the account "including her name and detailed
biographical material about her and her family, suggested that the *Times*
was challenging her account."

> No such challenge was intended, and the *Times* regrets that some parts of
> the article reinforced such inferences.
> The article should have explicitly asserted that nothing in the woman's
> known background could resolve the disputed testimony about her encounter
> with Mr. Smith. The *Times* regrets its failure to include such a clear statement
> of the article's limits and intent.

"Many readers inferred," of course, shifted the blame from the *Times* to
its readers. And the qualified nature of the editors' regrets carries a faint
echo of Richard Nixon's famous "mistakes were made." Still, the apology
is a remarkable admission from the powers that be of the nation's leading
newspaper. It occurred because the *Times* women made it happen.
"[F]rom beginning to end," wrote Nan Robertson, "it was the women of
the *Times* who pushed and prodded and called the paper and its most
powerful editors to account and insisted upon action."[72]

That these women's voices formed a powerful chorus was partly the
result of a class-action suit the *Times* settled in 1978, by agreeing to a new
affirmative action plan. "There's no doubt that they hired me as an
affirmative-action candidate," recalls Anna Quindlen.[73] At an anniversary event commemorating the suit, Quindlen, whose outraged voice
would condemn Bowman's treatment from the op-ed page of the paper
itself, thanked the women who had carried the suit through to settlement, "I'm not sure I would have had the guts to be one of them—who
made it possible for me to have eleven very good years at the *New York
Times*."[74]

A year before that suit was settled, and a decade and a half before the
Bowman/Smith case, Laura Foreman's sexual history was probed without
protest by reporters at the *Philadelphia Inquirer*. In September 1977, Foreman was forced out of her job in the Washington bureau of the *New York
Times* after revelations that she had an affair with and accepted gifts from a
source, Pennsylvania State Senator Henry J. (Buddy) Cianfrani, in an
earlier job covering Pennsylvania politics for the *Philadelphia Inquirer*. The
gifts came to light as part of a federal investigation of Cianfrani on charges
of racketeering, mail fraud, tax evasion, and obstruction of justice. Ultimately, Foreman would marry Cianfrani. Cianfrani would be imprisoned
after pleading guilty to federal charges. And Foreman, the first female

political reporter to be published by the *Inquirer* in its history, would leave journalism.

The *Inquirer* responded to news of its former reporter's affair by publishing a 17,000-word investigative report by its Pulitzer Prize-winning team of Donald Barlett and James Steele. The piece described Foreman as "a sultry, 31-year-old *Inquirer* reporter who speaks in a soft southern drawl and has an attachment for Faulkner."[75]

Where authors of the *Inquirer's* investigative report revealed the identity of Foreman's supposed lovers outside the *Inquirer,* they became strangely discreet about supposed inside liaisons. Other reporters on the *Inquirer* staff had been reluctant to report rumors of Foreman's affair with Cianfrani to *Inquirer* editors, said Barlett and Steele, because "They, like other *Inquirer* staff members, had heard a series of rumors linking Ms. Foreman romantically with several *Inquirer* reporters and editors."

Barlett and Steele seem to divide Foreman's sex life into "rumors," "apparent relationships," and relationships on "the record." "(As best as can be determined about such matters, Ms. Foreman apparently did have a romantic relationship with two *Inquirer* staff members—a married reporter and an editor, also married, who was in a position to pass judgment on her work. Both have since left the paper.)"[76]

"While most of the stories of Ms. Foreman's liaisons with Louisiana politicians [while on a former job] or public figures remain nothing more than rumors, there was one affair that became a matter of record both in New Orleans and later in Philadelphia. It was a relationship that developed between Ms. Foreman and a prominent New Orleans lawyer who had been an unsuccessful candidate for state and national political office. He was Benjamin Eugene Smith."[77]

The exposé raised the possibility of a double standard operating at the *Philadelphia Inquirer*. The *Inquirer's* national editor, Steve Seplow, said, "I had concluded very early that perhaps one or two male reporters were going out and (having affairs with) women on the Carter campaign staff . . . and I felt to pull her off this story because she was fooling around with a politician who has nothing to do with the campaign that there was just no reason to do it."[78] Although Seplow here presents an egalitarian stance, the *Inquirer* did not see problems posed by male reporters sleeping with sources in presidential campaigns as worthy of pursuit in print, nor apparently did it worry about editors sleeping with reporters whose conflicts of interest might as a result go unaddressed. Steele nevertheless insisted that "If there had been any comparable situation of a male reporter who had been personally involved with a public figure he was covering—we would have pursued it."[79] According to Steele, the women in the campaign involved with reporters had been secretaries or very low-level staff members.

The one reporter for the *Inquirer* who is quoted in non-*Inquirer* accounts seems to suggest that Foreman was guilty of violating gender stereotypes. "'She didn't have many female friends,' says *Inquirer* reporter Mary Walton. 'Very tough manner. I was a little afraid of her, and I don't feel much sympathy for her—there was plenty wrong with what she did.'"[80]

But women at other papers cried foul. Eleanor Randolph, then with the *Chicago Tribune,* said, "If they start opening up questions about people's lives, there's conflict everywhere. People covering the Hill are always screwing people. . . . There must be a lot of male reporters feeling pretty hypocritical talking about this."[81]

"Grievous though the experience has been," recalled Foreman, "it has at least afforded me a unique viewpoint. I doubt that any other reporter has ever been as close to what he covered as I have been, and I'm sure that no other reporter has ever been the target, rather than the author, of extended, deep, and painful probing into the most private crannies of life."[82]

A decade and a half later, internal and external critiques of the *Times*'s story on Smith's accuser would produce an apology; most commentary on the Foreman incident, however, centered not on the Barlett and Steele exposé, but on the justifiability of Foreman's conduct. The noteworthy exceptions were the *Washington Post*'s Richard Cohen, who explicitly decried the double standard he saw at work, and Foreman herself, who wrote an analysis for *The Washington Monthly.*

The presence of increasing numbers of women in the press corps may help to redress the balance, changing the sorts of questions asked of both men and women in the public sphere. NPR's Cokie Roberts believes that "had there not been so many women on the campaign bus, Gary Hart's womanizing wouldn't have been a story."[83] And, notably, it was female reporters who broke and pursued the claim that Anita Hill had been sexually harassed by Clarence Thomas.

Nor is political dirt the only issue. Observes Roberts, "when the budget negotiators from the Bush Administration and Congress were all holed up at Andrews Air Force Base, our male colleagues would ask them, 'How many MX's are there still left in the budget?' And I'd ask, 'Are mammograms still covered?'"

Women are changing behavior in the newsroom in other ways as well. In a misguided attempt to explain why women were less visible on the front page of the *New York Times* than on other papers, Max Frankel told a reporter from the *Washington Post* in 1990, "If you are covering the local teas, you've got more women than *The Wall Street Journal.*" The next day, he arrived to find protesting staffers wearing tea bags on their lapels.[84]

At a company retreat, Frankel welcomed recently appointed assisting managing editor Carolyn Lee to the staff members as the "latest 'adornment' to the paper's masthead," recalls Nan Robertson. "'Thank-you,' Carolyn said, 'but I have not worked so hard all these years to be called an *adornment.*' A woman editor who was there said, 'The men gasped, they were scandalized; the women were silently cheering.' Max apologized on the spot."[85]

One more force may counter institutional inertia in the press: the market. Where women were once more likely than men to read a newspaper, they are now less inclined to do so.[86] In 1991, a survey by the Knight-Ridder Women Readers Task Force found that women readers were unhappy with what they found in the nation's newspapers. Just as the gender gap is reshaping politics, so it may transform newspaper coverage as well.

9

The Stories We Tell

IT is difficult to write this sort of book without implying that someone other than women—society, organized religion, men—is imposing these binds. Women have, of course, been party to the process. But the women explicitly advocating women's rights usually assume that they are part of the solution, not the problem. While in the main I agree, there are times when a reasonable observer would surmise that those who identify with the movement (indeed occasionally in the name of the movement) inflict binds on other women with whom they disagree. They bring into play the bind of silence/shame.

Whether representatives of the movement should employ it is a question almost a century old. In January 1896, at the twenty-eighth annual meeting of the National American Woman Suffrage Association, an assigned committee presented a resolution to distance the association from Elizabeth Cady Stanton's *Woman's Bible*.

In the course of the debate, the association's president, Susan B. Anthony, turned over the gavel in order to oppose the resolution.

"The one distinct feature of our Association has been the right of individual opinion of every member," she said. "We have been beset at every step with the cry that somebody was injuring the cause by the expression of some sentiments that differed with those held by the majority of mankind. The religious persecution of the ages has been done under what was claimed to be the command of God. I distrust those people who know so well what God wants them to do to their fellows, because it always coincides with their own desires. . . . What you should do is to say to outsiders that a Christian has neither more nor less rights in our Association than an atheist. When our platform becomes too narrow for people of all creeds and of no creeds, I myself shall not stand upon it. Many things have been said and done by our orthodox friends that I have felt to be extremely harmful to our cause; but I should no more consent to a resolution denouncing them than I shall consent to this. Who is to draw the line? Who can tell now whether Mrs. Stanton's commentaries may not prove a great help to woman's emancipation from old superstitions that have barred her way? Lucretia Mott at first thought Mrs. Stanton had injured the cause of woman's other rights by insisting upon the demand for suffrage, but she had enough sense not to bring in a resolution against it. . . . You had better not begin resolving against individual action or you will find no limit. This year it is Mrs. Stanton; next year it may be me or one of yourselves who will be the victim."

Although the resolution was adopted by a vote of 53–41,[1] history has vindicated Anthony's dissent. Today, few who identify themselves as feminists find Stanton's work controversial.

In one unfortunate case, some in the women's movement repeated history. At issue was what a feminist scholar of history should have said in a court case. Rosalind Rosenberg, Professor of History at Barnard College, and Alice Kessler-Harris, Professor of History at Hofstra University, are both card-carrying feminists. Each is a respected historian. They differ in their explanation of why, between 1973 and 1980, women held three-quarters of the noncommissioned sales positions at Sears but were awarded only 40% of the promotions into the higher-paying commissioned sales jobs. As I noted in Chapter 5, Rosenberg found the explanation, at least in part, in women's acculturation and resulting preferences—women's difference. Kessler-Harris explained the disparity as evidence of discrimination by the giant retailer.

For her efforts, Rosenberg, in the words of the *New York Times,* was rewarded with "a debate that virtually isolated [her] from fellow feminists and historians, who accuse her of turning her scholarship against the women's movement."[2] Termed "immoral" by some, Rosenberg responded that "if the scholars allow their positions to drive their scholarship, they will be left with bad scholarship and misguided public policy."[3]

Like Susan B. Anthony's sisters ninety-nine years earlier, in December 1985, a committee of feminist historians passed a resolution affirming that "as feminists scholars we have a responsibility not to allow our scholarship to be used against the interests of women struggling for equity in our society."[4] An earlier version of the resolution had mentioned Rosenberg by name.

At issue is whether those who do not follow what is seen by some as feminist orthodoxy should be told that they ought to be silent rather than jeopardize the outcomes sought by the movement. Should those who perceive different truths in the history of women be ostracized, their professional credentials challenged? Should ideology—not the plausibility of the argument—be the test of a position? Should scholarship, one of the original tools used to break the binds, be subordinated to politics? Those who answer "Yes" perpetuate the silence/shame bind.

Rosenberg's critics said, in effect—and in the absence of identified victims of discrimination—that Sears was guilty until proven innocent, and that regardless of the facts of the case or the evidence of history, it was inappropriate for a feminist to argue the contrary.[5]

"The gauntlet that she [Rosenberg] has run since the trial," write two legal scholars, "goes beyond the ordinary scope of political disagreements in the academic world. It seems plainly designed to insure that no other historian, especially one without tenure, ever will dare to express similar views in court or in any other forum."[6]

Since Rosenberg is a tenured full professor, her economic and professional well-being are unlikely to be affected by such challenges. Their existence stands, however, as a warning to others that orthodoxy is not to be breached. The problem for those who wish to follow the dictum is identifying the orthodoxy. Rosenberg was condemned for arguing in court from a long-lived and respected tradition of scholarship by and about women. Her critics seem to say that there is a private sphere—within the academic journals and books—in which some sorts of scholarship are welcome but they become blameworthy and dangerous if revealed in public.

Kessler-Harris explains why Rosenberg should have demurred by asking,

> What was to be gained by such testimony? A successful argument would damage women who worked at Sears as well as past and future applicants. Worse, it would set a legal precedent that would inhibit affirmative action cases in the future. For if defendants could justify the absence of women in certain kinds of jobs on the grounds that insufficient numbers of women possessed any interest in them, one could foresee the resulting cycle. Expectations and aspirations conditioned by generations of socialization and labor market experience would now be used to justify continuing discrimination against women. The potential consequences were terrifying.[7]

Ultimately Kessler-Harris accepted the invitation to testify. "What then was a feminist doing in the courtroom at all? I had reacted viscerally to seeing my own work, badly distorted, put to the service of a politically destructive cause. I believed that the success of Sears' lawyers would undermine two decades of affirmative action efforts and exercise a chilling effect on women's history as a whole."[8]

But when Sears won, the consequences Kessler-Harris foresaw did not occur. As I indicated in Chapter 5, the next major discrimination case, based on clearer evidence and with victims of discrimination in court, was won by the female litigants. The prophesy was faulty.

For speaking in court, Rosenberg was shamed. The most effective document she offered the court was then suppressed by the influential and prestigious feminist journal *Signs*. Since Kessler-Harris rebutted Rosenberg's testimony, Rosenberg was given the opportunity by the court to respond in writing. She did so by documenting instance after instance in which Kessler-Harris's statements in court were at odds with Kessler-

Harris's published scholarship. *Signs* refused to publish this document. What it did provide its readers was Rosenberg's original offer of proof and Kessler-Harris's original testimony.

Since Rosenberg's written rebuttal to Kessler-Harris's weighed in the judge's ruling, eliminating it gives readers of the journal a distorted view of the exchanges that influenced the final decision. By giving her the final word, the editor's decision also sways the arguments in Kessler-Harris's favor. Nor is Rosenberg's rebuttal included in Kenneth Winston and Mary Jo Bane's *Gender and Public Policy: Cases and Comments.*[9]

Carl Degler, past president of the American Historical Association, and winner of the Pulitzer Prize, the Bancroft Prize, and the Beveridge Prizes in history, and author of *At Odds: Women and the Family in America from the Revolution to the Present,* agrees with both Rosenberg's decision to testify and with the "substance of her testimony."[10] And he sees the credibility of women's history at issue in the subsequent controversy. "I'm disturbed at the criticism of Rosenberg," he notes, "for having testified for Sears; I think criticism along those lines will hurt women's history, will make it seem to be simply a polemical subject and not a true historical subject. We still have to convince a lot of people that women's history is a real field of scholarly inquiry. If people have to follow a party line that will be fatal."[11]

The concerns raised by Rosenberg's treatment are not confined to discussions of labor history. Elizabeth Fox-Genovese opens the last chapter of her 1991 book *Feminism Without Illusions: A Critique of Individualism*[12] by noting,

> However much this book is intended as a feminist critique of individualism, it is bound to strike some—and perhaps many—as a critique of feminism. This risk is unavoidable. Feminism today has become so protean and wears so many faces that any notion of *a* feminist position has become utopian. Recognizing the danger that many other feminists, whose views differ from mine, may consider my critique disloyal, I can only hope that the scope of the movement permits the acceptance of divergent viewpoints as part of a continuing effort to make sense of what women need.

I share that hope.

Concerns about what can be said where are also occasionally voiced by some feminist literary theorists. In October 1989 three feminist scholars who had received their PhDs in the mid-1970s gathered to discuss "Criticizing Feminist Criticism." During the conversation, Marianne Hirsch, a professor at Dartmouth, observed, "I feel that 'the men' get a tremendous amount of pleasure and reassurance out of hearing feminists criticize each other. At Dartmouth we had a Women's Studies Conference in 1982. Lots of very established colleagues came to this conference and one person actually stood up and said, 'It is so gratifying to see that feminism has evolved to the point where feminists can criticize each other.' This is like the mark of maturity. But you could also see that the real pleasure for 'them' was to see that we are not a united front, that there are these splits, that in fact we're attackable."[13]

Using silence/shame against a scholar has the potential to eliminate from the repertoire of women's advocates one of the principle means of fracturing double binds. Since women have gained access to education, they have used scholarly evidence to establish that women's mental competence is not altered by menstruation, childbearing, and menopause; that use of male descriptors in job ads minimizes the likelihood of women seeing the jobs as applicable to them; that the measured aptitudes of women in science and math in all-girls and all-women's schools does not decline as they advance through the school cycle; and that girls and boys are treated differently when in the same math class.

When endocrinologist Dr. Estelle Ramey rose to rebut Dr. Edgar Berman's assertion that women could not aspire to national leadership because of their hormonal swings in 1970, her account was persuasive because it both summarized the available evidence and drew on her own considerable scholarship. In the more recent past, attempts to discredit the findings of Dr. Susan Fiske about stereotyping in *Hopkins v. Waterhouse* failed precisely because she was not presumed to be suppressing social and psychological evidence to advantage Hopkins. And law professor Wendy Williams's views about fetal protection shaped the ruling of the Supreme Court in *UAW v. Johnson Controls* precisely because her arguments pass the tests applied to scholarship: She grappled with the position of the other side, examined the precedents, drew on the available scientific data, and argued consistently.[14] So too did Inge Broverman who, with her colleagues, confirmed that we conflate supposedly masculine traits with mental health and supposedly feminine ones with disability. In each case, the female scholar proved credible in part because no one could persuasively argue that she had subordinated scholarship to political orthodoxy.

A final insidious legacy of an accumulated history of double binds is a tendency to regard the political world in the either/or of victim or victor, oppressed or oppressor. One result of the victim or victor mentality is a tendency to interpret and misinterpret available evidence through either lens, a move that becomes problematic when neither fits the facts. This is one of my complaints about Faludi's book—it rearranges the history of litigation on fetal protection and misstates women's proportional wages when it serves her ideological purposes.

But neither is Faludi an isolated instance nor is it the left only that is guilty of this form of filtering. So, for example, in the first edition of *The Beauty Myth*, Naomi Wolf contends that 150,000 women die annually from anorexia.[15] In *Who Stole Feminism? How Women Have Betrayed Women*, Christina Hoff Sommers exposes that figure as inaccurate and sets in place an alternative claim that "the correct figure is less than 100."[16] In her eager embrace of the posture of victim, Wolf has uncritically adopted a highly exaggerated figure. In her rush to unmask Wolf, Sommers asserts an erroneously low number. And Sommers then ignores the fact that Wolf corrected her error in her book's next edition.

Harold Goldstein and Harry Gwirtzman of the National Institute of Mental Health's Eating Disorders Program put the figure at about 1000 a year and add, "That is a far cry from 150,000, but it is substantially above the 'less than 100' Sommers proclaims. Too facile an acceptance of data in the service of ideology, from whichever end of the spectrum it comes, is ultimately a disservice to those women—and men—who suffer, and die, from eating disorders."[17]

And if Faludi should be called to question for re-ordering Court cases, Sommers is likewise culpable for injudicious deletion. Challenging the view that wife-beating was accepted in British common law, Sommers quotes Blackstone: "The husband was prohibited from using any violence to his wife. . . ." Lost to the ellipses is the Latin exception "aliter quam ad virum, ex causa regiminis et castigationis uxoris suae, licite et rationabiliter pertinet"—"other than that which appertains legally and reasonably to the husband for the governance and correction of his wife."[18]

As we reframe, recover, reclaim, recount, and shatter stereotypes, it is important that we not do unto others what has been done unto us. Thou Shalt Not Bind Thine Own Kind—Or Anybody Else. Sisterhood is a powerful metaphor; it ought not become a synonym for groupthink.

Double binds survive unscathed only so long as those caught in them grant their definitional power. One legacy of centuries of confronting and confounding binds is a persistent belief on the part of those who make up society, including women themselves, that we are outsiders. Even as women move into upper management and graduate from college in higher numbers than men, that image lingers. Often an albatross, the image is now, in some arenas, an asset. From the outsider position female candidates for elective office argue that, unlike those in power, they are not corrupted. And when a major organization requires reform, women are seen as candidates with unique qualifications for the job. Even Elizabeth Dole, a Washington insider by credential and marriage, carried that persona into her role as head of the American Red Cross. And when the head of United Way resigned in disgrace, that organization turned to Elaine Chao to clean the place up.

On the personal side, being an outsider can be liberating as well. Had she been an insider in the scientific community, Barbara McClintock would have been less likely to determine how transposable elements account for mutations in plants and animals—and in the process win the Nobel Prize. The same is true for Margie Profet who foreswore the PhD on the grounds that it would stifle her creativity. As I noted in Chapter 6, Profet last year took a scalpel to one vestige of the uterus/brain bind by demonstrating that the uterus is an active, not a passive organ. And it was an African-American woman, Hazel O'Leary, who did something no man—Republican or Democratic—had had the courage to do; open the files exposing the history of abuse in the United States's role as nuclear custodian.

Other images are not so useful. Those who grant the false dichotomies that undergird them permit others to define who they are and what they can and cannot do. Asserting control over the language through which we see ourselves is consequently an indispensable move in vanquishing the vestiges of the binds that tie. Reframing, recovering, reclaiming, recounting, and confounding the stereotypes are the tools being used to clear them away.

Reframing

The first step in overcoming a double bind is seeing it for what it is. Reframing invites an audience to view a set of options from a different perspective and confront the fact that the options offered are false—whether they present a no-choice-choice, a self-fulfilling prophecy, a no-win situation, a double standard, or an unrealizable expectation.

From its early days, the women's movement has used reframing against binds. Carrie Chapman Catt's remark in 1902 neatly reframes the situation then confronting women.

> This world taught woman nothing skillful and then said her work was valueless. It permitted her no opinions and then said she did not know how to think. It forbade her to speak in public, and said the sex had no orators. It denied her the schools, and said the sex had no genius. It robbed her of every vestige of responsibility, and then called her weak. It taught her that every pleasure must come as a favor from men, and when to gain it she decked herself in paint and fine feathers, as she had been taught to do, it called her vain.[19]

At its base, reframing meta-communicates. It steps back to critique the conventional rhetoric used to describe women's options. So, for example, at a 1993 fundraiser for the California Women's Political Caucus, state treasurer and gubernatorial candidate Kathleeen Brown invites her audience to see—and as a result reject as unfair—the dilemmas confronting a female candidate.

"If we are single, they say that we couldn't catch a man. If we are married, they say that we are neglecting him. If we are divorced, they say we couldn't keep him. If we are widowed, they say that we killed him."

At one time or another, most contemporary female politicians have recounted a version of the classic no-win situation that first identifies women in relation to men and then discredits any choice they can possibly make or any circumstance in which they can possibly find themselves. In the telling they help blunt the power of the bind.

Recovering

Recovery of the history of women's lives and accomplishments reminds us of the progress women have made and the means they have employed. Where historians once chronicled the life and times of our forefathers, the work of our foremothers is now valued as well. Their accounts have claimed not only shelf space in libraries but are also topics of study on their own. Just as important is the integration of half the human species into historical accounts not specifically focused on women's rights and roles.

Has women's place in the Pantheon been confirmed? In one instance, it may literally become so. When Marie Curie died ninety years ago, it was unthinkable that, as a tribute to her distinction (she won the Nobel Prize not once but twice), a woman would be interred in the Pantheon—the hallowed resting place of Voltaire, Victor Hugo, and other luminaries.

The only woman buried there at the time was Sophie Berthelot. An exception had been made for her not because of the accomplishments of her life, but the occasion of her death. When she and her scientist husband died together, their grieving friends argued that it was appropriate that they be buried together as well. If the president of France has his way, Sophie Berthelot will no longer be a token of one in the Pantheon. In March 1994, Francois Mitterand announced on International Women's Day that the government would ask her family for permission to rebury Marie Curie in one of France's more famous monuments.[20]

If justice is done by moving the remains of an accomplished woman to a place of honor, it is served as well by acknowledging the contributions of the living, regardless of their politics. Jeane Kirkpatrick is an accomplished scholar who authored a path-breaking book on women and leadership, and performed ably as U.S. representative to the UN (1981–85). In her role at the UN she confronted the femininity/competence bind head-on and her performance made it easier for her successor, Madeline Albright. In May 1994 when protesting students and faculty at Brandeis University opposed her receipt of an honorary degree, prompting her to refuse it, they did her and the cause of women an injustice.

Similarly, whether one embraces or rejects her politics, former British Prime Minister Margaret Thatcher was for a time the most powerful female leader in the world. In her exchanges with Parliament, she decisively dispatched the notion that a woman is incapable of thinking on her feet. In moments of crisis, she rose to eloquence. And by leading her country into war, she made it less likely that future female candidates for President of the United States will be asked whether they are tough enough to function as commander-in-chief. Those who bar Kirkpatrick and Thatcher from the Pantheon of female leaders do what the French did in keeping Marie Curie from France's most hallowed ground, and what the academy of the past did when it suppressed the work of female scholars and writers. They impoverish the range of models that future generations can draw from in seeking their own appropriate roles.

Reclaiming Language

As rhetorical critic Kenneth Burke argued, language can't do our drinking but language does do our thinking. Those who embrace the label "victim" should not be surprised to find themselves victimized.

Whether language is, in economists' terms, a leading or a lagging indicator, of course, depends on circumstance. When Sandra Day O'Connor joined the Supreme Court, those addressing the court could no longer refer to each assembled justice as they had in the past as Mr. Justice. The title "Justice" replaced its gender-identified predecessor when a woman took her place on the bench. In other instances, the power to name is the power to create and define. It is a sign of authority. And since Adam announced that he would call his mate "Eve" that power has been held by those presumably in charge. By assuming the power to name themselves, as Geraldine Ferraro and Kathleen Brown did when they retained their birth names in marriage, women confirm that they too control language. By insisting that they be identified as "pro-life," those opposing abortion demand the right to name themselves. So too do those favoring abortion rights when they insist on being identified as "pro-choice."

Language is a tool that changes our focus and our perceptions. When commentators observing the anniversary of D-Day praised servicewomen as well as servicemen, they were tacitly acknowledging that the women's movement had changed the way we think and speak. In so doing they had made a move toward rescuing important pieces of our past.

When we speak of reproductive choice rather than birth control, we again reflect a change in the language born of advocacy. Discussion of the realities of sexual assault took on an added dimension as the phrase "date rape" gained currency.

Centuries of pejoratives ridiculing older women fall aside when the identification changes to gray panthers. Patriarchy gives way to matriarchy if I pray to God the mother. And words such as "sexism," "agism," and "homophobia" condemn behaviors that might otherwise be tolerated if unlabeled.

What invited accusations of neglect elicits discussions about mutual commitment when I can replace the notion that a woman should spend full time with a child with the notion that a *parent* should spend *quality time*. When "feisty" shifted from the equivalent of a word ending in "ette" to mean strong and agile, women had again recovered some of the powers of language.

Recasting

The means of reclaiming language include recasting words used to disqualify into terms denoting qualification, and so transforming handcuffs into credentials. The suffragists argued that as mothers and homemakers they would use the vote to clean up politics. More recently, candidates such as Washington's Patty Murray have suggested that a "mom in tennis shoes" is just what the Senate needs if it is to be responsive to the needs of the people. The assumption is catching on. The May 1994 Democratic senatorial primary in Ohio pitted Joel Hyatt, the millionaire son-in-law of retiring Democratic Senator Howard Metzenbaum, against Cayahoga County Commissioner Mary Boyle. "The Senate doesn't need any more millionaire lawyers," she said in her TV ads. "What it needs is more moms." Outspent by Hyatt, Boyle lost. But arguing that motherhood is a qualification was an important move.

Equal Opportunity Alternatives

As I noted in Chapter 4, the silence/shame bind is perpetuated by a rich and varied range of pejorative words available to condemn the speech of women and the limited range of positive ones. Extinguishing "strident," "shrill," and their kind is one tack. Another turns the tables and applies them to men. In 1993, Kay Bailey Hutchison dexterously switched stereotypes usually used against women on her male opponent Democrat Bob Krueger, saying in response to his charges "I think he's getting hysterical. . . . This is crazy, I mean absolutely ludicrous. Three newspapers have not made those charges. He is out of control." Hutchison won.

A third move creates a range of condemnation applicable to men that mirrors that for women. This process of reversal is at work in feminist leader Gloria Steinem's essay "What If Freud Were Phyllis?," which envisions boys envying the "compactness, safety, and beauty" of the vagina. Hysteria becomes tysteria and penis envy, womb envy.[21] The reversal is predicated on a female instead of a male norm—female superiority, male inferiority. Each of these linguistic moves invites us to examine our presuppositions. Each makes it less likely that language penalizing women will be taken for granted in future exchanges.

In 1988, Dan Quayle felt insulted when a reporter identified him as a blond bimbo. The dig was also a slur on Quayle's masculinity. Suggesting someone who is brainless and sexually available, "bimbo" signals woman not man. In a forthcoming book titled *Manhood: The American Quest*, sociologist Michael Kimmel strikes back with the term "himbo." "Himbos" come in two forms: those created for women, such as model Fabio, and those constructed for men, such as Sylvester Stallone. A "himbo" might, in the words of reporter Susan Campbell, be described as a good looking man who is "two fries short of a Happy Meal."[22] Himbo can also be applied to men photographed as objects of sexual desire such as Marky Mark who appears in a larger-than-life billboard wearing Calvin Klein underwear. Similarly, to the phrase "trophy wife," pundits have added "trophy husband." To a history of ads that identified products with sexually attractive women, the sponsors of Diet Coke have added women ogling a bare-chested male construction worker downing a Coke on his break.

To bimbo-himbo and trophy husband-wife, toy manufacturers have added another pair. The card game, Old Maid, whose goal is to avoid being stuck with the Old Maid at all cost, now is accompanied on the toy-store shelves by Old Bachelor.

Language carries perspective in tow. A truckload of responsibility is shifted from men to women when our language is laden with unwed mothers but not unwed fathers, battered wives but not battering husbands. The trial of O. J. Simpson for the alleged murder of his ex-wife and a friend has been accompanied by a flood of media commentary asking why women do not leave and prosecute batterers. When we talk more about why men abuse women in the first place, we will have turned another corner.

If recontextualizing the words normally employed against women raises consciousness about the original uses of the offending language, so too does refusing to saddle men with similarly constraining assumptions. So, for example, Nancy Landon Kassebaum notes,

> Throughout my years here I have taken pride in the fact that I am a United States senator, not a woman senator. When some of my male colleagues have suggested that I know nothing about national defense issues because I am a woman, I have been offended. In the same vein, I have to assume that many of my male colleagues are offended by the notion that they cannot begin to understand the seriousness of sexual harassment or the anguish of its victims.[23]

Rewriting

The women's movement was launched in the United States with the Seneca Falls Declaration that rewrote the Declaration of Independence to assert that women too are created equal. In the backdrop of the revision stands the old order. Where reframing steps outside a piece of rhetoric, rewriting works from within.

This is the process at work in the recasting of a traditional prayer. A Jewish hymn drawn from Proverbs 31 is traditionally read on Friday evenings. The "Eishet Chayil" praises the woman who tends to her house, her husband, her children, and charity. In June 1994, author E. M. Broner paid tribute to the founder of the Israel Women's Network with a reconstructed version of the hymn that praises the "wise woman" who

> works alongside her husband,
> But outside the walls of the house,
> Outside the gates of her garden,
> She hears the cries of women in distress.
> She is their rescuer.
> She rises at dawn to organize. . . .[24]

Recounting

Across time, as Chapter 4 argues, storytelling has been a form open to and rewarded in women. Whether the process occurs in a boardroom or at the bedside of a child, a woman telling a story is, so to speak, in a traditional sphere. In an electronic age, television is drawn to the intimacy, self-disclosure, and natural unfolding of narratives. For this reason, as a woman moves into a place traditionally identified as male territory, she can use narrative—more effectively than other available means of persuasion—to explain who she is, why she belongs there, and what principles define her.

In the opening statement at her confirmation hearings, Attorney General Janet Reno did just that—rooting her principles in the life story of her mother who built a house from scratch. "One afternoon, mother picked us up at school," said Reno, "and she said, 'I'm going to build a house.' And we said. 'Well, what do you know about building a house?' And she says, 'I'm gonna learn.' And she went and talked to the brick mason and the electrician and the plumber, and she learned how to build a house. She dug the foundation with her own hands with a pick and shovel. She laid the blocks. She put in the wiring. She put in the plumbing. And daddy would help her with the heavy work when he got home from work at night. I have lived in that house ever since, and as I come down the driveway through the woods at night with a problem, an obstacle to overcome, that house is a symbol to me that you can do anything you really want to if it's the right thing to do and you put your mind to it."[25]

The narrative Justice Ruth Bader Ginsburg told at her confirmation hearings translated the lessons of her own life into an affirmation of the rights of women.

> I am, as you know, from my responses to your queries, a Brooklyn native, born and bred, a first-generation American on my father's side, barely second generation on my mother's. Neither of my parents had the means to attend college, but both taught me to love learning, to care about people, and to work hard for whatever I wanted or believed in. Their parents had the foresight to leave the old country when Jewish ancestry and faith meant exposure to pogroms and denigration of one's human worth.
>
> What has become of me could happen only in America. Like so many others, I owe so much to the entry this nation afforded to people yearning to breathe free. I have had the great good fortune to share life with a partner truly extraordinary for his generation, a man who believed at age 18 when we met, and who believes today, that a woman's work, whether at home or on the job, is as important as a man's.
>
> I became a lawyer in days when women were not wanted by most members of the legal profession. I became a lawyer because Marty and his parents supported that choice unreservedly. I have been deeply moved by the outpouring of good wishes received in recent weeks from family, neighbors, camp mates, classmates, students at Rutgers and Columbia Law Schools, law teaching colleagues, lawyers with whom I have worked, judges across the country, and many women and men who do not know me. That huge, spirit-lifting

collection shows that for many of our people, an individual's sex is no longer remarkable or even unusual with regard to his or her qualifications to serve on the Supreme Court.[26]

Each of their narratives subtly addresses the femininity/competence bind. Ginsburg and Reno both identify with their maternal heritage and employ the story form. While the narratives personalized the speaker, the principles each woman educed from them tied otherwise private experiences to the public role for which the nominee was being considered.

Confounding the Stereotypes

As women have campaigned for elective office, consultants have studied the behaviors that elicit negative and positive responses. So, for example, as I noted earlier, women are considered more credible when cast as uncorrupt outsiders, who are expert on matters associated in some way with a woman's traditional role: education, welfare, health. Voters are less likely to grant the credibility of a little known female candidate in such areas as defense.

But low expectations are not necessarily a liability. An exchange in one of the 1990 senatorial debates in New Jersey is illustrative. When Senator Bill Bradley asked his Republican opponent Christine Todd Whitman a tough, technical question about budgeting for a specific bomber, he was trying to capitalize on two areas of traditionally perceived male strength and female weakness. At the moment of questioning, focus groups I conducted responded to the move with strong approval of Bradley. Here was a moment in which Whitman's credibility was being tested. But when Whitman responded with a forceful, precise, knowledgeable answer, she gained *more* approval from the group than she did when she spoke in areas traditionally identified as within a woman's sphere. Her answer shattered stereotypes about the areas a woman can be competent in and elicited a strong positive response.

Women working at high levels in university administration use the same principle when they demonstrate a facility in working through university budget figures. Demonstrating unexpected competence carries a bigger bonus than showing that one can do what is expected. The corollary, of course, comes to the fore when a women seems inadequate in an area of perceived natural strength; such incompetence carries bigger penalties.

Marshaling nonverbal cues in unexpected ways also fractures stereotypes. A female candidate can, for example, use "warm" nonverbal cues to counter tough questions. In the 1993 Texas senatorial race, after Republican contender Kay Bailey Hutchison "boasted that she had completed a Spanish course the previous year as part of her effort to gain an undergraduate degree, reporters asked her to say a few words in Spanish. 'Nolo contendere,' she replied with a laugh, using the Latin term that defendants use when they do not contest charges lodged against them. Asked again, she sputtered, ungrammatically, 'Me no sabe.' But she kept grinning through the whole exchange, and it never developed as a major campaign issue."[27]

Observing such tactics, Molly Ivins, columnist for the *Fort Worth Star-Telegram,* commented, "In the future women who would be candidates would do well to study her nigh-flawless combination of saccharine and steel. Hutchison has always dressed in a style that is close to a parody of I-am-a-nice-girl fashion. . . . At the same time, in debate after debate against Krueger [her Democratic senatorial opponent], whenever she was pressed, out came the steel, the edge to the voice, the don't-mess-with-me-you-despicable-worm-how-dare-you-suggest-that form. She was tough."

The need for female candidates to establish that they are both tough and caring can be satisfied by sending tough cues in verbal form and caring cues nonverbally or vice versa—an art former Texas Governor Ann Richards has mastered. Attacked by her opponent Clayton Williams in a candidate debate, she smiled throughout an answer in which she chided him for not releasing his taxes and accused him of lying. Respondents in focus groups who were asked to evaluate the exchange used none of the words ordinarily elicited when a woman attacks.

Another way of fracturing stereotypes is use of a tactic I call "Speaking softly while carrying a big statistic." Indeed, consultants have learned that opening a female candidate's campaign with data-driven material and following it up with warmer, more empathetic content enables their candidates to mute the question *is she tough enough* while communicating that she is indeed *caring enough* to lead. These moves are designed to overcome the "bitch" factor, shorthand for the femininity/competence bind.

The nature of the times is reflected in the stories that are told about women and the stories that, as women, we tell. To recap the issues central to this book, let me recount some of them. I call the first set "The bad old days," the second "The times they are a changin'," the third, "Women will be seen and heard," the fourth, "We have overcome."

The Bad Old Days

Not too long ago, the stories we told spoke not of misidentification but of exclusion and rejection. Few professions have done justice to their fore-mothers. The scientific establishment in France found it difficult even to recognize Madam Curie. Confronted by her nomination to the Academie des Sciences, members of the Institut de France voted 90–52 "that no woman should ever be elected to its membership."[28]

Until women were finally elected to full membership in the Royal Society in 1945, the only woman inside this prestigious British academic society was "a skeleton preserved in the society's anatomical collection."[29] The history of women in science can be written as an ongoing struggle for a lab of her own. "Many of [the female Nobel Prize winners] faced enormous obstacles," writes Sharon Bertsch McGrayne. "They were confined to basement laboratories and attic offices. They crawled behind furniture to attend science lectures. They worked in universities for decades without pay as volunteers—in the United States as late as the 1950s."[30]

The conditions under which Nobel winner Lise Meitner worked were representative. "Using a private entrance, Lise Meitner entered her basement laboratory—and stayed there. A former carpentry shop, it was the only room in Berlin's chemistry institute that she was permitted to enter. No females—except, of course, cleaning women—were allowed upstairs with the men. Prohibited even from using a rest room in the chemistry building, she had to use facilities in a hotel up the street."[31]

Feelings of exclusion persist among female scientists. "Maybe it's because I am a woman, but I have never felt like one of the boys," former NIH head Bernadine Healy says. "I have never really been in the inner circle. It doesn't matter that I was a professor, it doesn't matter that I was president of the Cleveland Clinic. I have always been on the edge. . . . It doesn't matter to me that the club is angry with me, because I've never been a member."[32]

To succeed against such odds, women had to be exceptional. In the 1960s this fact would be expressed in an angry slogan: "To make it, a woman has to be twice as good as a man; fortunately, that's not difficult." It was a notion repeated by female leaders. Ottawa's former mayor, Charlotte Whitton, phrased it a bit differently, saying that to be thought half as good as a man a woman had to do twice as well. But, she added, "Luckily, this isn't difficult."[33]

Female reporters had it no better. When reporter Mary Ellen Leary joined her male colleagues at the San Francisco Commonwealth Club in the 1950s for a briefing on a forthcoming piece of legislation, "the Commonwealth Club secretary came and motioned me with his finger outside. And I left my purse and my raincoat and everything and walked out with him. And he said, 'Mary Ellen, I'm afraid we're not allowed—we cannot have you in here. We don't allow women at any of our meetings and you can't be here.' And I said, 'I'm not here as a woman, I'm here as a reporter.' Well, I lost on that, I couldn't stay."

"Until 1956, the [National Press Club's] prestigious power lunches with political figures were not open to female members of the press. After 1956, women reporters were allowed into the club's balcony—'purdah,' [UPI's Helen] Thomas called it—to watch and listen to the speeches, but they were not allowed to ask questions or to sit and eat with male colleagues at tables on the floor."[34]

Since the State Department booked most major leaders into speaking engagements at the Club, the Women's Press Club protested that women were being denied the opportunity to do their jobs. Members of the Women's Press Club carried their case to the foreign embassies as well as to the State Department. In 1959, the organization got its first break when Soviet head of state Nikita Khrushchev insisted that he would not speak unless the female reporters were given access.

The result: the National Press Club, on a one-time only basis, invited thirty women reporters to join their male colleagues for the luncheon and the speech. The event was important. In that speech the Soviet leader explained what he had meant by his widely quoted claim "We will bury you." It took twelve more years of pressure to finally open the Club to women; the vote to admit occurred in 1971.[35]

Sooner or later, state associations fell in line. "Albany has the oldest Legislative Correspondents Association in the country," recalls the *New York Times'* Kit Seelye. "They put on a show every year, the LCA show. It was an extremely important event. Legislators, the governor, the senators and everybody goes to the LCA show. But women weren't allowed. It was a two-night event. . . . It is like the gridiron. It wasn't until '74 that women were allowed in."[36]

Nor was the law a profession that welcomed women. When Sandra Day O'Connor was nominated to the Supreme Court, reporters dug out the fact that "Fresh out of law school, engaged to fellow law-review editor John Jay O'Connor III, she found that none of the large California firms would hire a woman, so she went to work as deputy county attorney for San Mateo County."[37]

Ruth Bader Ginsburg's nomination prompted a similar string of stories. In 1956 Harvard Law School Dean Erwin Griswold held a tea for the nine women in one of Harvard's three cohorts of would-be lawyers. At that event, the dean asked the women how they felt "about taking places earmarked for men." "Fearful of seeming too assertive," Ginsburg "mumbled that studying law would help her better understand her husband's work, and could possibly lead to part-time employment for herself."[38]

At a Social Security office where she worked while her husband completed military service, the 21-year-old Ginsburg failed to hide her pregnancy. She was demoted. Nine years later, while teaching at Rutger's Law School, she hid her second pregnancy under oversized clothing borrowed from her mother-in law.[39] When she applied for a clerkship with Justice Felix Frankfurter, he asked, "Does she wear skirts? . . . I can't stand girls in pants." Despite testimony that she met his dress code, Ginsburg didn't get the post. While arguing a sex discrimination case before the Supreme Court in the 1970s, she was asked by Justice Rehnquist "why women were not satisfied with their station now that the face of Susan B. Anthony was on a coin."[40] And when she urged that luncheon meetings of the American Law Institute be moved from a club closed to women, an A.L.I. member and law professor at the University of Texas Austin wrote her a letter full of admiration for qualities he would have found unnoteworthy in a man. "You never hesitated in making your own position known with force and candor. . . . What I have marveled at, however, is the complete serenity with which you take part in these debates. No lock of hair is ever askew. No hint of anger or elation ever appears on your face or in your voice. Yours is always the position of calm reason and confidence."[41]

Learning of Ginsburg's nomination to the High Court, a lawyer in Bernardsville, New Jersey, used his formal remarks after a Rotary Club dinner to recall that in law school he and his friends referred to Ginsburg by the nickname "Bitch." When told of this, Ginsburg replied, "Better bitch than mouse." When retold in *The New Republic,* her response became a resource for others and fodder for further discussion.[42]

Other women now in positions of political power recall similar stories. "I remember looking for a job before the 1964 Civil Rights bill was passed," reports California Senator Dianne Feinstein. "I remember being often passed over by men with less of a resume, less education, than I had. [And yet, until] Title VII of the 1964 Civil Rights Act, it was commonplace to discriminate against a woman for employment."[43]

Most of us have mothers and grandmothers who recall similar situations. "Many years ago, before 1935, I took a test for this job in the museum," remembers the grandmother of one of the researchers for this book, "and I came out number one and I did not get the job because they said flatly, 'We do not want a woman.'"[44]

Exclusion was not the only problem. Women working behind the lines often found their work credited to someone else. Each profession seems to offer one capsulizing story more memorable and more often repeated than others about the usurpation of a woman's significant work. Female scientists return time and again to one glaring instance of intellectual theft and a subsequent campaign to denigrate its victim.

"Had [Rosalind] Franklin not had her work secretly taken from her and had she thus been allowed enough time to use her data to solve her puzzle, there is hardly any doubt that she would have unraveled the helix—perhaps even before Crick and Watson," writes G. Kass-Simon in *Women of Science: Righting the Record.*[45] Anthony Serafini observes that "James Watson made use of Rosalind Franklin's data without crediting her in the famed DNA race. . . . Certainly Watson and Crick would not have gotten the Nobel Prize had they not stolen her data."[46]

Unauthorized use of Franklin's work is not the only charge against Watson: He went on to add insult to injury. "In Watson's book, Franklin plays the role of 'Rosy,' the wicked stepmother," historian Sharon Bertsch McGrayne points out. "She is both Watson's central rival and the stereotypical old maid who keeps the plot line moving. Besides denigrating her personality, Watson attacked her scientific abilities, accusing her of being categorically 'anti-helical' and opposed to model-building."[47] He also described her in terms that called up visions of ill-tempered school marms. "There was not a trace of warmth or frivolity in her words," he complained. Franklin never wore glasses, but, in his imagination, he put them on her and wondered "how she would look if she took off her glasses and did something novel with her hair." He called her "Rosy . . . Maurice's assistant," and portrayed her behavior with male colleagues as bringing back unpleasant memories of lower school.[48]

The Times, They Are a Changin'

Every man my age I know can recount a slide show on a technical topic or speech to a Lodge or all-male civic club that for comic relief or to spike lagging attention included pictures of nude women. Such titillation was apparently a male lecturer's stock in trade. They were the equivalent of pin-up calendars hung on the wall of the local garage.

Where my middle-aged husband can recount such stories, our teenage son cannot. Nor can his older brother. The reason: Audiences who once were comfortably assumed to be male have been infiltrated. Women tell stories about moments of transition. "I . . . had a job as a policy analyst, but the policies were in transportation and hazardous materials. Very male oriented, male dominated field," recalls Kathy Dalton, now a federal liaison to the New York State Department of Social Services. "I remember going to a state police training course on hazardous material transport. The trainer didn't expect women in the room. Some of the slides showed women dressed provocatively. He'd say, 'Now here is a hazardous material, but we wouldn't caution you against touching it.' Or he'd say, this was before AIDS and AIDS education, 'You definitely don't need to be protected when approaching this hazardous material.'

"I was sitting in the front row and he just looked at me all of a sudden and said, 'Oh my God! I didn't expect you to be here.' I said, 'Surprise!' and smiled. I didn't mean to say more than that except to acknowledge that 'Yes I was here' and that 'yes I noticed that what he did was not appropriate.' When he took his break, I saw him shifting through the slides really quickly to get other pictures out."

The presence of powerful women is changing other forms of entertainment as well. In June 1993, Colorado state legislator Michelle Lawrence accepted an invitation to play in a charity golf tournament to benefit a children's hospital. "When our threesome arrived at the 11th green," she recalls, [the tournament sponsor] "was there to take our picture. Nearby, sitting in a golf cart, was a young woman, probably in her early 20s and dressed in a bikini, who we later learned was available to have her photo taken, with or without the top to her bikini, with the golfers that day. . . . Later . . . we noticed that this same woman was, by then, completely nude."

Lawrence confronted the sponsor and was told that "a topless woman has been available each year since the tournament began. . . . He said that one year a golfer put the topless woman on his shoulders and carried her to the portion of the course bordering Colorado Boulevard, which in his words 'stopped traffic.'"[49]

The legislator blew the whistle. A subsequent interview with her on a local talk radio show included the following exchange:

HOST: "Was that a picture of the stripper in the paper or is that you cuz you're not bad looking."

SECOND HOST: "What if they had nude men? Would you feel better about that?"
LAWRENCE: "Not at all. I think that would be rather hypocritical, don't you think?"[50]

But the retrogrades at the radio station didn't have community backing. Lawrence's revelation prompted the hospital to refuse the funds raised at the tournament and the host to declare that he would not include such "entertainment" in the future.

Women are unmasking gender bias and inappropriate behavior in more sober surroundings as well. During the confirmation hearings of Judge David Souter, Senator Strom Thurmond appeared to think he was complimenting the panel of women seated before him when he said, "Mr. Chairman, we have a group of lovely ladies here. We thank you for your presence. I have no questions." The "lovely ladies"—Molly Yard, president of NOW, and Eleanor Smeal, head of the Fund for the Feminist Majority—grimaced. When Senator Simpson disapproved of their reaction to his colleague, Yard responded, "Maybe you could explain to him that we would like to be treated the way you treat everybody else. You don't say to men, 'Gentlemen, you all look so lovely.'"[51]

From the beginning, advocates of equality have also included men. Their contributions—from those of W. E. B. DuBois to John Dewey—merit memorials as do the small but nonetheless significant actions of those less often mentioned. In 1987, sports reporter Suzyn Waldman of WFAN entered the Toronto Blue Jays locker room after a major game. There she joined the male reporters who were interviewing left fielder George Bell. Seeing Waldman approach, Bell began shouting obscenities in Spanish and English and ended the interview. "I didn't want anyone to see me break down," recalls the reporter. "I started to go for the door—the room was still dead quiet—when I heard someone call me. It was (right fielder) Jesse Barfield." Watching the scene, Barfield had asked a teammate the reporter's name. "Hey Suzyn," he called out, "I went 3 for 4 today. Don't you want to talk to me?" Moving to the corner in which he was standing, the reporter conducted an interview. Why did he respond? "She was just trying to do her job just like I'm trying to do mine," said Barfield. "And I only did what I'd want done to me. As long as they respect us, they should be respected as well."[52]

Women: Seen and Heard

Increasingly, women are telling other women how they used speech to transform a situation. "My male opponent for the post of County Commissioner of Allegheny County, the City of Pittsburgh, had spent a lot of money to win. I'd spent fifteen thousand dollars," recalls Pennsylvania's Auditor General Barbara Hafer. "Near the end of the race, the press asked him, 'What do you think of her?' He said, 'What does she know? She's just a little nurse from Elizabeth.' The next day, we quoted that statement in letters we sent to twenty-five thousand registered nurses. I won."

At a summer 1993 conference on gender bias in the courts, a male judge opened a session with a joke. "Bob Packwood and Dan Quayle were competing in a spelling bee. The word to be spelled was harass. Quayle got it right. Packwood lost. He thought harass was two words."

"There were groans and looks of disbelief throughout our group," recalls a male judge who was there. A women scheduled to speak on behalf of one of the attending groups altered her remarks to respond. "Looking right at him and speaking in a measured, firm manner, she said that one of the obstacles to the implementation of gender-bias programs was the kind of inappropriate joke we had just heard. 'If it can happen here, it can happen anywhere,' she said. 'We have to talk about this.' There was immediate applause from all corners of the room."

Later in the day, the judge apologized for the joke. "As the judge described his state's report, he mentioned a chapter on sexist jokes. He paused, then added, 'I guess I didn't read that chapter closely enough.' As we all laughed, the conference chair left her seat at the head table, walked up to the podium, and hugged her colleague with obvious affection. That hug was a message for us all: 'My friend blew it, but you should forgive him as I do. He's a good man. We need him for this important work.'"[53]

But what is one to do when one is seen but not heard? When the Family Medical Leave Act came to the floor of the U.S. House of Representatives, first-term Congresswoman Lynn Woolsey spoke for it. She had never before addressed the floor of the chamber. Later that day the chair of one of the committees approached her and said, "I saw you give your talk. I was in my office in a meeting and I saw you on television." The Congresswoman recalls readying herself for a compliment. Instead he said, "I couldn't hear what you said because we didn't have the sound up, but we do that a lot. . . . But one of those men on my committee said, 'Who's that good looking babe?'" "You know," Woolsey observed, "He thought he was complimenting me and what I really wanted to hear was what he thought of my words."

So Woolsey devised a strategy to direct his attention to the content of her messages. "Since then," she says, "I have learned to ask my chair, 'How am I doing?' What did I say? Did I say the right things?' So they would start listening instead of looking."[54]

There are various ways to say "Enough." In the early 1980s, "I was in a packed courtroom prosecuting a big death-penalty case," recalls Melba Marsh, now Judge of Cincinnati's Hamilton County Municipal Court. "The opposing lawyer said, 'Your Honor, let's pick a trial date when the prosecutor doesn't have PMS'—and the judge said *nothing*. So I replied, 'I only have PMS a few days of the month. There's *no* part of the month when *you're* not an asshole.' Only then did the judge get his tongue back. He yelled at both of us,"[55]

We Have Overcome

Every woman I know tells a story of mistaken identity. The female doctor mistaken for a nurse, her male nurse assumed to be the doctor. The female executive treated as her secretary.

Sometimes, this game of Clue involves women mistaking the professional identity of men. "I've certainly been guilty of making an assumption based on looks, and then am shocked to find how off-base I've been," reports *Lear*'s Fashion Consultant Vicki Path.[56] "Once I tipped the head of a record company—I thought he was the parking valet." But more often the story is told about those not expected to hold positions of authority, particularly women, blacks, and Hispanics.

Not even Nobel Prize winners are immune. Among the traditions of the awarding surrounding the Nobel Prize, a Swedish student is given the honor of escorting the prize winner from the banquet table to the dias. Instructed to escort Dr. Yalow, the student faced a problem. Seated at the table were two Yalows, one male, one female. "Proceeding down the wrong side of the table, he carefully came to attention behind Aaron Yalow. Across the table, Rosalyn [Sussman Yalow] realized what had happened. Presented with the choice of two Dr. Yalows, the student had bet on the man. Standing straight and tall in her long blue gown, Rosalyn Yalow marched herself up to the podium as the hapless aide accompanied her along the opposite side of the table."[57]

En route to the July 1993 meeting of the National Women's Political Caucus in Los Angeles, the Senior Vice President for Medical Affairs of the R. W. Johnson Pharmacetical Research Institute, Dr. Itri, was traveling first class. Over the intercom she heard a call asking whether there was a physician or nurse on board. Dr. Loretta Itri rang her call button to ask the attendant the nature of the problem. "Well, there's a young woman in the back of the plane who is having abdominal pain," came the reply. "And then," Dr. Itri recalled at the plenary session of the National Women's Political Caucus, "she said to me in a very conspiratorial tone, 'She has her period.'"

"I have learned," Itri told the Caucus, "that that's a loaded remark. . . . But, at any rate, I went back to the back of the plane with her [where] . . . I was confronted by another stewardess. And she looked at me, head to toe, and said in a very stern voice, 'You know we'll have to see some identification.'"

While the first attendant returned to first class to locate the pieces of plastic that would confirm that Itri is a doctor, the surrounding passengers played Robert Young a.k.a. "I'm not a doctor but I play a doctor on TV." Their conclusion: The passenger was probably "hysterical."

"Now, finally they decided I was who I said I was," recalled Itri, "and so I got to examine the patient"—a young black woman, in her mid-thirties, who was a second-year pharmacy student. "She is clearly in severe pain, lower abdominal pain. After eliciting a history and doing a cursory physical exam, it becomes apparent that her problem has nothing to do with her menstrual cycle. She's probably having an episode of renal colic. Now for any of you who have never seen anybody with a kidney stone, let me tell you this is pain. And given the circumstances, this woman was handling herself with a great deal of poise."

In Los Angeles, Dr. Itri accompanied the passenger to a local emergency room. "But as I sped towards the hotel in the back of a car I had a little time to ruminate on the events that had transpired. And two questions came to mind . . . if a well-dressed man came out of first class identifying himself as a physician, would he have been asked for identification? I'm from New York and I know there are a lot of crazies out there, but . . . my honest answer to that question is "No." The second . . . more disturbing . . . question [is] whether [if] a man in economy class doubled over in pain, would anyone have thought he was hysterical? I don't think so."[58]

During my early years as a teacher at the University of Maryland, a faculty member in the Speech department could earn an extra $1500 a semester by teaching a night course. With our second child on the way, and an extra bedroom being built as a buffer against sibling rivalry, I took on a night class for cash. A single line in the University's night school catalogue announced: Dr. K. Jamieson. "Introduction to Public Speaking." 3 credits.

The class attracted adults from the Maryland community, and most were the age of my parents. I was past morning sickness but also past the point at which disguising a pregnancy is possible. So I strode expectantly in my dress-for-success maternity best into the first class. At the end of the first lecture, a middle-aged man approached. He was the head of the trash collectors for the county, he reported. The class was a means of professional advancement and he shared the anxiety that dogs most beginning speakers. Signing up and coming to campus had not been easy, but he wanted to learn what he thought the class would teach: how to defend his budget before his supervisors. I addressed his concerns in turn by offering the standard assurances. Experience in speaking reduces anxiety about speaking, I recall telling him.

"Dr. Jamieson," he said at the end of our conversation. "You weren't what I expected." Assuming that he was taken aback by my impending motherhood, or by what Ronald Reagan would have termed "my comparative youth," I uttered a syllable that sounded, I suspect, like a cross between an observation, a question, and an apology: "Oh." After a pause, he explained, "I thought you'd have a beard."

Wearing a beard would have solved one problem. In the early 1970s, the number of obviously pregnant faculty members on college campuses was small. As I ambled around campus, people stared at my abdomen with an unbecoming curiosity. Women and men I hardly knew assumed the right to pat my stomach as they inquired about the name, sex, and projected time of arrival of the presence we now know as our second son Patrick. The attention was annoying. Who, I mused, would pay any attention to my expanding middle if I were sporting a Snydley Whiplash mustache or a full-chinned Henry the Eighth?

A similar encounter occurred when Assistant Secretary of Defense Bill Perry was sworn in at the Pentagon. Since he was a constituent of hers, Congresswoman Anna Eshoo was invited. There was standing-room only with the Congresswoman positioned near the front of the room. Once there, she was approached by a general who "bumped into me and I mean I rocked back on my heels to gain my composure," she remembers. "I thought he was going to look down and say, 'Excuse me.' [Instead] he looked at me and he said, 'All the staff is over there.' I got to look up and say, 'With all due respect to you sir, I am the secretary's Congress-woman."[59]

These stories contain a moment of vindication. Whether the custodians of the ID cards think she looks like a member of Congress or not, it is Eshoo who casts the votes that affect legislation on the floor of Congress. Bearded or not, I was the teacher; the student did not flee. By semester's end, his competence as a speaker had demonstrably improved. The Nobel Prize was given to Rosyln Sussman Yalow as was the money and fame that accompany it. Loretta Itri did diagnose the problem, escort the patient to the hospital, and arrived as scheduled to deliver her speech to the National Women's Political Caucus.

As the 2000 members applauded her story, they and she demonstrated that women have come a long way and at the same time have a long way to go. Women are distinguishing themselves in professions once closed to them. And as men and women come in contact with doctors, lawyers, engineers, professors, and senators in skirts, the surprise at finding us in such positions should subside. It hasn't yet, but it will.

With increasing numbers of women in power, space has been created to which women have unique access. Women are telling stories that establish that doors once closed are now ajar. For decades, women were denied jobs as reporters and lobbyists on the grounds that they couldn't follow their male sources into the men's room. With women now in legislative positions, there are deals to be made in the women's room as well. And female sports reporters are now permitted in the locker rooms of both men's and women's teams.

"The usual thing [is that] the deals are made in the men's room," reports Sunny Jones, Western Government Affairs Manager for the Miller Brewing Company. "One time up in Sacramento, we were in a conference with a lot of different legislators. I had to go to the bathroom and so I went in the bathroom and there were three women and their legislators. I said 'Finally.'"[60]

Similarly, *Washington Post* reporter Molly Moore notes that during the Gulf War "in Saudi Arabia the senior military officers seemed to be taking both the women in uniform and the women journalists more seriously."

"There was good reason on both counts," she explains. "Army Lieutenant Lori Fanning, a 22-year-old postal officer from Columbia, South Carolina, summarized it best when I interviewed her at the Dhahran air base, where she was sorting huge orange sacks of mail: 'We have an equal opportunity to die for our country. You live with the possibility of somebody dropping a bomb on you or some terrorist blowing you up.'"61

Those in power are changing its face. And with this change are different endings to the story. Increasingly, stories of exclusion are being turned into stories of access. And where in the 1950s women were admitted to all-male preserves by gracious men, by the 1990s they were hosting the meetings. The contrast is plain in two stories of access—one from the 1950s, the other from 1992.

"Oh, I've been kicked out of some of the best places," recalls reporter Fran Harris. Stan Kresge, of the family that founded the department store chain, was, she recalls, a family friend. A donor to Albion College, Kresge had called a press conference in the 1950s to introduce the Detroit press to a new college president whom he had helped recruit. The location of the breakfast announcement was the Detroit Club, an establishment closed to women except during lunch and dinner. Women "had to go in a separate door. When I went there, the separate door was not open so of course I went in the front door, naturally." She was turned away. "'What are you doing here?' asked a manager. And I told him that I was Mr. Kresge's guest. 'Oh. Well, he's upstairs. You can't go upstairs. You shouldn't even be here.' I said, 'I don't think Mr. Kresge would like that very well because he invited me.'

"So the little man took me down the stairs out the front door and went around and unlocked the women's entrance door, unlocked the women's entrance elevator, and up we went. He kept his eye on me, took me to the actual room where Kresge was holding his little soiree. He was stunned because when Stan Kresge saw me, he said, 'Fran, hello! I'm so glad you're here.' And the little man just diminished."

A story Representative Maxine Waters tells shows a different playing field in 1992. Where Harris found locked doors that could be opened only by men, Waters employed rhetoric to assert an authority that opened doors. "[W]hen President Bush called a meeting of government leaders last May to discuss urban concerns in the wake of the Los Angeles riots, Maxine Waters very much minded not being invited, and said so," notes the December 1992 issue of *Working Woman*.62 "Ignoring protocol, the black congresswoman from the violence-torn district just showed up issuing a statement to the press that she should have been invited and would attend. She did—thereby making national news."

At the July 1993 meeting of the National Women's Political Caucus, Waters retells the story to the applause of an audience of women's rights activists. Her telling consists of a series of confrontations with individuals who give way and grant her right to access. First she is informed by House Speaker Tom Foley that the president controlled the invitation list. She told Foley that she planned to attend. At the gate of the White House she simply informed the guard that she was there for the meeting, flashed her credentials, and was admitted. The same scene occurred at the door of the White House. And once in the meeting room, she asked where at the table she was supposed to sit and was given a place. Where Harris's story might be tagged "Eve's pal protects her right to tea," Waters's could be captioned "Eve storms the Garden."

The times are changing because women are being accommodated in places that once required only men's rooms. And as women move into positions of influence, the risk in giving them offense mounts. For the past four years, women in politics have been relishing the story I would title "Respect or Consequences" or "The revenge of the donor."

For giving the Republican party $100,000 in 1988, California developer Kathryn G. Thompson became a member of the party's "Team 100." To her surprise, at a party function then-Republican National Committee Chairman Clayton Yeutter asked the newly honored Thompson, "And who do you belong to, little lady?" In the past, the story would have ended here. Chances are Thompson would have been the spouse of a donor. And she wouldn't have spoken her distress at the assumption that she "belonged" to anyone. But times have changed and with them the stories we are able to tell.

"I used to belong to Team 100," she responded.

This story has a sequel. In 1992, Thompson's money shifted parties. The Democrats' response to her support was likely to elicit additional support: She was among those invited to Clinton's December 1992 economic summit.[63]

The lesson: Women belong to themselves. And in the process of being themselves, women are changing the workplace. By simultaneously lactating and dictating, Patricia Diaz Dennis, who went on to serve as a FCC commissioner, established that the womb/brain bind was a thing of the past. "When I graduated from law school, I was the first woman and first Latino hired by a major downtown Los Angeles law firm. After about a year of practice I became pregnant and had my first child [in 1974] and it was a very difficult juggling act, as most women will tell you. I was determined to nurse. . . . We hired a nanny who would bring her at lunch time because I went back to work after my oldest daughter was three months old, after I established her nursing schedule. . . . When I was nursing Ashley one evening, I had turned on this little red sign on my door that was essentially a 'Do Not Disturb' sign and I was sitting there, very indelicately and indiscreetly nursing her . . . dictating up a memorandum with one hand, at that time they had these Dictaphones, and you had to hold this piece in your hand, and on the other arm I had Ashley nursing and, as I said, I was not being very discreet, and I am sitting there dictating away and the door flung open and one of the senior partners from the labor department burst in and he turned beet red . . . and he blurted out, 'Well, you don't have to come to the labor department meeting,' and I said, 'I hadn't planned to,' and he slammed the door shut and went back to the labor meeting and he said, 'You know, we have the only associate in town who lactates and dictates at the same time.' It was a real education process for them as well as for me."[64]

Recalling Different Treatment While Confirming We Have Overcome

The anecdotes that emerged about Ruth Bader Ginsburg prompted the *New York Times* to ask other prominent women about the difficulties they had faced. Because these women have achieved prominence, the telling is a form of vindication, an affirmation that things are now better.

"Dr. Maria Iandolo New, who is 64, can still recite the letter she received from the dean of the University of Rochester's medical school in 1950 after her husband had been invited for an interview, and she had not. She had written the dean asking that he not reject her just because she was married and pointing out that she actually had the better academic record.

'You are an impertinent young lady,' the dean replied, 'and I am more sure than ever that we do not want you in our medical school.'"[65]

"When Constance Baker Motley, a young lawyer working on school desegregation cases in the early 1950s for the NAACP Legal Defense Fund, tried to enter the New York City Bar Association's library, the doorman looked her up and down. She assured him she was in fact, a member, and told him her name.

'I carefully said Mrs. Motley, because black women were always addressed by their first name,' she said.

'Oh, right this way, Constance,'"[66] he replied.

Notes a reporter for the *New York Times*, "Judge Motley later became the first woman to serve a full term in the New York State Senate and in 1966, the first black woman appointed as a Federal judge to the Federal District Court in Manhattan, where she still presides at 72. But she says that for years she was not given the same opportunities as men to serve on committees or speak at conferences, activities that build judges' careers."[67]

These stories form a chorus that invites outrage at the affronts these women suffered and relief that they nonetheless persevered and triumphed.

"If I did it, you can too" is a message women in positions of power are now communicating to future generations. "I sat where you sat in about 1950," Attorney General Janet Reno tells an elementary school class in Washington, D.C. "When I was your age, they said ladies didn't become lawyers. Ladies certainly didn't become attorneys general." She was one of sixteen women in her Harvard Law School class of 500, she reports. But a law firm that denied her a summer job because she was a woman years later offered to make her a partner. And now she is the nation's first female Attorney General. The lesson: "You can be what you want to be . . . if it's the right thing to do, and you put your mind to it, and you study hard." The story is then transmitted to a mass audience in the August 1993 issue of *Vogue*.[68]

Some female leaders recall that they themselves once defined their current roles according to a male model. "I always thought senators had to be tall and rich and male," five-foot tall Senator Patty Murray told the Senate in the midst of the debate over the Clinton economic plan. "But my parents always told me that the little guy was just as important as the big guy. And I believe that."[69] Since she is making the statement from the well of the Senate, those listening are invited to believe it as well.

Other problems confront women in power. One is fine but two's a crowd seems to be an unspoken rule when the one wears a skirt. And those in authority have found ways to reward women for excluding others of their kind. "I went through the generation of women told that there was going to be one woman in the Board of Supervisors and we already have her," recalls Michigan's Democratic gubernatorial candidate Debbie Stabinow. "A woman got a lot of attention for being the one and only and there are two kinds of traps. One is that we enjoy having attention as the one and the only. As a result, women would play against each other. . . . Even now I get press attention that my male colleagues in the Democratic primary won't because I am a woman. The other is then telling women, we must pick one of you. When I came in through the House, there were no women in the leadership. I decided to run for Assistant Majority Floor Leader in the early '80's and there were other women who were interested and we were told that we had to agree on one. We couldn't all run."[70] Stabinow ran nonetheless—and won.

These are stories of women within the circle of power, not those staring in from the outside. Because they are the repository of the possible, the stories that women tell about the binds that tie matter. Often, however, they are stories of well-educated women, members of Congress or the Court, MBAs, M.D.s, and PhDs, who have attained positions of power unusual for both men and women. We don't often hear the stories of the women who won the case against Lucky Stores or lost the case at Sears. A contribution of Faludi's book is its focus on working-class women. My complaint is that she tells the stories of their failure, but not their success.

By misrepresenting the world as it is, these tales thwart the desire to create the world as it ought to be. The story of the women who fought the battery-making company Johnson Controls all the way to the Supreme Court is indicative of the difference between the world of backlash and the world of bind-breaking.

As described in Chapter 3, Johnson Controls made batteries in which lead is a key ingredient. Like American Cyanamid, Johnson Controls required medical documentation of women employees' inability to bear children as a condition of employment in an area in which lead levels were thought to endanger a fetus. Since those areas paid higher salaries, the affected women were faced with a loss of wages, possible loss of a job, or sterilization.

Here was a matter of unequal treatment, of treating every female as potentially pregnant. Even if the company policy was motivated strictly by health and safety concerns, other factors—in addition to matters of privacy—made the Johnson Controls policy troublesome. One is the fact that high levels of lead are injurious to the workers themselves, both male and female, with the company's own literature stating that lead could damage the reproductive capacities of males. The other is that the lower-paying jobs to which women were moved appeared, in many instances, to expose them to levels of lead and other toxins at rates that were just as high as those encountered in higher-paying positions.

In impassioned detail Faludi recounts the stories of the women who were told by American Cyanamid that they had to be sterilized if they wanted to keep their jobs. When their lawyers miscalculated and took their case into court on OSHA grounds, they lost. Their sex discrimination claims, however, had previously been settled out of court.

As Faludi describes it, the "fetal protection" plans of the very corporations that had once denied women access to high-paying jobs[71] were a "choice" "like so many of the other options the backlash magnanimously granted women . . . framed as a clear-cut and forward-looking development—it represented progress for women. . . . [But] There was never anything straightforward, helpful, or enlightened about the options presented the Cyanamid women. In fact, their alternatives were paradoxical, harmful, and regressive—and rigged against them from the start."[72]

They were rigged. And rather than perpetuating them, the Supreme Court in the Johnson Controls ruling banned them. Where the American Cyanamid decision reinforced the womb/brain bind, Johnson Controls vanquished it. But Faludi doesn't tell her readers that. Instead, by treating American Cyanamid *after* Johnson Controls, she invites the false conclusion that women can still be bound by the no-choice-choice of so-called "fetal protection policies."[73]

International Union, UAW v. Johnson Controls, Inc.[74] was moreover, won in 1991, before Faludi's book was published and at the climax of what she sees as the backlash decade of the 1980s, with the Bush administration and a Reagan Court still firmly in place.

The importance of the Johnson Controls case was forecast by Judge Frank H. Easterbrook who, in his dissent in the Seventh Circuit, said it "is likely the most important sex-discrimination case in any court since 1964, when Congress enacted Title VII." Why? Because if the Circuit Court's support of Johnson Controls' position held in the High Court "by one estimate 20 million industrial jobs could be closed to women."[75]

Let me tell you the story of the women affected by what Easterbrook considered "the most important sex-discrimination case in the court since 1964." Each worked at one of Johnson's fourteen battery plants from Bennington, Vermont, to Fullerton, California. In 1982, the company decided that no fertile women could work in a high lead area. Representing six women and one man, the United Autoworkers of America sued the company for sex discrimination.

After graduating from Southern High School in Louisville, Kentucky, Patricia Briner did office work. In November 1977, she was hired to work in the shipping department at the Johnson Controls factory in Louisville. When a better job came open in the loading area she applied. It required working overnight from 11 P.M. until 7 A.M., which was difficult for the recently divorced mother of two small children. Wanting to spend more time in the day with her children, she watched for an opening.

When a higher-paying daytime job was posted, she applied. But "the next morning," she recalls, "the foreman called me in the office and told me I couldn't have the job because it was in a leaded area and I was a fertile female."

"I was dumbfounded," Ms. Briner notes. "I said, 'I am a what? What is that?' 'Oh,' he said, 'that's the new company policy.' Needless to say I didn't get my first shift job. A junior man in the plant took it."

The children were eight and nine at the time, she recalls. "I couldn't afford (to have someone stay overnight with them while I was at work) . . . I could barely hang on . . . because I kept the house when I went through my divorce and I had all the bills . . . it was just a real struggle to make ends meet. So they were there by themselves. I developed an ulcer. I had to go to the doctor. I had an ulcer from worrying about leaving the house at night and leaving my kids there. Of course they were in bed but everything would run through my mind. Fire, you know . . . it was really hard. That is where I felt cheated, not the money . . . but I was denied a lot of first shift [day] jobs because of the (fertility issue)."

She stayed on the night shift until a day job in an unleaded area came open about five years later. "I read the OSHA book and it stated that any male or female that intended on having a family should not work in a leaded area and I thought that if they are worried about the safety, then why are they letting the men [work there]?"

"I just felt like it was their way of keeping women in the lower paying jobs or the ones that weren't as skilled." Most of the other women "went ahead and had their tubes tied. . . . [But I didn't want to do that.] At the time I had just gone through a divorce and I had two small children and I didn't know what lay ahead or my future held. And I thought well I could lose a kid and what if I decided to have one ten years down the road?"

"I figured I was smart enough that if I wanted to have a child that I wouldn't work in a leaded area to begin with because I wouldn't want to take a risk. It was just like they were denying me any intelligence."

"When the union finally . . . asked if any of the women there felt like they had been discriminated against, I said, 'Give me that piece of paper.'"

Deborah Moore, whose pregnancy instigated the Fetal Protection Policy, also joined the suit. After graduating from high school, she followed her father and a number of uncles and cousins working in the Middletown Delaware Plant of Johnson Controls in 1981. "I loaded plates," she recalls, for about $11 an hour. "I would insert the plates into a machine which would put the post on top of them as it would go around the machine and come out the other side. They were loaded into the cases on the other side."

The process exposed her to a level of lead that in the judgment of her employer endangered a fetus. When she became pregnant, she was moved to a janitorial job at lower pay. The Fetal Protection Policy was then put in place for all fertile women.

During and after the pregnancy "They just put me on whatever kind of job they had. . . . I did anything they came up with. I didn't just work in shipping. I did painting. I painted walls. I did landscaping. In the formation area they have these pipes which the air went through. I cleaned them with a steam jenny and I did whatever it is they could find for me to do. I had no position."

In fact cleaning the pipes exposed her to high levels of airborn lead and other contaminants. "Well we knew we were being discriminated against. I mean it is not fair. Everybody comes in there knowing that there is lead and there could be a problem. We do have common sense."

"[N]ot only is there lead in there, but they have acid with the formation of the batteries and you breathe the fumes from that. . . . When the other women started to be pulled, Lois Sweeten went to the union about what we could do and that is when the union initiated the suit."

Joanne Leard began working at the Johnson Controls factory in Bennington, Vermont, in 1973 at the age of 34. While there she was named health and safety representative from the union.

"I had gotten appointed to that position because my background had been in Health and Safety . . . so I was the most likely candidate."

"We were told that we would not be able to work in [the high lead areas] as long as we were fertile females, but I was not actually removed from my job because at that time my job was not considered a high lead job. Virginia's [Green, her friend who worked in the pasting dept.] was." The 50-year-old Green had been demoted on the assumption that the level of lead in her blood might endanger a fetus if she were to conceive. "So she was automatically removed and . . . when that happened, we said, 'No. This isn't right.' So we filed a grievance. The women filed a grievance with Ginny named as the actual grievant at that time because she was the one that was losing her job. So they told her that they were going to put her in . . . what we call the laundry. She washed the respirators and stuff like that which [were] covered with lead no less. That was where they put Virginia and she has been there ever since and like I said we filed a grievance."

What was clear was that Johnson Controls was selective in what it considered "high-risk" and "low-risk" work. High-risk jobs that paid well appeared to be riskier than high-risk jobs that paid less. Leard said that "I was denied that opportunity to bid on a job in a high lead area. . . . I wasn't really affected like Virginia was but I was affected in another sense."

When the company asked for proof of infertility, "I told them that I didn't feel that that was right. That was my personal business. I don't have to let everybody know that I am fertile or not fertile. You might have just as well put it over the loudspeaker in the plant. . . . [T]here are only like twelve women working here and the rest are all men. It is just like announcing to them, 'Okay guys. I am not a fertile female anymore.' You know, it is none of their business.

"[M]y family was already there. I was not about to go having any more children, but that was the whole point. It should be my decision and this is what I told them. I said, 'Well who do you think you are that you can tell me whether or not I can work in this job?'

"I have got enough common sense to know that if I am going to work on this job . . . and if I were to get pregnant that it could harm a child. . . . If I take the job it is because I know that I am not going to get pregnant, but it is my decision and not the company's. So I resented that and they knew that I resented that. I never let up on it.

"The company naturally protested, saying that it was for the protection of the fetus and all this. So then it went to the next stage. When you file a grievance you have so many stages you got to go through and then it finally went to the third step where the national comes in and argues a grievance. Then the company still said, 'No. We were going to keep this policy.'"

Then "the international got involved and the legal department took over and so it has been all these years that we have been trying to get this thing settled until finally . . . we took it to Supreme Court, and we finally won it.

"So I went running down to the cage, which is where Ginny works. We call it the cage. I went running down to Virginia [Green] and screaming, 'Ginny! Ginny! We won it, we won it!' And the two of us were like a couple of idiots, standing out in the hallway and all of the people in the office were all looking there wondering what the screaming was about. They looked down and I told them what happened, and they were all excited too. They were all, 'Yeah! Yeah! Yeah!' Like I said, they were behind us also. It was just that they were told to do this from the headquarters. So it was well worth the wait, believe me, to win the case."

As I noted in Chapter 3, the Supreme Court's majority opinion is unequivocal. Johnson Controls "has chosen to treat all its female employees as potentially pregnant," said the Court. "[T]hat choice evinces discrimination on the basis of sex. . . . Johnson Controls has attempted to exclude women because of their reproductive capacity. Title VII and the PDA (Pregnancy Discrimination Act) simply do not allow a woman's dismissal because of her failure to submit to sterilization."[76]

Margaret Thatcher, Hillary Clinton, Janet Reno, Hazel O'Leary, Carol Moseley Braun, Barbara Mikulski, Nancy Kassebaum, Alice Rivlin, Jeane Kirkpatrick, Lesley Stahl, Carole Simpson, Cokie Roberts, and Ann Hopkins are not the only harbingers of the future. So too are Deborah Moore, Joanne Leard, Patricia Briner, Virginia Green, and the others at Johnson Controls and Lucky Stores who, as surely as their more famous sisters, have shown us how to break the binds that tie.

Notes

Chapter 1 The Binds That Tie

1. Friedrich Spee, *Cautio Criminalis Seu De Processibus Contra Sagas Liber Ad Magistratus* (Sumptibus Ioannis Gronaei, Austria, 1632). See especially corollaries xiii and xiv, pp. 61ff.

2. "Why the gay ban hurts *all* military women." *Glamour* (August 1993), p. 98.

3. Rosalyn Carter, *First Lady from Plains* (Boston: Houghton Mifflin, 1984), p. 175.

4. Nancy Reagan, *My Turn: The Memories of Nancy Reagan* (New York: Dell, 1990, p. ix.

5. Sharon Bertsch McGrayne, *Nobel Prize Women in Science, Their Lives, Struggles and Momentous Discoveries* (New York: Carol Publishing, 1993), p. 148.

6. Gregory Bateson et al., "Toward a Theory of Schizophrenia." *Behavioral Science* (1956), Vol. I, pp. 251–64.

7. Rosabeth Moss Kanter, *Men and Women of the Corporation* (New York: Basic Books, 1977), p. 10.

8. Nancy A. Nichols, "Whatever Happened to Rosie the Riveter?" *Harvard Business Review* (July/August 1993), p. 60.

9. Jane M. Ussher, *Women's Madness: Misogyny or Mental Illness* (Amherst: The University of Massachusetts Press, 1991), pp. 73–74.

10. *A Very Serious Thing: Women's Humor and American Culture* (Minneapolis: University of Minnesota Press, 1988), p. 142.

11. "Portia in a Different Voice: Speculations on a Woman's Lawyering Process." *Berkeley Women's Law Journal* (Fall 1985), vol. 1, no. 1, pp. 39–63 at p. 53.

12. Senator Barbara Boxer with Nicole Boxer, *Stranger in the Senate: Politics and the New Revolution of Women in America* (Washington, D.C.: National Press Books, 1994), p. 73.

13. Paula J. Caplan, "Don't Blame Mother." *Ms.* (September/October, 1993), p. 96.

14. Marguerite Holloway, "A Lab of Her Own." *Scientific American* (November 1993), pp. 94–103, at p. 103.

15. Linda Witt, Karen M. Paget, and Glenna Matthews, *Running as a Woman: Gender and Power in American Politics* (New York: The Free Press, 1994), p. 119.

16. Karen Lehrman, "The Feminist Mystique." *New Republic* (March 16, 1992), p. 31.

17. Susan Faludi, *Backlash: The Undeclared War Against American Women* (New York: Crown, 1991), p. 46.

18. Ibid., p. 48.

19. Ibid., p. 58.

20. In Faludi's universe, women "kick against the backlash barricades"; they struggle to "gain sufficient momentum to crash its [the backlash agenda's] steel reinforced gates." And in the 1980s, when "women tried to drive privately against the antifeminist forces . . . they most often found their wheels spinning, frustration and disappointment building as they sank deeper in the same old ruts." The metaphors are those of mechanics and physics; the path is linear, the barriers external and obstructionist.

21. Betty Friedan, *The Second Stage* (New York: Dell, 1981), p. 28.

22. Urvashi Vaid, Naomi Wolf, Gloria Steinem, and bell hooks, "Let's Get Real About Feminism: The Backlash, the Myths, the Movement." *Ms.* (September/October 1993), pp. 34–43.

23. Faludi, op. cit., p. 398.

24. *Cannon v. University of Chicago,* 441 U.S. 677.

25. 465 U.S. 555 (1984).

26. Faludi, op. cit., p. 47.

27. *Time* (March 9, 1992), p. 54.

28. Saint [John] Chrysostom, "Homily IX," in *A Select Library* of *The Nicene and Post-Nicene Fathers,* Ed. Philip Schaff, XIII (New York: The Christian Lit. Co., 1889), p. 436.

29. Constantia Munda, "The Worming of a Mad Dog," Katharine Usher Henderson and Barbara F. McManus, Eds., *Half Humankind* (Urbana: University of Illinois Press, 1985), p. 253.

30. Mary Tattlewell and Joan Hit-Him-Home, "The Women's Sharp Revenge," Reprinted in Usher Henderson and McManus, op. cit., p. 313.

31. Londa Schiebinger, *The Mind Has No Sex? Women in the Origins of Modern Science* (Cambridge, Mass.: Harvard University Press, 1989), p. 274.

32. Ibid., p. 37.

33. Sharon Bertsch McGrayne, op. cit., p. 71.

34. Ibid., p. 72.

35. Virginia Woolf, *A Room of One's Own* (New York: Harcourt, Brace and World, 1929), p. 65.

36. Londa Schiebinger, op. cit., p. 250.

37. Ibid., p. 247.

38. *The Works of Anne Bradstreet,* ed. Jeannine Hensley (Cambridge: Harvard University Press, 1967), p. 16.

39. Londa Schiebinger, op. cit., p. 270.

40. Joanna Russ, *How to Suppress Women's Writing* (Austin: University of Texas Press, 1983), p. 76.

41. Barbara Miller Solomon, *In the Company of Educated Women* (New Haven: Yale University Press, 1985), p. 61.

42. *The Nation* (February 18, 1925), vol. cxx, no. 3111, p. 173.

43. Quoted by Barbara Miller Solomon, op. cit., p. 61.

44. Ibid., 253–54.

45. (New York: Harper, 1975), pp. 61–62.

46. Bateson, op. cit., p. 261.

47. E. K. Smith, "The Effect of Double-Bind Communications Upon the State Anxiety of Normals," *Dissertation Abstracts International* (University Microfilms, 1973, 34, 427B, No. 73–16, 583).

48. Carroll Smith-Rosenberg, *Disorderly Conduct: Visions of Gender in Victorian America* (New York: Oxford University Press, 1985), p. 192.

49. *Commentaries on the Laws of England* (1765), Vol. 1 (Philadelphia: Geo. T. Bisel Co., 1922), p. 4111.

50. *Los Angeles Times,* April 21, 1981.

51. Florynce Kennedy, *Color Me Flo* (Englewood Cliffs, N.J.: Prentice-Hall 1976), p. 8.

52. *Johnsonian Miscellanies,* ed. George Birkbeck Hill, Vol. II (New York: Barnes and Noble, Inc., 1966 [1784]), p. 11. Johnson went on to add, "My old friend, Mrs. Carter . . . could make a pudding, as well as translate Epictetus from the Greek, and work a handkerchief as well as compose a poem."

53. Carry A. Nation, *The Use and Need of the Life of Carry A. Nation* (Topeka: F. M. Steves & Sons, 1905), p. 188.

54. Robert K. Merton, "The Self-Fulfilling Prophecy." *The Antioch Review* VIII (June 1948), pp. 193–210, at p. 195.

55. Dorothy Cantor and Toni Bernay with Jean Stoess, *Women in Power* (Boston: Houghton Mifflin, 1992), p. 61.

56. See Ellen DuBois, in "Feminist Discourse." *Buffalo Law Review* (1985), p. 68.

57. Carl N. Degler, *At Odds: Women and the Family in America from the Revolution to the Present* (New York: Oxford University Press, 1980), p. 403.

58. Faludi, op. cit., p. 399.

59. *UAW v. Johnson Controls, Inc.,* 499 U.S. 187 (1991).

60. "A Mother's Work Never Seems to End," *The Wall Street Journal,* July 29, 1994, p. B1.

61. Faludi, op. cit., p. 449.

62. Paula Ries and Anne J. Stone, eds., *The American Woman, 1992–93: A Status Report* (New York: W. W. Norton and Co., 1992), p. 355, citing U.S. Bureau of Labor Statistics, *Handbook of Labor Statistics,* August 1989, Table 41. For a discussion of the factors including occupational segregation that account for lower wages, see Chapter 6 "Why Are Women's Wages Low?" in Barbara R. Bergmann, *The Economic Emergence of Women* (New York: Basic Books, 1986). Also see June O'Neill and Solomon Polachek, "Why the Gender Gap in Wages Narrowed in the 1980s." *Journal of Labor Economics* (January 1993), Vol. 11, No. 1, Part 1, pp. 205–228; and Robert G. Wood, Mary E. Corcoran, and Paul N. Courant, "Pay Differences among the Highly Paid: The Male-Female Earnings Gap in Lawyers' Salaries." *Journal of Labor Economics* (July 1993), Vol. 11, No. 3, pp. 417–440.

63. *Reed v. Reed,* 404 U.S. 71 (1971).

64. *Frontiero v. Richardson,* 441 U.S. 677 (1973).

65. *Weingler v. Druggists Mutual Insurance Co.,* 446 U.S. 142 (1980).

66. *Orr v. Orr,* 440 U.S. 268 (1979).

67. Pat Parker, *Movement in Black* (Ithaca, N.Y.: Firebrand Books, 1989), p. 68.

68. Wendy Kaminer, "Feminism's Identity Crisis." *The Atlantic* (October 1993), Vol. 272, No. 4, pp. 51–68 at p. 68.

Chapter 2 Hillary Clinton as Rorschach Test

1. "Gnotes." *Glamour* (June 1993), p. 133.

2. "The Governor: On His Music, His Marriage and His Future," in Landon Y. Jones, "Road Warriors." *People* (July 20, 1992), pp. 68ff, at p. 72.

3. "Just as Ross Perot isn't only about an election but about national Angst, so the squirming over Hillary Clinton isn't so much about a First Lady as about ambivalence over women, power, work and marriage," noted a profile in the *Los Angeles Times* (July 14, 1992, p. A1). "She is a lightning rod for the mixed emotions we have about work and motherhood, dreams and accommodation, smart women and men's worlds (*Los Angeles Times*, July 15, 1992, p. A21).

4. "Campaign's New Rules for Wives of Hopefuls," *Newsday*, March 24, 1992, p. 86.

5. Letty Cottin Pogrebin, "Hillary Clinton and the Year of the Woman," June 16, 1992, p. 7A. All translations from Spanish were done by Annenberg graduate student Isabel Molina Gerzman.

6. Hadassah Lieberman, identified by *Nightline* as "a career woman and wife of Senator Joseph Lieberman," argued that like "most political wives" Hillary Clinton faced "a no-win situation"—what I will call a double bind. "When they say nothing, there's a problem. When they say something, there's an even greater problem. And it sort of—what it does is, it reinforces the notion that a political wife should just step into the background and [not] say anything that's controversial."

7. (New York: The Free Press, 1994), pp. 193–94.

8. (*CBS This Morning*), January 27, 1992.

9. Patt Morrison, "Time for a Feminist First Lady?," *Los Angeles Times*, July 14, 1992, p. A1.

10. "Macho Feminism, R.I.P.," *New York Times*, January 27, 1992, p. A21.

11. Quoted in Judith Warner, *Hillary Clinton: The Inside Story* (New York: Penguin Books, 1993), p. 185.

12. Historical revisionism would have it that Hillary clearly tied cookies and tea to a woman's right to choose her own role. The time-coded-transcript-from-tape I am working from establishes that is not the case. The transcript was provided by a network source who wishes to remain anonymous. By September 1992, revisionists, abetted by Clinton spin-control artists, had begun reconstructing "cookies and tea." First in *Time* in September 1992, then in *Vanity Fair* in June 1993, *Time*'s Margaret B. Carlson would mistakenly recall a sure-footed answer by Hillary Clinton that directly tied "cookies and tea" to a career dedicated to choices for women. In this revision, "When he [Jerry Brown] attacked her law practice as a conflict of interest in a state where her husband was governor," wrote Margaret Carlson in *Vanity Fair*, "she shot back, 'I suppose I could have stayed home and baked cookies and had teas.' Less reported was her next sentence: 'The work that I have done . . . has been aimed . . . to assure that women can make choices . . . whether it's full-time career, full-time motherhood, or some combination." Margaret Carlson, "A Hundred Days of Hillary." *Vanity Fair* (June 1993), p. 169.

13. "The Hillary Problem," *New York Times*, March 26, 1992, p. A23.

14. Karen Ball, "Debate Erupts into Vicious Feud Before Midwestern Votes," AP, May 16, 1992, 15:07 EST.

15. That evening, CBS retained the statement about being careful and used it to anchor a story on conflict of interest. ABC dropped the "careful" claim and tied the cookies and tea quote to Clinton's implication that Jerry Brown is a sexist. "The

Clintons counterattacked on two fronts: painting Jerry Brown as a bully, and a sexist one at that."

16. Interview, December 21, 1993.

17. *Nightline* did make this point on March 26, reporting that 59% of U.S. families have dual incomes.

18. Cited by Elaine S. Povich, "Millions Left Out of GOP Pitch to Moms at Home," *Chicago Tribune,* August 23, 1992, p. D1.

19. Election debriefing, Annenberg School, University of Pennsylvania, December 1992.

20. *NBC News at Sunrise,* April 15, 1992.

21. *NBC Today,* September 7, 1992.

22. "Essay: The Hillary Problem," *New York Times,* March 26, 1992, p. A23.

23. World News Tonight with Peter Jennings, *ABC News,* August 18, 1992.

24. Marjorie Williams, "First Ladies," *Washington Post Magazine,* November 1, 1992, p. W11.

25. Ronald Brownstein, "Age Issue Would Likely Shape Bush-Clinton Debate," *Los Angeles Times,* April 24, 1992, p. A1.

26. John Broder, "First Lady Takes the Gloves Off," *Los Angeles Times,* August 19, 1992, p. A1.

27. "Oven Mitts Are Off in Clash of the First Ladies," *Los Angeles Times,* July 24, 1992, p. 1E.

28. In early July, The *Washington Post*'s Lloyd Grove characterized Hillary Clinton as "the governor's talented, outspoken and controversial wife" ("The Short List for Spouses," July 8, 1992, p. C1).

29. Editorial: "Vote for My Wife: The Candidates' Women Carry More Political Weight Each Time," *El Mundo,* September 2, 1992, p. 14.

30. Dan Balz and David Broder, "Fighting Doubts and Each Other," The *Washington Post,* March 22, 1992, p. 1.

31. *Los Angeles Times,* August 19, 1992, p. A1.

32. "113 Days Is Enough," *New York Times,* March 12, 1987, p. A31.

33. "The First Lady Stages a Coup," *New York Times,* March 2, 1987, p. A17.

34. Maureen Dowd, "The 1992 Campaign Trail: From Nixon, Predictions on the Presidential Race," *New York Times,* February 6, 1992, p. A18.

35. Ibid.

36. Edward Klein, "Masako's Sacrifice," *Vanity Fair* (June 1993) p. 76.

37. March 26, 1992.

38. Gail Sheehy, "What Hillary Wants." *Vanity Fair* (May 1992), p. 140.

39. Sally Quinn, "Tabloid Politics, the Clintons, and the Way We Now Scrutinize Our Potential Presidents." *Washington Post,* January 26, 1992, p. C1.

40. Tom Shales, "Campaign '92: The Muck Starts Here." *Washington Post,* January 28, 1992, p. E1.

41. Editorial: "Leers, Smears, and Governor Clinton," *New York Times,* January 28, 1992, p. A20.

42. Judy Keen, "Leading the Crusade: First Lady Takes Role as Saleswoman," *USA Today,* September 28, 1993, p. 1A.

43. Conducted between September 29 and 30, 1993. My thanks to Tom Hannon of CNN for providing a copy.

44. Patt Morrison, op. cit.

45. William Safire, "The Hillary Problem," *New York Times,* March 26, 1992, p. A23.

46. Judith Warner, *Hillary Clinton: The Inside Story* (New York: Penguin Books, 1993), p. 182.

47. Lloyd Grove, "The Limited Life of a Political Wife," *Washington Post*, March 19, 1992, p. C1.

48. Patt Morrison, op. cit.

49. "Excerpts from an Interview with Clinton After the Air Strikes," *New York Times*, January 14, 1993, p. A10.

50. Mary McGrory, "Putting Women to Work," *Washington Post*, August 21, 1992, p. A2.

51. Dan Balz, "Clinton's Wife Finds She's Become Issue," *Washington Post*, March 17, 1992, p. 1.

52. Marjorie Williams, "First Ladies," *The Washington Post Magazine*, November 1, 1992, p. 12.

53. February 10, 1992.

54. Susanne Braun Levine, "News-speak and 'Genderlect'—It's Only News If You Can Sell It." *Media Studies Journal* (1993), Vol. 7, Nos. 1–2, p. 118.

55. Judith Warner, op. cit.

56. Thomas B. Edsall, "G.O.P. Plans a 'Family Values' Offensive Pitch to White Middle Class, Casts Bush as Defender of Social Norms," *Washington Post*, August 19, 1922, p. A1.

57. Anna Quindlen, "GOP Fairy Tale Has a Sleeping Barbara and Wicked Hillary," *Chicago Tribune*, August 18, 1992, p. C13.

58. Andrew Rosenthal, "Women Are Triumphant in Their Gains, But Cautious," *New York Times*, July 14, 1992, p. A9.

59. Stephanie Grace, "Hillary Clinton Stresses Children's Issues," *Los Angeles Times*, July 16, 1992, p. A6.

60. Joyce Purnick, "Let Hillary Be Hillary," *New York Times*, July 15, 1992, p. A20.

61. Sally Friedman, "Keeping a Close Watch on First Ladies," *New York Times*, November 1, 1992, p. 13NJ-3.

62. Patt Morrison, op. cit.

63. Andrew Rosenthal, op. cit.

64. Mary Voboril, "Elect Him, You Get Hillary," *Philadelphia Inquirer*, May 4, 1992, p. C1.

65. James Carville, *CBS Evening News*, September 25, 1992.

66. Alessandra Stanley, "Family Values and Women: Is G.O.P. a House Divided?," *New York Times*, August 21, 1992, p. A1.

67. Patt Morrison, op. cit.

68. John Balzar, "Hillary's Role Resurrects an Old Arkansas Dilemma," *Los Angeles Times*, October 4, 1992, p. 26.

69. Anna Quindlen, "Public and Private: The Two Faces of Eve," *New York Times*, July 15, 1992, p. A21.

70. "The General Election Project—Interim Report."

71. Alessandra Stanley, "A Softer Image for Hillary Clinton," *New York Times*, July 13, 1992, p. B1.

72. John Kass and Elaine Popovich, "Democratic Convention '92: As Women Step Up, Hillary Steps Back," *Chicago Tribune*, July 14, 1992, p. C1.

73. *New York Times*, July 13, 1992, p. B1.

74. Joan Beck, "Other Women Shine as Hillary Becomes a Stepford Mother," *Chicago Tribune*, July 16, 1992, p. C27.

75. Maureen Dowd, "The Faces Behind the Face That Masks Clinton's Smile," *New York Times,* October 25, 1992, p. Al.

76. World News Tonight with Peter Jennings, *ABC News,* August 19, 1992.

77. *NBC Nightly News,* November 2, 1992.

78. Kathey O'Malley and Dorothy Collin, "O'Malley and Collin," *Chicago Tribune,* July 15, 1992, p. C18.

79. Anthony Lewis, "Abroad at Home; Merchants of Hate," *New York Times,* August 21, 1992, p. A25.

80. Dan Balz, "Values Night Looks Like Family Feud," *Washington Post,* August 20, 1992, p. Al.

81. Robin Toner, "Republicans Send Bush into the Campaign Under a Banner Stressing 'Family Values': Bestow Nomination," *New York Times,* August 20, 1992, p. Al.

82. David Maraniss, "Clinton Camp Sizes Up the Rhetoric," *Washington Post,* August 19, 1992, p. A26.

83. Alan Miller, "Marilyn Quayle Has Unconventional Style," *Los Angeles Times,* August 19, 1992, p. A4.

84. Howard Kurtz, "TV Coverage Ignores G.O.P. Script: Viewers Miss Fiery Speech," *Washington Post,* August 18, 1992, p. A15.

85. Charles Madigan, "Reagan to Nation: Trust Bush, Republicans Begin with Bow to Right," *Chicago Tribune,* August 18, 1992, p. Cl.

86. Alessandra Stanley, "Family Values and Women," op. cit.

87. World News Tonight with Peter Jennings, *ABC News,* August 19, 1992.

88. *NBC Nightly News,* August 19, 1992.

89. Howard Kurtz, op. cit., pp. A15–A16.

90. David Maraniss, op. cit.

91. Ruth Marcus, "Republicans Aim Barbs at Hillary Clinton," *Washington Post,* August 19, 1992, p. A21.

92. Colbert I. King, "Once This Was the Party of Frederick Douglass," *Washington Post,* August 19, 1992, p. A19.

93. Anthony Lewis, op. cit.

94. Editorial: "Elephant in The Tent: Who's a Republican?," *Los Angeles Times,* August 19, 1992, p. B6.

95. John Kass and Elaine S. Povich, op. cit.

96. "Republicans Say She's a Radical," *Los Angeles Times,* August 25, 1992, p. El.

97. Anna Maria Chavez, "Turn-Off for the 90s: Educated and Career-Minded Young Women," *Los Angeles Times,* September 13, 1992, p. 3M.

98. R. W. Apple, Jr., "No Choice But to Confront the Economy," *New York Times,* August 21, 1992, p. Al.

99. Anna Quindlen, "Public and Private," op. cit.

100. Lloyd Grove, op. cit.

101. *Public Man, Private Woman: Women in Social and Political Thought* (Princeton, NJ: Princeton University Press, 1981), pp. 333–34.

102. "Sex, Society, and the Female Dilemma: A Dialogue Between Simone de Beauvoir and Betty Friedan." *Saturday Review* (June 14, 1975), p. 18.

103. Marilyn Quayle, "Workers, Wives, and Mothers," *New York Times,* September 11, 1992, p. A35.

104. CNN, October 2, 1993.

105. Mark Landsbaum, "First Lady Backs Sen. Seymour," *Los Angeles Times* (Orange County Edition), March 25, 1992, p. 7B.

106. *New York Times,* November 20, 1992, p. Al.

107. Interview, December 21, 1993.

108. "Salute America's First Ladies," *Good Housekeeping,* April, 1992, p. 2.

109. Adam Clymer, *New York Times,* September 29, 1993, p. Al.

110. Ibid., p. A19.

111. Ibid.

112. Interview, December 13, 1993.

113. Gwen Ifill, "Role in Health Expands Hillary Clinton's Power," *New York Times,* September 22, 1993, p. A24.

114. Ibid.

115. "Hillary's Work on Health Care Has Silenced All Those Bashers," *Philadelphia Inquirer,* September 24, 1993, p. A23.

116. CNN/USA Today Poll, September 29, 1993–September 30, 1993.

117. Margaret Carlson, "A Hundred Days of Hillary," *Vanity Fair* (June 1993), p. 170.

118. Judy Keen, "Leading the Crusade: First Lady Takes Role as Saleswoman," *USA Today,* September 28, 1993, p. lA.

119. Larry King, "With Hillary on the Home Front," *USA Today,* October 4, 1993, p. 2D.

120. Judy Keen, op. cit.

121. Interview with Katie Couric on *The Today Show,* February 17, 1993. Appreciation to Sara Pines for providing a transcript.

122. Maureen Dowd, "Hillary Rodham Clinton Strikes a New Pose and Multiplies Her Images," *New York Times,* December 12, 1992, p. E3.

123. Ibid.

124. October 1, 1993.

Chapter 3 Double Bind Number One: Womb/Brain

1. Christopher Lyndon, "Role of Women Sparks Debate by Congresswoman and Doctor," *New York Times,* July 26, 1970, p. 35.

2. Karen Tumulty, " 'Pillow Talk' Taught Colleagues Health Issues, Rep. Stark Says." *Los Angeles Times,* March 17, 1994, p. A4.

3. Quoted in Blanche Wiesen Cook, *Eleanor Roosevelt* (New York: Penguin Books, 1992), p. 300.

4. John Darnton, "With a Mace, Madam Rules a Most Unruly House," *New York Times,* September 27, 1993, p. A4.

5. John Winthrop, *The Short Story of the Rise, Reign, and Ruine of the Antinomians . . . . (London, 1644).* Reprinted in *The Female Experience: An American Documentary,* Ed. Gerda Lerner (New York: Oxford University Press, 1977), p. 466.

6. Ibid., pp. 470–71.

7. Michael Levin, "Exalted by Their Inferiority." *Research in Social Movements, Conflicts and Change (1991),* Vol. 13, pp. 199–220.

8. Susan E. Marshall, "In Defense of Separate Spheres: Class and Status Politics in the Antisuffrage Movement," *Social Forces,* 65:2 (December 1986), p. 334.

9. George Burrows, *Commentaries on the Causes, Forms, Symptoms, and Treatment, Moral and Medical, of Insanity* (London, 1828); reprinted New York: Arno Press, 1976), p. 146.

10. Edward H. Clarke, *Sex in Education; or, A Fair Chance for the Girls* (Boston: J. R. Osgood, 1873; reprinted New York: Arno Press & The New York Times, 1972), p. 39.

11. Ibid., p. 94.

12. *Research in Social Movements, Conflicts and Change,* op. cit.

13. (Philadelphia: P. Blakiston, 1895 [Eighth Edition]), pp. 209–10.

14. Quoted in Carolyn Marvin, "The Body of the Text: Literacy's Corporeal Constant" *Quarterly Journal of Speech* 80 (2) (May 1994), pp. 129–49.

15. *Research in Social Movements, Conflicts and Change,* op. cit.

16. Deborah Howell, "True Confessions—My Life as a White Male." *Media Studies Journal* (Winter/Spring 1993), pp. 197–203 at p. 199. Ms. Howell, Washington Bureau chief for Newhouse Newspapers, is a former editor of the *St. Paul Pioneer Press.*

17. Sharon Bertsch McGrayne, *Nobel Prize Women in Science, Their Lives, Struggles and Momentous Discoveries* (New York, Carol Publishing, 1993), p. 76.

18. Barbara J. Harris, *Beyond Her Sphere: Women and the Professions in American History* (Westport, Conn.: Greenwood Press, 1942).

19. Christopher Lyndon, op. cit., p. 95.

20. Quoted by Carl N. Degler, *In Search of Human Nature: The Decline and Revival of Darwinism in American Social Thought* (New York: Oxford University Press, 1991), p. 108.

21. Barbara Miller Solomon, *In The Company of Educated Women* (New Haven: Yale University Press, 1985), p.63.

22. Carl Degler, op. cit., p. 124.

23. Charlotte Perkins Gilman, *Women and Economics: A Study of the Economic Relation Between Men and Women as a Factor in Social Evolution* (Boston: Small, Maynard and Co., 1898; reprinted New York: Source Book Press, 1970), p. 159.

24. Carl N. Degler, *At Odds: Women and the Family in America from the Revolution to the Present* (New York: Oxford University Press, 1980), p. 55.

25. Ibid., p. 181. See also Janet Farrell Brodie, *Contraception and Abortion in 19th-Century America* (Ithaca and London: Cornell University Press, 1994), p. 2.

26. Carl Degler, *At Odds,* p. 187.

27. Margarete J. Sandelowski, "Failures of Volition: Female Agency and Infertility in Historical Perspective," in *Ties That Bind,* Eds. Jean F. O'Barr, Deborah Pope, and Mary Wyer (Chicago: The University of Chicago Press, 1990), pp. 35–59 at p. 43.

28. Carl Degler, *At Odds,* p. 237.

29. William Acton, op. cit., pp. 185–86.

30. William Acton, op. cit., p. 140.

31. Carol Hymowitz and Michaela Weissman, *A History of Women in America* (New York: Bantam, 1978), p. 342.

32. 97 F.2d 510 (2nd Cir. 1938).

33. 381 U.S. 479 (1965).

34. *Eisenstadt v. Baird,* 405 U.S. 438 (1972).

35. *Roe v. Wade,* 410 U.S. 113 (1973).

36. 42 U.S.C. at 2000e(k).

37. David J. Garrow, "Justice Souter Emerges," *The New York Times Magazine,* September 25, 1994, p. 39.

38. Bella Abzug, *Gender Gap* (Boston: Houghton Mifflin, 1984), p. 168.

39. Marion Burros, "Even Women at Top Still Have Floors to Do," *New York Times,* May 31, 1993, p. 11.

40. Rosalind C. Barnett and Grace K. Baruch, "Correlates of Fathers' Participation in Family Work," in *Fatherhood Today: Men's Changing Role in the Family,* ed. Phyllis Bronstein and Carolyn P. Cowen (New York: John Wiley and Sons, 1988), p. 72; see also Arlie Hochschild, *The Second Shift: Working Parents and Revolution* (New York: Viking Books, 1989).

41. Michael E. Lamb, "The Changing Role of Fathers," in *The Father's Role: Applied Perspectives,* ed. Michael E. Lamb (New York: John Wiley and Sons, 1986), pp. 9ff.

42. Felice N. Schwartz, "Management Women and the New Facts of Life." *Harvard Business Review* (January–February 1989), p. 68.

43. Marion Burros, op. cit.

44. Colin Harrison, "Here's Baby. Dad Stays Home. Dad Gets Antsy," *New York Times,* August 31, 1993, p. A17.

45. Thomas Birch, *The Works of Mrs. Catherine Cockburn, Theological, Moral, Dramatic, and Poetical with an Account of the Life of the Author,* Vol. 1 (London: J. and P. Knapstoa, 1751), p. xl.

46. (London: Routledge and Kegan Paul), pp. 153–54.

47. Shirley M. Tilghman, "Science vs. Women—A Radical Solution," *New York Times,* January 26, 1993, p. A23.

48. Lester Thurow, "62 Cents to the Dollar: The Earnings Gap Doesn't Go Away." *Working Mother* (October 1984), p. 42.

49. The professional groups whose breast cancer rates were elevated included nuns, clergywomen, teachers, librarians, white counselors, mathematicians, computer scientists, secretaries, finance officers, pharmacists, supervisors, bank tellers, clerks, lawyers, judges, managers, administrators, and nurses but not white women physicians. Black women professionals were at higher risk than their white counterparts. Carol Hogfoss Rubin, et al., "Occupation as a Risk Identifier for Breast Cancer." *American Journal of Public Health,* Vol. 83 (9), p. 1311.

50. Kay Mills, *A Place in the News: From the Women's Pages to the Front Page* (New York: Dodd, Mead and Co., 1988), p. 41.

51. Ibid., p. 67.

52. Lindsy Van Gelder, "Countdown to Motherhood, When Should You Have a Baby?" *Ms.* (December 1986), p. 37.

53. Cited in Joyce Gelb and Marian Lief Palley, *Women and Public Policies* (Princeton, N.J.: Princeton University Press, 1982; revised, 1987), p. 163.

54. 414 U.S. 632 (1974).

55. 474 U.S. 395 (1973).

56. Lucinda M. Finley, "Transcending Equality Theory: A Way Out of the Maternity and the Workplace Debate," *Columbia Law Review* (October 1986), Vol. 86, No. 6, p. 1135.

57. Op. cit. at p. 645.55.

58. *Sprogis v. United Airlines, Inc.,* 444 F.2d 1194, cert. denied 404 U.S. 991.

59. *General Electric v. Gilbert,* 429 U.S. 125 (1976).

60. Cited by Marguerite Holloway, "A Lab of Her Own." *Scientific American* (November 1993), p. 103.

61. Bruno Bettelheim, "The Problem of Generations." *Daedalus* 91, pp. 68–96 at p. 84.

62. Elizabeth Wyman and Mary E. McLaughlin, "Traditional Wives and Mothers." *Counseling Psychologist* 8 (1979), p. 25.

63. Paul Anderson, *Janet Reno: Doing the Right Thing* (New York: John Wiley and Sons, 1994), p. 207.

64. Antoinette Martin, "Governor Who?," *Detroit Free Press Magazine,* July 4, 1993, pp. 6–17 at p. 10.

65. Walt Harrington, "A Woman Between Two Worlds," *The Washington Post Magazine,* July 9, 1989, p. 35.

66. Penelope Gilliatt quoted by James Childs in "Rebirth," *The Hollywood Screenwriters,* Richard Corliss, ed. (New York: Discus Books, 1972), p. 237.

67. Interviewed by author. Legislator requested anonymity.

68. Martin Walker, "Britain's Uncommon Leader," *The Washington Post Book World,* October 31, 1993, p. 2.

69. *The Girls in the Balcony* (New York: Random House, 1992), p. 183.

70. Interview, June 2, 1993.

71. Malcolm Gladwell, "The Healy Experiment," *The Washington Post Magazine,* June 21, 1992, p. 9.

72. Ibid.

73. *Los Angeles Times,* July 3, 1992, p. A1.

74. Robin Young, "The Selling of Motherhood." *Newsweek* (October 1, 1990), p. 12.

75. Interview with Debbie Stabenow at National Women's Political Caucus Convention, Los Angeles, July 8–11, 1993.

76. Charles Trueheart, "Woman to Succeed Mulroney," *Washington Post,* June 14, 1993, pp. 1 and A15 at p. A15.

77. *Political Babble,* Ed. David Olive (New York: John Wiley and Sons, 1992), p. 232.

78. Sam Roberts, "Just Casually Clenched, Holtzman Looks Ahead," *New York Times,* September 30, 1993, p. B7.

79. Timothy M. Phelps and Helen Winternitz, *Capitol Games* (New York: Harper Perennial, 1992), p. 367.

80. David Brock, "The Real Anita Hill," *The American Spectator* (March 1992), p. 27.

81. Ibid.

82. Timothy M. Phelps and Helen Winternitz; op. cit., p. 269.

83. Ibid., p. 323.

84. Ibid., p. 366.

85. National Women's Political Caucus, Friday, July 9, 1993, 9:00–10:30 A.M. Issues Forum: "The Religious Right: A View From the Field," Kathy Frasca, Director, Mainstream Voters Project, San Diego.

86. Lavinia Edmunds, "Barbara Mikulski," *Ms.* (January 1987), p. 63.

87. Quoted by John B. Judis, "Voice from the Right," *The Washington Post Book World,* September 5, 1993, p. x2.

88. Eleanor Clift, "Not the Year of the Women." *Newsweek* (October 25, 1993), p. 31.

89. Karen Schneider, "Women Seeking Office Say Rumors Hint of Double Standard," *Philadelphia Inquirer,* September 20, 1992, p. A8.
Of course tagging a woman a lesbian is only a liability if the audience assumes

that lesbianism is bad, unnatural, or inappropriate in leaders. Otherwise the charge may instead advantage the identified person. As the gay rights movement succeeds in making "gay," "homosexual," and "lesbian" neutral descriptors or positive words, those who want the claim to be an indictment are increasingly freighting it with presumably pejorative adjectives such as "radical," "mean," and "militant." So, for example, Jesse Helms challenged the nomination of a gay rights activist as deputy at the Department of Housing and Urban Development, claiming that "She's a militant-activist-mean lesbian working her whole career to advance the homosexual agenda." (AP, "Helms Calls Nominee to HUD a 'Mean' Lesbian," *Philadelphia Inquirer,* May 8, 1993, p. A3.

90. Mitchell Landsberg, "Reno Is Strictly No Frills," *Philadelphia Inquirer,* February 14, 1993, p. A12.

91. Elizabeth Gleick, "General Janny Baby." *People* (March 29, 1993), pp. 40–41.

92. Wendy Webster, *Not a Man to Match Her: The Marketing of a Prime Minister* (London: The Women's Press, 1990), pp. 4-136.

93. *Condit v. United Air Lines, Inc.,* 558 F.2d 1176 (4th Cir. 1977); *EEOC v. Delta Air Lines, Inc.,* 441 F. Supp. 626 (S.D. Tex. 1977).

94. *Maclennan v. American Airlines, Inc.,* 440 F. Supp. 466 (E.D. Va. 1977).

95. *Burwell v. Eastern Air Lines,* 633 F.2d 361 (4th Cir. 1980).

96. Susan Faludi, *Backlash* (New York: Crown, 1991), p. 440.

97. *Oil, Chemical and Atomic Workers v. American Cyanamid Co.,* 741 F.2d 444 (D.C. Cir. 1984).

98. Interview, December 3, 1993, University of Pennsylvania.

99. *Hayes v. Shelby Memorial Hosp.,* 726 F.2d 1543 (11th Cir. 1984). Rehearing denied.

100. 499 U.S. 187.

Chapter 4 Double Bind Number Two: Silence/Shame

An earlier version of this chapter appeared in my *Eloquence in an Electronic Age* (Oxford, 1988).

1. Carey A. Moore, "Susanna: A Case of Sexual Harassment in Ancient Babylon." *Bible Review* (June 1992), pp. 21–29 and 52 at p. 28.

2. Jews and Protestants find the story in the Apocrypha, Catholics in the thirteenth chapter of the Book of Daniel.

3. Carey A. Moore, op. cit., p. 29.

4. "The History of Susanna," in *The Apocrypha* (New York: Tudor Publishing Co., 1937), p. 246.

5. Carey A. Moore, op. cit., p. 22.

6. Margaret R. Miles, *Carnal Knowing: Female Nakedness and Religious Meaning in the Christian West* (Boston: Beacon Press, 1989), p. 122.

7. Mary D. Garrard, "Artemesia and Susanna," *Feminism and Art History* (New York: Harper and Row, 1982), pp. 147–72 at p. 150. See also Rozsika Parker and Griselda Pollock, *Old Mistresses: Women, Art and Ideology* (New York: Pantheon Books, 1981), pp. 20–25.

8. Mary D. Garrard, op. cit., p. 163.

9. Ibid., pp. 148–49.

10. *Countertraditions in the Bible: A Feminist Approach* (Cambridge: Harvard University Press, 1992), p. 10.

11. Ibid.

12. 1 Timothy 2:9–15.

13. *The New Dictionary of Thoughts,* ed. Tryon Edwards (New York: Standard Book Co., 1966), p. 735.

14. *Politics* 1, 12. 260a: 20–30; III. 4.1277b: 21–25.

15. Hunter College Women's Studies Collective, *Women's Realities, Women's Choices* (New York: Oxford University Press, 1983), p. 406.

16. *The Concept of Anxiety,* trans. Reidar Thomte (Princeton, N.J.: Princeton University Press, 1980), p. 66.

17. Thomas Wilson, *The Art of Rhetoric* (University Park: Pennsylvania State University Press, 1994), p. 226.

18. Gerda Lerner, *The Female Experience* (Indianapolis: Bobbs-Merrill Educational Pub., 1977), p. 417.

19. Susan Brownmiller, *Femininity* (New York: Fawcett Columbine, 1984), p. 112.

20. Ibid., p. 113.

21. Robin Morgan, *The Word of a Woman: Feminist Dispatches 1968–1992* (New York: W.W. Norton, 1993), p. 192.

22. David E. Sanger, "Silent Empress, Irate Nation (And Contrite Press)," *New York Times,* December 24, 1993, p. A4.

23. Ibid.

24. M. P. Sadker and D. M. Sadker, "Sexism in the Classroom of the '80's." *Psychology Today* (March 1985), pp. 54–57.

25. Dominick A. Infante, "Motivation To Speak on a Controversial Topic." *Central States Speech Journal* 34 (1983), pp. 96–103.

26. AAUW, *How Schools Shortchange Girls,* prepared by The Wellesley College for Research on Women (Washington, D.C.: AAUW Educational Foundation, 1992). These conclusions are consistent with those of Myra Sadker and David Sadker, "Sexism in the Schoolroom of the '80s," *Psychology Today* (March 1985), pp. 54–57.

27. Susan M. Barbieri, "Office Politics in the Nicer '90s." *Working Woman* (August 1993), p. 36.

28. Timothy Phelps and Helen Winternitz, *Capitol Games* (New York: Hyperion, 1992), p. 369.

29. "The Worming of a Mad Dogge," (London, 1617), reprinted in Katherine Usher Henderson and Barbara F. McManus eds., *Half Humankind: Contexts and Texts of the Controversy About Women in England, 1540–1640* (Urbana: University of Illinois Press, 1985), p. 253.

30. Joseph Swetnam," The Arraignment of Lewd, Idle, Forward, and Unconstant Women . . ." (London, 1615), in ibid., p. 209.

31. "Ester hath hanged Haman." (London, 1617), in ibid., p. 242.

32. Hunter College Women's Studies Collective, op. cit., p. 417.

33. Gerda Lerner, ed. *The Female Experience: An American Documentary* (New York: Oxford University Press, 1977), p. 214.

34. Brian Harrison, *Separate Spheres: The Opposition to Women's Suffrage in Britain* (New York: Holmes & Meier Publishers, Inc., 1978), p. 113.

35. Gerda Lerner, op. cit., p. 492.

36. Sarah Moore Grimke, *Letters on the Equality of the Sexes and the Condition of Woman, Addressed to Mary Parker, President of the Boston Female Anti-Slavery Society* (Boston: Isaac Knapp, 1838), p. 11.

37. Benjamin F. Butler, *Butler's Book* (Boston: A. M. Thayer and Co., 1892), p. 418.

38. Ibid., p. 421.

39. Ibid., p. 419.

40. Ibid.

41. Ibid., p. 420.

42. Linda Kerber, *Women of the Republic* (Chapel Hill: University of North Carolina Press, 1980), pp. 212–13.

43. Julia O'Faolain, and Lauro Martines, eds., *Not in God's Image* (New York: Harper Torchbooks, 1973), pp. 191–92.

44. Quoted by Cheris Kramarae, *Women and Men Speaking* (Rowley, Mass.: Newbury House, 1980), p. xvii.

45. Antonia Fraser, *The Weaker Vessel* (New York: Alfred Knopf, 1984), p. 246.

46. Stephen E. Wessley, "The Guglielmites: Salvation Through Women," in *Medieval Women,* ed. Derek Baker (Oxford: Basil Blackwell, 1978), pp. 289–303.

47. Christine Berg and Philippa Berry, "Spiritual Whoredom: An Essay on Female Prophets in the Seventeenth Century," in *Feminist Literary Theory,* ed. Mary Eagleton (Oxford: Basil Blackwell, 1986), p. 125.

48. Quoted by Gerda Lerner, *The Creation of Feminist Consciousness: From the Middle Ages to Eighteen-Seventy* (New York: Oxford University Press, 1993), p. 71.

49. Elizabeth Cady Stanton, *The Woman's Bible* (1895); reprinted Boston: Northeastern University Press, 1993), pp. 7–8.

50. Barbara Sapinsley, *The Private War of Mrs. Packard* (New York: Paragon, 1991), p. 5.

51. Elizabeth Parsons Ware Packard, *Modern Persecution or Insane Asylums Unveiled as Demonstrated by the Report of the Investigating Committee of the Legislature of Illinois,* Vol. 1 (Hartford, CT: Case, Lakewood and Brainard, 1875), pp. 99–100.

52. Thomas Szasz, *The Manufacture of Madness* (New York: Harper & Row, 1970), p. 130.

53. Barbara Sapinsley, op. cit., pp. 16–17.

54. See Jane M. Friedman, *America's First Woman Lawyer: The Biography of Myra Bradwell* (Buffalo, N.Y.: Prometheus Books, 1993), p. 205.

55. John Putnam Demos, *Entertaining Satan* (New York: Oxford University Press, 1982), p. 125.

56. The quote from Riddle is found in Boyce Rensberger, "Contraception the Natural Way: Herbs Have Played Role From Ancient Greece to Modern-Day Appalachia," *Washington Post,* July 25, 1994, p. A3. Riddle is the author of *Contraception and Abortion from the Ancient World to the Renaissance* (Cambridge, Mass. and London: Harvard University Press, 1992).

57. John Putnam Demos, op. cit., p. 95.

58. Ibid., p. 93.

59. Winthrop's Journals, Vol. II, p. 225; cited by Edward S. Morgan, *The Puritan Family: Religion and Domestic Relations in Seventeenth Century New England* (New York: Harper & Row, 1966), p. 44.

60. Ellen Bassuk, "The Rest Cure: Repetition or Resolution of Victorian Women's Conflicts?" in *The Female Body,* ed. Susan Rubin Suleiman (Cambridge, Mass.: Harvard University Press, 1986), pp. 139–51.

61. Charlotte Perkins Gilman, *The Living of Charlotte Perkins Gilman. An Auto-biography* (New York: D. Appleton-Century Co., 1935), pp. 72ff.

62. Ibid., p. 96.

63. (London: John Churchill, 1853), p. 119. Cited by Carroll Smith Rosen-berg, *Disorderly Conduct*, (New York: Oxford University Press, 1985), pp. 210–11.

64. Carroll Smith-Rosenberg, *Disorderly Conduct*, op. cit., p. 208.

65. "Managing Female Minds," in *Women's Studies: Essential Readings*, ed. Stevi Jackson (New York: New York University Press, 1993), pp. 378–80 at p. 378.

66. Carol McPhee and Ann Fitzgerald, comp. *Feminist Quotations* (New York: Thomas Y. Crowell, 1979), p. 59.

67. E. S. Valenstein, "The Practice of Psychosurgery: A Survey of the Literature (1971–1976)," in *Psychosurgery* (Washington, D.C.: Government Printing Office, 1977), pp. 3–11.

68. *Politics* 1, 12, 1260b, 28–31.

69. *History of Animals*, 608a32–619.

70. Thomas Wilson, (University Park, Pa.: Pennsylvania State University Press, 1994), p. 204.

71. *Institutes*, XI. III, 19.

72. Antonia Fraser, op. cit., p. 4.

73. Sarah B. Pomeroy, *Goddesses, Whores, Wives, and Slaves* (New York: Schoc-ken Books, 1975), p. 175.

74. Alice S. Rossi, ed. *The Feminist Papers: From Adams to de Beauvoir* (New York: Columbia University Press, 1973), p. 166.

75. Speech to the Troops at Tilbury on the approach of the Armada in 1588. Maria Theresa of Austria made a comparable claim on her ascent to the throne in 1741. "I am only a woman," she said, "but I have the heart of a King." See C. A. Macartney, *Maria Theresa and the House of Austria* (London: English Univer-sities Press, 1969), p. 47.

76. Sarah B. Pomeroy, op. cit., p. 172.

77. Ibid., p. 175.

78. Mary Putnam-Jacobi, *Common Sense Applied to Woman Suffrage* (New York: G. P. Putnam's Sons, 1894), p. 94.

79. *Institutes*, x, 11, 12.

80. Daniel Webster, *The Writings and Speeches of Daniel Webster* (Boston: Little, Brown and Co., 1903).

81. Robert Raymond, *The Patriotic Speaker* (New York: A. S. Barnes and Burr, 1864).

82. Faith Middleton, "She's Prepared." *Ms.* (November, 1986), p. 29.

83. Focus on the rhetoric of women and women's liberation has produced a rich body of scholarly research.

Karlyn K. Campbell, "The Rhetoric of Women's Liberation: An Oxymoron." *Quarterly Journal of Speech* (1973), 59, pp. 74–86.

Karlyn K. Campbell, "Elizabeth Cady Stanton," in H. R. Ryan, ed., *American Orators Before 1900: Critical Studies and Sources* (New York: Greenwood, 1987), pp. 340–49.

Karlyn K. Campbell, "What Really Distinguishes and/or Ought to Distinguish Feminist Scholarship in Communication Studies?" *Women's Studies in Communica-tion* (1988), 11, pp. 4–5.

Karlyn K. Campbell, *Man Cannot Speak For Her: A Critical Study of Early Feminist Rhetoric*, Vol. 1 (New York: Greenwood Press, 1989).

Karlyn K. Campbell, "Hearing Women's Voices." *Communication Education* (1991), 40, pp. 33–48.

Karlyn K. Campbell, and E. C. Jerry, "Woman and Speaker: A Conflict in Roles," in S. Brehm, ed., *Seeing Female: Social Roles and Personal Lives* (New York: Greenwood Press, 1988), pp. 123–33.

Celeste Condit, "What Makes Our Scholarship Feminist? A Radical Liberal View." *Women's Studies in Communication* (1988), 11, pp. 6–8.

Charles Conrad, "Agon and Rhetorical Form: The Essence of Old Feminist Rhetoric." *Central States Speech Journal* (1981), 32, pp. 45–53.

Charles Conrad, "The Transformation of the Old Feminist Movement." *Quarterly Journal of Speech* (1981), 67, pp. 284–97.

Bonnie J. Dow, "The 'Womanhood' Rationale in the Woman Suffrage Rhetoric of Frances E. Willard." *Southern Communication Journal* (1991), 56, pp. 298–307.

Phyllis Japp, "Esther or Isaiah?: The Abolitionist-Feminist Rhetoric of Angelina Grimke." *Quarterly Journal of Speech* (1985), 71, pp. 335–48.

Cheris Kramarae, *Women and Men Speaking: Frameworks for Analysis* (Rowley, Mass.: Newbury House, 1981).

84. Karlyn Campbell, *Man Cannot Speak For Her* (New York: Greenwood Press, 1989), p. 52.

85. Cf. Nancy Chodorow, *The Reproduction of Mothering* (Berkeley: University of California Press, 1978), pp. 6–7.

86. W. Arkin and L. R. Dobrofsky. "Military Socialization and Masculinity." *Journal of Social Issues* 34 (1978), pp. 151–68; P. J. Stein and S. Hoffman, "Sports and Male Role Strains." *Journal of Social Issues* 34 (1978), pp. 136–50.

87. Jesse Bernard, "Talk, Conversation, Listening, Silence," *The Sex Game* (New York: Atheneum, 1972), pp. 135–64.

88. H. M. Gilley and C. S. Summers, "Sex Differences in Use of Hostile Verbs." *Journal of Psychology* 76 (1970), pp. 33–37.

89. Claudia Mitchell-Kernan, "Signifying," in *Mother Wit from the Laughing Barrell*, Ed. Alan Dundes (Englewood Cliffs, N.J.: Prentice-Hall, 1973), p. 328.

90. Michael Burgoon, J. P. Dillard, and N. E. Doran, "Friendly or Unfriendly Persuasion." *Human Communication Research* 10 (1983), pp. 283–94.

91. In "Feminist Discourse," in *Buffalo Law Review* (*1985*), p. 50.

92. Edgar F. Borgatta and J. Stimson, "Sex Differences in Interaction Characteristics." *Journal of Social Psychology* 60 (1963), pp. 89–100; Judith A. Hall, *Nonverbal Sex Differences* (Baltimore: Johns Hopkins University Press, 1984), p. 145.

93. David H. Dosser, et. al., "Male Inexpressiveness and Relationships." *Journal of Social and Personal Relationships* 3 (1986), pp. 241–58.

94. Chris Greene, "Women: The Emotional or the Expressive Sex?" *Psychology Today* (June 1986), p. 12.

95. Hall, op. cit., p. 143.

96. R. Exline, D. Gray, and D. Schutte, "Visual Behavior in a Dyad as Affected by Interview Control and Sex of Respondent." *Journal of Personality and Social Psychology* 1 (1965), pp. 201–9.

97. D. Feshbacker, "Sex Differences in Empathy and Social Behavior in Children," in *The Development of Prosocial Behavior*, ed. N. Eissenberg (New York: Academic Press, 1982), pp. 315–38.

98. A. Vaux, "Variations in Social Support Associated with Gender, Ethnicity, and Age." *Journal of Social Issues* (1983), 41, pp. 89–110.

99. The distinction gained currency in the 1950s with the publication of Talcott Parsons and Robert F. Bales, *Family, Socialization and Interaction Process* (Glencoe, Il: Free Press, 1955).

100. Lea P. Stewart, Pamela J. Cooper, and Sheryl A. Friedley, *Communication Between the Sexes* (Scottsdale, Ariz.: Gorsuch Scarisbrick, 1986), p. 114. The same distinction wears different language in the literature on personality. There males are cast as agency oriented and females as tending toward communion. Some attribute the task orientation of men to their more aggressive personalities; by this logic, women's more nurturant personalities dispose toward the maintenence of group well-being. Also see R. Carlson. "Understanding Women: Implications for Personality Theory and Research." *Journal of Social Issues* 28 (1972), pp. 17–32.

101. Marjorie Swacker, "The Sex of the Speaker as a Sociolinguistic Variable," in *Language and Sex*, eds. Barrie Thorne and Nancy Henley (Rowley, Mass.: Newbury House, 1975), pp. 76–83.

102. Judith Kegan Gardiner, "On Female Identity and Writing by Women," in *Writing and Sexual Differences*, ed. Elizabeth Abel (Chicago: University of Chicago Press, 1982), p. 185.

103. Josephine Donovan, "The Silence Is Broken," in *The Feminist Critique of Language*, Ed. Deborah Cameron (London: Routledge, 1990), pp. 41–56 at p. 47.

104. Ibid., p. 48.

105. Gerda Lerner, *The Creation of Feminist Consciousness: From the Middle Ages to Eighteen-Seventy*, op. cit., p. 233.

106. Quoted by Mary Kinnear, *Daughters of Time: Women in the Western Tradition* (Ann Arbor: The University of Michigan Press, 1982), p. 95.

107. Bruno Bettleheim, *The Uses of Enchantment* (New York: Alfred Knopf, 1977), p. 153.

108. Meg Bogan, *The Women Troubadours* (New York: W. W. Norton Co., 1976), p. 13.

109. K. K. Ruthven, "Feminist Literary Studies," in *Feminist Liberary Theory: A Reader*, Ed. Mary Eagleton (London: Basil Blackwell, 1980), p. 93.

110. Steeven Guazzo, *The Civile Conversation of M. Steeven Guazzo*, Trans. George Pettie (1581) (London: Constable, 1925), p. I. 69.

111. Ibid.

112. Joseph Addison, *Spectator*, ed. George A. Aitkin (London, John C. Nimmo, 1898), 3, 376.

113. JoAnn Bromberg-Ross, "Storying and Changing: An Examination of the Consciousness-Raising Process." *Folklore Feminists Communication* 6 (1975), pp. 9–11; Meg Bogan, op. cit.; Susan Harding, "Women and Words in a Spanish Village," in *Toward An Anthropology of Women*, Ed. Rayne Reiter (New York: Monthly Review Press, 1975); Susan Kalcik, "'. . . Like Ann's Gynecologist or the Time I Was Almost Raped.': Personal Narratives in Women's Rap Groups." *Journal of American Folklore*, 88 (1975), pp. 3–11.

114. T. W. Smith, "The Polls: Gender and Attitudes Toward Violence." *Public Opinion Quarterly* 48 (1984), pp. 384–96.

115. Cf. Sandra Baxter and Marjorie Lansing, *Women and Politics* (Ann Arbor: University of Michigan Press, 1981), p. 57.

116. Ibid., p. 50.

Chapter 5 Double Bind Number Three: Sameness/Difference

1. Bruce Lambert, "Richard Salant, 78, Who Headed CBC News In Expansion, Is Dead," *New York Times,* February 17, 1993, p. B8.

2. *Bradwell v. Illinois,* 83 U.S. (16 Wall) 130, 141-2 (1872).

3. See Jane M. Friedman, *America's First Woman Lawyer: The Biography of Myra Bradwell* (Buffalo, N.Y.: Prometheus Books, 1993), p. 142.

4. *State v. Hall,* 187 So. 2nd 861, 863 (Miss.), appeal dismissed, 385 U.S. 98 (1966).

5. Sue Tolleson Rinehart, *Gender Consciousness and Politics* (New York: Routledge, 1992), pp. 71–72.

6. See Carleton Mabee's analysis in *Sojourner Truth: Slave, Prophet, Legend* (New York: New York University Press, 1993), pp. 74–82.

7. Gage's version can be found in Elizabeth Cady Stanton, Susan B. Anthony, and Matilda Joslyn Gage, *History of Woman Suffrage,* 3 vols. (1881–87) in vol. 1, (New York: Source Book, 1970), pp. 115–17.

8. Salem, OH, *Anti-Slavery Bugle,* June 21, 1951. Quoted in Carleton Mabee, op. cit., p. 81.

9. Quoted by Jacqueline Bobo, "The Color Purple: Black Women as Cultural Readers," in *Female Spectators: Looking at Film and Television,* ed. E. Diedre Probram (London: Verso, 1988), p. 106.

10. Cf. Elizabeth Spelman, *Inessential Woman: Problems of Exclusion in Feminist Thought* (Boston: Beacon Press, 1988).

11. Charlotte Perkins Gilman, *The Man-Made World; Or, Our Androcentric Culture* (New York: Johnson Reprint, 1911/1971), pp. 20ff.

12. Either choice was problemmatic. "[I]f women claim they are the same as men in order to secure the rights of men," writes legal theorist Martha Minow, "any sign of difference can be used to deny those rights; and if women claim they are different from men in order to secure special rights, those very differences can be cited to exclude women from the rights that men enjoy."
Since the beginnings of the suffrage debate, the sameness-difference bind has haunted women's rights activists and divided their allies. "When equality and difference are paired dichotomously," argues Joan Wallach Scott, "they structure an impossible choice. If one opts for equality, one is forced to accept the notion that difference is antithetical to it. If one opts for difference, one admits that equality is unattainable."*Gender and the Politics of History* (New York: Columbia University Press, 1988), p. 172.

13. For a discussion, see B. C. Gelpi, "The Politics of Androgyny." *Women's Studies* 2 (1974), pp. 151–60; C. Secor, "Androgyny: An Early Reappraisal." *Women's Studies* 2 (1974), pp. 161–69; C. R. Stimpson, "The Androgyne and the Homosexual." *Women's Studies* 2 (1974), pp. 237–48; E. E. Sampson, "Psychology and the American Ideal," *Journal of Personality and Social Psychology,* 35 (1977), pp. 767–82; Sandra Lipsitz Bem, *The Lenses of Gender* (New Haven: Yale University Press, 1993).

14. (Cambridge: Harvard University Press, 1982).

15. Cf. Linda K. Kerber, Catherine G. Green, Eleanor Maccoby, Zella Luria, Carol B. Stack, and Carol Gilligan, "On *In a Different Voice:* An Interdisciplinary Forum." *Signs* (Winter 1986), Vol. 11, No. 2, pp. 304–33; John M. Broughton, "Women's Rationality and Men's Virtues: A Critique of Gender Dual-

ism in Gilligan's Theory of A Moral Development," *Social Research* (Autumn 1983), pp. 597–642.

16. "Feminist Discourse," *Buffalo Law Review* (1985), p. 74.

17. The text is included in *Man Cannot Speak For Her: Key Texts of the Earlier Feminists,* Vol. II, comp. Karlyn Kohrs Campbell (New York: Greenwood Press, 1989), p. 34. Hereafter cited as K. K. Campbell, *Texts.*

18. Ibid., p. 810.

19. J. S. Mill, *The Subjection of Women* (London: Longman, Green, Reader, and Dyer, 1869), p. 1.

20. K. K. Campbell, *Texts,* op.cit. pp. 37–38.

21. Ibid.

22. Ibid., p. 38.

23. *Women's Studies International Forum* (1984), Vol.7 No.6, pp. 455–65. Printed in Great Britain.

24. *Jane Addams: A Centennial Reader* (New York: Macmillan, 1960), p. 115.

25. Sharon Bertsch McGrayne, *Nobel Prize Women in Science, Their Lives, Struggles and Momentous Discoveries* (New York: Carol Publishing, 1993), pp. 34–35.

26. Linda K. Kerber, *Women of the Republic: Intellect and Ideology in Revolutionary America* (Chapel Hill: University of North Carolina Press, 1980), p. xii.

27. Anne Firor Scott, *Natural Allies: Women's Associations in American History* (Urbana and Chicago: University of Illinois Press, 1993), p. 15.

28. (New York: Wiley, 1960), p. 448.

29. Ibid., p. 490.

30. Ibid.

31. Reprinted in Sara Evans, *Personal Politics: The Roots of Women's Liberation in the Civil Rights Movement and The New Left* (New York: Vintage Books, 1979), pp. 233–34 at p. 233.

32. Edward Constantini and Kenneth H. Craik, "The Social Background, Personality and Political Careers of Female Party Leaders." *Journal of Social Issues* 28 (1972), p. 235.

33. Celia Morris, "Changing the Rules and the Roles: Five Women in Public Office," in *The American Woman 1992–93,* ed. Paula Ries and Anne Stone (New York: W. W. Norton, 1992), pp. 95–126, at 108.

34. *Winning with Women,* A survey commissioned by EMILY's List, the National Women's Political Caucus and the Women's Campaign Fund (Washington, DC: 1991), p. 5.

35. "The World's Experience Upon Which Legislation Limiting the Hours of Labor for Women Is Based," in Josephine Goldmark, *Fatigue and Efficiency: A Study in Industry* (New York: Charities Publication Committee, 1912).

36. *Quong Wing v. Kirkendall,* 223 U.S. 59 (1912), 63. Indeed, working women presumably needed protection because, as the court held in *Muller v. Oregon,* the "two sexes differ in structure of body, in the functions to be performed by each, in the amount of physical strength, in the capacity for long-continued labor . . . . (*Muller v. Oregon* 208 U.S. 412, 422 [1908]).

37. For discussion see Norma Nasch, "The Emerging Legal History of Women in the United States: Property, Divorce, and the Constitution." *Signs* 12 (Autumn 1986), pp. 112–15.

38. Elizabeth Faulkner Baker, *Protective Labor Legislation: With Special Reference to Women in the State of New York* (1925; reprinted New York: Columbia Uni-

versity's Studies in History, Economics, and Public Law, AMS Press, 1969), p. 432.

39. Quoted by Alice Kessler-Harris, *Out to Work: A History of Wage-Earning Women in the United States* (New York: Oxford University Press, 1982), p. 208.

40. Donald G. Mathews and Jane Sherron DeHart, *Sex, Gender, and the Politics of ERA* (New York: Oxford University Press, 1990), pp. 169–70.

41. For a discussion of the role and rhetoric of Ervin, see ibid., pp. 48ff.

42. Ibid., p. 179.

43. Phyllis Shlafly (New York: Harcourt Brace Jovanovich, 1977), pp. 109–10.

44. Joanne Meyerowitz, *Women Adrift: Independent Wage Earners in Chicago, 1880–1930* (Chicago: University of Chicago Press, 1988), p. 33.

45. Deborah L. Rhode, *Justice and Gender* (Cambridge: Harvard University Press, 1989), p. 158.

46. Cynthia Fuchs Epstein, *Women in Law* (Urbana and Chicago: University of Illinois, 1993), p. 443.

47. Ibid.

48. Alice Kessler-Harris, *A Woman's Wage* (Lexington: University Press of Kentucky, 1990), p. 111.

49. Ibid.

50. Donald Mathews and Jane Sherron DeHart, op.cit., p. 34.

51. Samuel Gompers, "Women's Work, Rights and Progress." *American Federationist,* 20 (August 1913), p. 625; quoted by Kessler-Harris, *A Woman's Wages,* op. cit., p. 19.

52. The Honorable Ruth Bader Ginsburg, "American University Commencement Address." *The American University Law Review* (Summer 1982), pp. 891–902 at p. 895.

53. Cited by Joan W. Scott, "Deconstructing Equality-Versus-Difference," in *Conflicts in Feminism,* Eds. Marianne Hirsch and Evelyn Fox Keller (New York: Routledge, 1990), p. 143.

54. "Deconstructing Equality-Versus-Difference: Or, the Uses of Poststructuralist Theory for Feminism, in ibid.

55. Samuel Freedman, "Of History and Politics: Bitter Feminist Debate," *New York Times* June 6, 1986, pp. B1/B4; Carol Sternhell, "Life in the Mainstream." *Ms.* (July 1986), pp. 48–51, 86–89.

56. *EEOC v. Sears, Roebuck & Co,* 628 F. Supp. 1264 (1986), at page 1308.

57. Offer of Proof Concerning the Testimony of Dr. Rosalind Rosenberg, *EEOC v. Sears* (No 79-C-4373) at para. 1.

58. (Berkeley: University of California Press, 1978), cited by Rosenberg in footnote 19a of her "Proof." Gilligan is cited in footnotes 19a and 19c.

59. *American Couples: Money, Work, and Sex* (New York: William Morrow, 1983).

60. Offer of Proof, op.cit., para. 19a.

61. Written Testimony of Alice Kessler-Harris, *EEOC v. Sears* (No.79-C-4373) at para. 2a.

62. Ibid. at para. 11.

63. Ibid. at para. 13. Judge Nordberg explicitly rejected Kessler-Harris's claim that the discrepancy could *only* be accounted for by discrimination. See footnote 63 of his opinion.

64. Op. cit., at p. 1314.

65. Jane Gross, "Big Grocery Chain Reaches Landmark Sex-Bias Accord," *New York Times,* December 17, 1993, p. 1/B10.

66. Carol Mueller, "Nurturance and Mastery: Competing Qualifications for Women's Access to High Public Office," in *Research in Politics and Society,* ed. Gwenn Moore and Glenna Spitze Vol. 2 (Greenwich, Conn.: JAI Press, 1986), pp. 211–32 at p. 219.

67. "Confidence in Institutions." *The American Enterprise* (November/December 1993), pp. 94–95 at 94. Survey conducted by the National Opinion Research Center, February/April 1993.

68. Voter Research & Surveys (VRS), Election day exit polls, 1992. CBS, NBC, ABC, and CNN pool their resouces to create this common exit poll.

69. *Winning with Women* op. cit., p. 3.

70. Interview, October 1993.

71. A panel discussion sponsored by the National Women's Political Caucus, "Gender and the 1993 Gubernatorial Campaigns," National Press Club, November 10, 1993. I moderated the session.

72. National Women's Political Caucus, "Gender and the 1993 Gubernatorial Campaigns," November 10, 1993, National Press Club.

73. *Frontiero v. Richardson* 411 U.S. 677, 684 (1973).

74. Ibid., at 686–87.

75. Martha Albertson Fineman, *The Illusion of Equality: The Rhetoric and Reality of Divorce Reform* (Chicago: The University of Chicago Press, 1991), p. 3.

76. 511 U.S. (1994).

77. *Dothard v. Rawlinson,* 433 U.S. 321.

78. 442 F.2d 385 (5th Cir. 1971), cert denied 404 U.S. 950.

79. Diaz 311 F. Supp. 559, 565–66 reprinted in *Sex-Based Discrimination: Text Cases and Materials,* 3rd ed., ed. Herma Hill Kay (St. Paul, Minn.: West Publishing Co., 1988), p. 612.

80. *Wilson v. Southwest Airlines,* 517 F. Supp. 292 (N.D Tex. 1981). For discussion, see Lucinda M. Finley, "Transcending Equality Theory: A Way Out of the Maternity and the Workplace Debate." *Columbia Law Review,* (October 1986), Vol. 86, No. 6, pp. 1134–1135.

81. *Harris v. Forklift Systems Inc.,* 114 S.Ct. 367 (1993).

82. Barbara Presley Noble, "Little Discord on Harassment Ruling," *New York Times,* November 14, 1993, p. 25.

83. Ibid.

84. Ibid.

Chapter 6 Double Bind Number Four: Femininity/Competence

1. Inge K. Broverman, Donald M. Broverman, et al., "Sex-Role Stereotypes and Clinical Judgments of Mental Health." *Journal of Consulting and Clinical Psychology* 34 (1970), pp. 1–7.

2. ABA Commission Summary, (Chicago: American Bar Association, 1988), p. 2.

3. Robin Lakoff, *Language and Woman's Place* (New York: Harper Colophon Books, 1975), p. 6.

4. Francis Olsen, *The Sex of Law,* unpublished, pp. 1–4. Cited by Christine A. Littleton, "Reconstructing Sexual Equality." *California Law Review* (July 1987), Vol. 75, No. 4, pp. 1279–1337 at p. 1332.

5. Betty Friedan, *The Second Stage* (New York: Bantam, 1981), p. 154.

6. Nancy A. Nichols, "Whatever Happened to Rosie the Riveter?" *Harvard Business Review* (July/August, 1993), p. 60. "In this era of opening opportunity, women are beginning to move into positions of authority," writes linguist Deborah Tannen in her best selling book *You Don't Understand* (New York: Morrow, 1990). "At first we assumed they could simply talk the way they always had, but this often doesn't work. Another logical step is that they should change their styles and talk like men. Apart from the repugnance of women's having to do all the changing, this doesn't work either, because women who talk like men are judged differently—and harshly."

7. Interview with Rebecca Stafford, President, Monmouth College. June 4, 1993, Washington, D.C.

8. Malcolm Gladwell, "The Healy Experiment," *The Washington Post Magazine* (June 21, 1992), p. 25.

9. Robert K. Merton, "The Self-Fulfilling Prophecy." *The Antioch Review,* VIII (June 1948), p. 198.

10. *Men and Women of the Corporation* (New York: Basic Books, nc. 1977), p. 214.

11. Simone de Beauvoir, *The Second Sex,* trans and edited by H. M. Parshley (New York: Alfred A. Knopf, 1952; reprinted New York: Vintage Books, 1989), p. xxii. As if to confirm the author's claim, responses from some readers stigmatized her with classic stereotypes of Other. "Unsatisfied, frigid, priapic, nymphomaniac, lesbian, a hundred times aborted," recalled de Beauvoir in her memoirs. "I was everything, even an unmarried mother. People offered to cure me of my frigidity, or to temper my labial appetites."

Yet even de Beauvoir embraces the notion that the male specifies the norm. "To gain the supreme victory," she writes in the closing sentence of *The Second Sex,* "it is necessary, for one thing, that by and through their natural differentiation men and women unequivocally affirm their brotherhood." In 1952, sisterhood had not become powerful.

12. *The Second Sex,* op.cit., p. 701.

13. *The Second Sex,* op.cit., pp. 701–2.

14. Junda Woo, "Widespread Sexual Bias Found in Courts," *The Wall Street Journal,* August 20, 1992, pp. B-1, B-3.

15. M. E. Lockheed and K. P. Hall, "Conceptualizing Sex as a Status Characteristic: Applications to Leadership Training Strategies." *Journal of Social Issues* 32 (1976), pp. 111–24; B. F. Meeker and P. A. Weitzel-O'Neill, "Sex Roles and Interpersonal Behavior in Task-Oriented Groups." *American Sociological Review* 42 (1977), pp. 91–105.

16. Dorothy W. Cantor and Toni Bernay with Jean Stoess, *Women in Power* (New York: Houghton Mifflin, 1992), p. 167.

17. Kay Mills, *A Place in the News* (New York: Dodd, Mead, 1988), p. 61.

18. Interview July 10, 1993, National Women's Political Caucus, Los Angeles.

19. Michele A. Paludi and Lisa A. Strayer, "What's in an Author's Name? Differential Evaluations of Performance as a Function of Author's Name. *Sex Roles* 12:3/4 (1985), pp. 353–61.

20. Norma Costrich, Joan Feinstein, Louise Kidder, Jeanne Marecek, and Linda Pascale, "When Stereotypes Hurt: Three Studies of Penalties for Sex-Role Reversals." 11 *Journal of Experimental Social Psychology* (1975), p. 520.

21. M. L. MacDonald, "Assertion Training for Women," in *Social Skills Training,* Eds. J. P. Curran and P. M. Monti (New York: Guilford, 1981).

22. Madeline E. Heilman and Melanie H. Stopeck, "Attractiveness and Corporate Success: Different Causal Attributes for Males and Females." 70 *Journal of Applied Psychology* (1985), 70:2, pp. 379–88.

23. P. H. Bradley, "The Folk-Linguistics of Women's Speech: An Empirical Examination." *Communication Monographs* (1981), 48, pp. 73–90.

24. A. Bridges and H. Hartman, "Pedagogy by the Oppressed." *Review of Radical Political Economics* (Winter 1975) Vol. 6, No. 4, pp. 75–79 at p. 77. See also Norma Wikler, "Sexism in the Classroom," paper presented at the American Sociological Association (1976); cited by Arlie Russell Hochschild, *The Managed Heart* (Berkeley: University of California Press, 1983), p. 168.

25. Dorothy W. Cantor and Toni Bernay with Jean Stoess, op.cit., pp. 85–86.

26. Sandra Lipsitz Bem, "Probing the Promise of Androgyny," in *The Psychology of Women,* Ed. Mary Roth Walsh (New Haven: Yale University Press, 1987), pp. 206–25 at p. 211.

27. Not all the concepts associated with femininity are negative, however. Masculinity has been associated with task-oriented or problem-solving focus and on instrumental action, action as a means to an end; by contrast, femininity has been associated with a concern for harmony and process, an expressive orientation. There has been a tendency to see these approaches as mutually exclusive. One can either be instrumental or expressive, focused on task or on process, focused on relationships or on outcome.

28. M. D. Storms, "Sex Role Identity and Its Relationship to Sex Role Attributes and Sex Role Stereotypes." *Journal of Personality and Social Psychology* 37 (1979), pp. 1779–89.

29. J. T. Spence and R. L. Helmreich, *Masculinity and Femininity: Their Psychological Dimensions, Correlates, and Antecedents* (Austin: University of Texas Press, 1978).

30. Katherine Usher Henderson and Barbara F. McManus, *Half Humankind* (Urbana and Chicago: University of Illinois Press, 1985), p. 3.

31. Charles Trueheart, "Woman Set to Succeed Mulroney," *Washington Post,* June 14, 1993, pp. 1 and A15 at p. A15.

32. "Judge Ginsburg's Life, A Trial by Adversity," *New York Times,* June 25, 1993, pp. 1 and A19 at p. A19.

33. Georgette Mosbacher, *Feminine Force: Release the Power Within to Create the Life You Deserve* (New York: Simon and Schuster, 1993), p. 189.

34. Kristin Clark Taylor, *The First to Speak: A Woman of Color Inside the White House* (New York: Doubleday, 1993), p. 59.

35. Joyce Sterling, "Women on the Bench," quoted by Cynthia Fuchs Epstein, *Women in Law,* 2nd ed. (Urbana: University of Illinois Press, 1993), p. 454.

36. Camille Paglia, *Sex, Art, and American Culture* (New York: Vintage Books, 1992), p. 250.

37. Bernice Lott, "The Devaluation of Women's Competence," in *Seldom Seen, Rarely Heard: Women's Place in Psychology,* ed. Janis S. Bohan (Boulder: Westview Press, 1992), pp. 171–91 at pp. 179–80.

38. Anne Stibbs, ed., *A Women's Place: Quotations About Women* (New York: Avon Books, 1992), p. 209.

39. Letter to Eleanor Roosevelt, July 25, 1942.

40. "Indira Gandhi's Guiding Star," *London Observer,* November 4, 1984.

41. Sharon Bertsch McGrayne, *Nobel Prize Women in Science, Their Lives, Struggles and Momentous Discoveries* (New York: Carol Publishing, 1993), p. 43.

42. Golda Meir, *My Life* (New York: G. P. Putnam's Sons, 1975) p. 114.

43. Kathleen Hall Jamieson, *Eloquence in an Electronic Age* (New York: Oxford University Press, 1988), p. 87.

44. Bella Abzug, *Gender Gap* (Boston: Houghton Mifflin, 1984), p. 171.

45. Jerry Gray, "And Now, Whoppers vs. Waffles," *New York Times* October 7, 1993, p. B8.

46. Susan T. Fiske and Steven T. Neuberg, "A Continuum of Impression Formation from Category-Based to Individuating Processes: Influences of Information and Motivation on Attention and Interpretation," in M. Zanna, ed., *Advances in Experimental Social Psychology,* 23, pp. 1–74.

47. "If 'Good Managers' are Masculine, What Are 'Bad Managers'?" *Sex Roles* 10 (1984), nos. 7/8, pp. 477–84.

48. Ibid., p. 477.

49. "Probing the Promise of Androgyny," in *The Psychology of Women: Ongoing Debates,* ed. Mary Roth Walsh (New Haven: Yale University Press, 1987), pp. 206–25 at pp. 207–8.

50. Joel R. Heerboth and Nervella V. Ramanaiah, "Evaluation of the BSRI Masculine and Feminine Items Using Desirability and Stereotype Ratings." *Journal of Personality Assessment* (1985), 49, 3, pp. 264–70.

51. Ibid., p. 267.

52. Mary R. Beard, *Women as a Force in History* (New York: The Macmillan Co., 1946), p. 49.

53. "Feminist Discourse," *Buffalo Law Review* XXXIV (1985), p. 39.

54. Susan Moller Okin, *Women in Western Political Thought* (Princeton, NJ: Princeton University Press, 1979), pp. 99–100.

55. Ruth Herschberger, *Adam's Rib* (1948; reprinted New York: Harper and Row, 1970), p. 72. See also Emily Martin, "The Egg and the Sperm: How Science Has Constructed a Romance Based on Stereotypical Male-Female Roles." *Signs: Journal of Women in Culture and Society* (1991), Vol. 16, No. 31, pp. 485–501.

56. Ruth Herschberger, *Adam's Rib,* op.cit., p. 82.

57. See Emily Martin, *The Woman in the Body: A Cultural Analysis of Reproduction* (Boston: Beacon, 1987).

58. Ibid., p. 50.

59. Natalie Angier, "Radical New View of Role of Menstruation," *New York Times* (September 21, 1993), pp. C1/C10.

60. Margie Profet, "Menstruation as a Defense Against Pathogens Transported by Sperm." *The Quarterly Review of Biology* 68, no. 3 (September 1993), pp. 335–81 at p.336.

61. *Language: Its Nature, Development and Origin* (London: Allen and Unwin, 1922), pp. 245–50.

62. *Language and Woman's Place* (New York: Harper and Row, 1975).

63. Ibid., p. 7.

64. Ibid., p. 19.

65. Deborah Cameron, Fiona McAlinden, and Kathy O'Leary, "Lakoff in Con-

text: The Social and Linguistic Functions of Tag Questions," in *Women's Studies: Essential Readings,* ed. by Stevi Jackson et al. (New York: New York University Press, 1993), pp. 421–26 at p. 424.

66. Jennifer Coates, "Gossip Revisited," in ibid., pp. 426–29 at p. 426.

67. *Man Made Language* (London: Routledge and Kegan Paul, 1980), p. 8. "The focus on female difference, of course, emphasizes the underlying assumption that the female is a deviant while the male is 'normal' and speaks 'the language'." Nancy Henley and Cheris Kramarae, "Gender, Power and Miscommunication," p. 419 in Stevi Jackson, op. cit.

68. E. Diedre Pribram, *Female Spectators: Looking at Film and Television* (London: Verso, 1988), p. 1.

69. Jackie Byars, "Gazes/Voices/Power: Expanding Psychoanalysis for Feminist Film and Television Theory," in E. Pribram, op.cit., pp. 110–31 at p. 111.

70. *Fire with Fire: The New Female Power and How it Will Change the 21st Century* (New York: Random House, 1993).

71. Naomi Wolf, *The Beauty Myth* (New York: Doubleday, 1992).

72. Ibid., p. 19.

73. Tamar Lewin, "Feminists Wonder If It Was Progress to Become 'Victims,'" *New York Times,* May 10, 1992, p. E6.

74. 477 U.S. 57 (1986).

75. I am following the account provided by the Court in *Hopkins v. Price Waterhouse,* U.S. Court of Appeals, District of Columbia Circuit 1987, 825 F2d 458; certiorari granted 485 U.S. (1988).

76. In the Supreme Court of the United States, October Term, 1987, *Price Waterhouse v. Ann B. Hopkins,* on writ of Certiorari to the United States Court of Appeals for the District of Columbia Circuit. Brief for Amicus Curiae American Psychological Association in Support of Respondent, June 18, 1988.

77. Ibid.

78. *Price Waterhouse v. Hopkins,* 490 U.S. 228 (1989).

79. *Hopkins v. Price Waterhouse,* 737 F. Supp. 1202 (D.D.C.) affirmed 920 F 2d 967 (D.C. Cir. 1990).

80. Brief for Amicus Curiae American Psychological Association in Support of Respondent, June 18, 1988, 28, 1–30.

81. U.S. Department of Labor, *A Report on the Glass Ceiling Initiative* (Washington, D.C., 1991) 6.

82. *The New Diversity: Women and Minorities on Corporate Boards* (Chicago: Hedrick and Struggles, 1993), pp. 2–6.

83. Ibid., p. 3.

84. Martha Fay Africa, "Forum on Hiring," in *Lawyer Hiring and Training Report* (May 1993), 13, no. 5, p. 2.

85. Ibid., pp. 6–7.

86. *The New Diversity,* op.cit., p. 8.

87. Madeline E. Heilman, "The Impact of Situational Factors on Personnel Decisions Concerning Women: Varying the Sex Composition of the Applicant Pool." 26 *Or. Behavior and Human Performance* 174 (1984), pp. 386–95.

88. Op.cit., p. 29.

89. Jennifer Crocker and Kathleen M. McGraw, "What's Good for the Goose Is Not Good for the Gander: Solo Status as an Obstacle to Occupational Achievement for Males and Females." 27 *American Behavioral Scientist* (1984), pp. 357–70.

90. Robin J. Ely, "Organizational Demographics and Women's Gender Identity at Work," Working Paper, John F. Kennedy School of Government. Harvard University, 1992.
91. Linda Tarr-Whelan, "Realigning Priorities." *Social Policy* (Summer 1993), Vol. 23, No. 4, pp. 8–13 at p. 13.
92. Patrice Duggan Samuels, "Enterprise." *Lear's* (January 1994), pp. 17–18 at p. 18.
93. Angie Cannon, "Clinton Has Named Women to 40% of Top Government Jobs," *Philadelphia Inquirer,* September 28, 1993, p. A7.
94. The Honorable Ruth Bader Ginsburg, "Speech: American University Commencement Address, May 10, 1981," in *The American University Law Review* (Summer, 1981), p. 896.
95. Karen Schmidt and Colleen Collins, "Showdown at Gender Gap." *American Journalism Review* (July/August 1993), p. 42.
96. David Weaver, Roy W. Howard, G. Cleveland Wilhoit, *The American Journalist in the 1990s.* The Freedom Forum World Center, Arlington, Va., November 17, 1992. Funded by The Freedom Forum.
97. Karen Schmidt and Colleen Collins, op. cit., p. 42.
98. Ibid., p. 8.
99. Ibid., p. 9.
100. James Rose, "Women Are No Longer the Second Strings in Classical-Music Circles," *Philadelphia Inquirer,* May 10, 1994, p. F5.
101. Center for the American Woman and Politics, Rutgers University.
102. National Women's Political Caucus, "Factsheet on Women's Political Progress" (June 1993), p. 1.
103. Ibid., p. 4.
104. Ibid., p. 7.
105. Jody Newman, "Perception and Reality: A Study Comparing the Success of Men and Women Candidates" (Washington, D.C.: National Women's Political Caucus, 1994) p. 2.

Chapter 7 Double Bind Number Five: Aging/Invisibility

1. *Rhetoric* II, ch. 13.
2. P. Baltes and K. Schaie, "The Myth of the Twilight Years," *Psychology Today,* (March 1974), p. 37.
3. Simone de Beauvoir, *The Coming of Age* (New York: G. P. Putnam's Sons, 1972), p. 3.
4. F. M. Deutsch, C. M. Zalenski, and M. E. Clark, "Is There a Double Standard of Aging." *Journal of Applied Social Psychology* 16 (1986), pp. 771–85.
5. H. Zepplin, R. A. Sills, and M. W. Halth, "Is Age Becoming Irrelevant? An Exploratory Study of Perceived Age Norms." *International Journal of Aging and Human Development* 24 (1987), pp. 241–56.
6. J. Drevenstedt, "Perceptions of Onsets of Young Adulthood, Middle Age, and Old Age." *Journal of Gerontology* 31 (1976), pp. 53–57.
7. H. Zepplin et al., op. cit.
8. Margot Jeffreys, "The Elderly in Society," *Textbook of Geriatric Medicine and Gerontology,* 4th ed., ed., J. C. Brocklehurst, R. C. Tallis, H. M. Fillot, (Edinburgh: Churchill Livingstone, 1992), pp. 971–79 at p. 976.

9. W. H. Crockett and M. L. Humbert, "Perceptions of Aging and the Elderly," in C. Eisdorfer, ed., *Annual Review of Gerontology and Geriatrics,* Vol. 7 (New York: Springer, 1987), pp. 217–41.

10. Fay Lomax Cook, "Ageism: Rhetoric and Reality." *The Gerontologist* 32:3 (1992), pp. 292–95 at p. 292.

11. For an argument that a shift has occurred see R. H. Binstock, "The Aged as Scapegoat." *The Gerontologist* 23 (1983), pp. 136–43.

12. Louis Harris and Associates, *Survey of Legislators and Regulators, 1988,* Table 9-1.

13. E. B. Palmore, *Ageism: Negative and Positive* (New York: Springer, 1990).

14. Robert N. Butler, "Care of the Aged in the United States of America," in *Textbook of Geriatric Medicine and Gerontology,* 4th ed., Eds. J. C. Brocklehurst, R. C. Tallis, and H. M. Fillit (Edinburgh: Churchill Livingstone, 1992), pp. 993–99 at p. 995.

15. B. L. Neugarten, "Age Groups in American Society and the Rise of the Young-Old," *The Annals of the American Academy of Political and Social Science* 415 (1974), pp. 187–98.

16. Meredith Minkler, "Gold in Gray: Reflections on Business' Discovery of the Elderly Market." *The Gerontologist* 29:1 (1989), pp. 17–23 at p. 21.

17. S. Sontag, "The Double Standard of Aging," in P. Bart, ed., *No Longer Young: The Older Woman in America* (Detroit, Mich.: Wayne State Institute of Gerontology, 1975).

18. Edith M. Lederer, "Birth of a Controversy: Is New Mother Too Old?," *Philadelphia Inquirer,* December 29, 1993, p. 1.

19. Ibid.

20. Susan Chira, "Of a Certain Age, and in a Family Way," *New York Times,* January 2, 1994, p. 5.

21. Carroll Smith-Rosenberg, "Puberty to Menopause: The Cycle of Femininity in Nineteenth-Century America." *Feminist Studies,* 1 (1973), pp. 58–73 at p. 59.

22. See C. Smith-Rosenberg and C. Rosenberg, "The Female Animal: Medical and Biological Views of Woman and her Role in Nineteenth Century America," in *Women and Health in America,* ed. J. W. Leavitt (Madison: University of Wisconsin Press, 1984), pp. 12–27.

23. M. Anderson, *The Menopause* (London: Faber and Faber, 1983), p. 58.

24. *Feminine Forever* (New York: M. Evans and Co., 1966), pp. 97–98.

25. Ibid., p. 16.

26. Ibid., p. 44.

27. Arlie Russell Hochschild, *The Managed Heart: Commercialization of Human Feeling* (Berkeley: University of California Press, 1983), p. 180.

28. F. M. Deutsch, C. M. Zalenski, and M. E. Clark, "Is There a Double Standard of Aging?," *Journal of Applied Social Psychology* 16 (1986), pp. 771–85.

29. Satires, trans. by Niall Rudd (Oxford: Clarendon Press, 1991), p. 41.

30. *Witches, Devils and Doctors in the Renaissance,* ed. George Mora (Binghamton, N.Y.: Medieval and Renaissance Texts and Studies, 1991) , p. 498.

31. Lois W. Banner, *In Full Flower: Aging Women, Power, and Sexuality* (New York: Vintage Books, 1993), p. 171.

32. Boccaccio, *The Corbaccio,* trans. Anthony K. Cassell (Urbana: University of Illinois Press, 1975), p. 46; Desiderius Erasmus, *The Praise of Folly,* trans. Clarence H. Miller (New Haven, CT.: Yale University Press, 1979); pp. 48–49. With regard

to older men Erasmus also wrote: "It is my doing [Dame Folly] that you see everywhere men as old as Nestor . . . fall[ing] head over heels in love with some young girl and outdo[ing] any beardless youth in amorous idiocy" (p. 48).

33. Anne Stibbs, *A Woman's Place: Quotations About Women,* (New York: Avon Books, 1992), p. 333.

34. "Estranged Bedfellows," May 9, 1993, pp. 12–15 at p. 14.

35. Sharon Barman, "The Business of Beauty." *Working Woman* (August 1993), p. 50.

36. T. Virginia Atkins, Martha C. Jenkins, and Mishelle H. Perkins, "Portrayal of Persons in Television Commercials Age Fifty and Older." *Psychology* (1991), Vol. 27, No. 4 and Vol. 28, No. 1, pp. 30–37 at p. 30.

37. See also R. H. David and J. A. David, *TV's Image of the Elderly: A Practical Guide for Change* (Lexington, Mass.: Lexington, 1986).

38. R. Seidenburg, "Drug Advertising and Perceptions of Mental Illness," *Mental Hygiene* 55 (1971), p. 21; J. Prather and L. S. Fiddell, "Sex Differences in the Content and Style of Medical Advertisements," *Soc. Sci. Med.* 9. (1975), p. 23.

39. Selina Redman, Gloria R. Webb, Deborah J. Hennrikus, Jill J. Gordon, and Robert W. Sanson-Fisher, "The Effects of Gender on Diagnosis of Psychological Disturbance." *Journal of Behavioral Medicine* 14:5 (1991), pp. 527–40 at p. 527.

40. G. Klerman and M. Weissman, "Depressions Among Women," in ed. Juanita Williams, *Psychology of Women: Selected Readings* (New York: W. W. Norton and Co., 1985), pp. 484–513.

41. Robert N. Butler, op.cit., p. 994.

42. *Women and Minorities on Television: A Study in Casting and Fate.* A report to the Screen Actors Guild and the American Federation of Radio and Television Artists, June 1993, by George Gerbner, Annenberg School for Communication, University of Pennsylvania.

43. John Bell, "In Search of a Discourse on Aging: The Elderly on Television." *The Gerontologist* 32:3 (1992), pp. 305–11 at p. 309.

44. Ibid., p. 310.

45. B. S. Greenberg and D. D'Alessio, "Quantity and Quality of Sex in the Soaps." *Journal of Broadcasting and Electronic Media* (Summer 1985), 29, pp. 309–21.

46. Amy Dawes, "SAG: Women Shortshrifted." *Variety,* August 8, 1990, p. 3.

47. Robert N. Butler, op.cit., p. 993.

48. Cited by Germaine Greer, *The Change: Women, Aging and the Menopause* (New York: Fawcett Columbine, 1991), p. 315.

49. Jim Jerome, "Saved by Love," *Redbook* (February 1994), pp. 88–91, 138–40 at 140.

50. David Weaver, Roy W. Howard, and G. Cleveland Wilhoit, *The American Journalist in the 1990s.* The Freedom Forum World Center, Arlington, Va., November 17, 1992. Funded by The Freedom Forum.

51. Interview with Ruth Ashton Taylor by Shirley Biagi, *Women in Journalism,* oral history project of the Washington Press Club Foundation, January 11, 1992, p. 82, in the Oral History Collection of Columbia University and other collections.

52. "Briefs." *Media Report to Women,* (Fall 1993), Vol. 21, No. 4, p. 8.

53. Interview, February 9, 1994.

54. Interview, February 17, 1994.

55. Speech by Dr. Susan Love on "Changing the Face of American Politics."

Health Issues Affecting Older Women. July 10, 1993. National Women's Political Caucus. Los Angeles, Calif.

56. Simone de Beauvoir, *The Second Sex,* trans. and ed. H. M. Parshley (New York: The Modern Library, 1968; originally published 1952), p. 575.

57. *Force of Circumstance,* trans. R. Howard (New York: G. P. Putnam's Sons, 1965), pp. 655ff.

58. Muriel Oberleder, "Study Shows Mindset About Aging Influences Longevity," *Washington Post,* September 15, 1982. See also Gordon L. Bultena and Edward A. Powers, "Denial of Aging: Age Identification and Reference Group Orientations." *Journal of Gerontology* 33, 5 (1978), pp. 748–54.

59. Betty Friedan, *The Fountain of Age* (New York: Simon and Schuster, 1993), p. 66.

60. Germaine Greer, op.cit., p. 125.

61. Gail Sheehy, *The Silent Passage* (New York: Pocket Books, 1993), p. 9.

62. Cited by Pauline B. Bart and Marlyn Grossman, "Menopause," in *The Woman Patient: Medical and Psychological Interfaces,* eds. Malkah T. Notman and Carol C. Nadelson (New York: Plenum Press, 1978), Vol. 1, pp. 337–54 at p. 337.

63. This is the title of Germaine Greer's book (New York: Fawcett Columbine, 1992).

64. Gail Sheehy, op.cit.

65. Edited by Joan C. Callahan (Bloomington: Indiana University Press, 1993).

66. Ralph Linton, *The Study of Man* (New York: Appleton-Century-Crofts, 1936), p. 119.

67. Lois W. Banner, op.cit., p. 116.

68. See Ellen McGrath's 1990 report for the American Psychological Association titled *Women and Depression.*

69. John B. McKinlay, Sonja M. McKinlay, and Donald Brambilla, "The Relative Contributions of Endocrine Changes and Social Circumstances to Depression in Mid-Aged Women." *Journal of Health and Social Behavior* 28 (1987), pp. 345–63.

70. Paula Brown Doress and Diana Laskin Siegal, co-authors of *Ourselves, Growing Older* in Foreword to Ivan K. Strausz, M.D., *You Don't Need a Hysterectomy* (Reading, Mass.: Addison-Wesley, 1993), p. xi.

71. Ivan K. Strausz, *You Don't Need a Hysterectomy* (Reading, Mass.: Addison-Wesley, 1993), p. xv.

72. Judith Rodin and Ellen Langer, "Aging Labels: The Decline of Control and the Fall of Self-Esteem." *Journal of Social Issues,* 36 (1980), pp. 12–29 at p. 15.

73. Lynn Darling, "Age, Beauty and Truth," *New York Times,* January 23, 1994, pp. 1, 5 at p. 5.

74. Susan Antilla, "Workplace Discrimination? Don't Try It Around Her," *New York Times,* February 13, 1994, p. F7.

75. ABC's "Good Morning America," July 20, 1994. Transcript from tape.

76. September 8, 1977, 12.

77. M. B. Brewer, V. Dull, and L. Lui, "Perception of the Elderly: Stereotypes as Prototypes." *Journal of Personality and Social Psychology* 41 (1981), pp. 656–70.

78. Don Terry, "Swift Change of Image for Illinois Candidate," *New York Times,* March 21, 1994, p. A12.

79. Wendy Webster, *Not a Man to Match Her: The Marketing of a Prime Minister* (London: The Women's Press, 1990), pp. 4–136 at p. 117.

80. Ibid., p. 118.

81. Gerda Lerner, *The Grimke Sisters from South Carolina: Pioneers for Woman's Rights and Abolition* (New York: Schocken Books, 1971), p. 366.

82. Susan B. Anthony and Ida Harper, eds., *The History of Woman Suffrage*, Vol. IV (Rochester, N.Y.: Susan B. Anthony, 1902), pp. 350–51.

83. Lois W. Banner, op.cit., p. 282; Lois W. Banner, *Elizabeth Cady Stanton; A Radical for Woman's Rights* (Boston: Little, Brown, 1979), pp. 109–10.

84. Wendy Webster, op.cit., p. 95.

85. Helen E. Fisher, "Mighty Menopause," *New York Times*, October 21, 1992, p. 19.

86. S. M. Shappert, "Office Visits to Psychiatrists: United States, 1989–90." *Advanced Data*, No. 237 (Hyattsville, MD: National Center for Health Statistics, 1993).

87. Betty Friedan, op.cit., p. 70.

88. Gloria Steinem, *Revolution from Within* (Boston: Little, Brown, 1993), p. 248.

89. Patricia Aburdene and John Naisbitt, *Magatrends for Women* (New York: Villard Books, 1992), p. 160.

90. Sue V. Rosser, "Is There Androcentric Bias in Psychiatric Diagnosis?" *The Journal of Medicine and Philosophy*, 17,2 (April 1992), pp. 215–32 at p. 227.

91. Nicole Lurie, Jonathan Slater, Paul McGovern, Jacqueline Ekstrum, Lois Quam, and Karen Margolis, "Preventive Care for Women." *The New England Journal of Medicine* (August 12, 1993), pp. 478–82 at p. 478.

Chapter 8 Newsbinds

1. W. Speers, "A Blessing for the Movie Biz: Cardinal Rejects Censorship," *Philadelphia Inquirer*, October 2, 1992, p. C2.

2. *United Press International Stylebook* (New York: UPI, 1977), p. 48.

3. *The Associated Press Stylebook and Libel Manual* (New York: AP, 1980), p. 59.

4. James M. Perry and Jeffrey H. Birnbaum, "We the President," *The Wall Street Journal*, January 28, 1993, p. 1.

5. Ellen Goodman, "Cheers for the First Volunteer," *Philadelphia Inquirer*, February 27, 1993, p. A9.

6. Senator Barbara Mikulski, speech at the University of Pennsylvania, February 28, 1994.

7. *New York Times*, June 7, 1993, p. A15.

8. Robert B. Gunnison, "A Senate Sweep for Feinstein, Boxer 'Washington, Ready or Not, Here We Come,'" *San Francisco Chronicle*, November 4, 1992, p. Al.

9. Loe Cannon, "2 California Democrats Diverge in Senate Races, Feinstein and Boxer React to Foes' Campaigns," *Washington Post*, October 19, 1992, p. Al.

10. John Wildermuth and Harriet Chiang, *San Francisco Chronicle*, November 4, 1992, p. A3.

11. Susan Gilmore, "The Big Winners: Women. Murray's Win: Clincher Was Her Gender," *Seattle Times*, September 16, 1992, p. Al.

12. On the weekend of October 2–4, 1992, for example, *Women are Good News*, a San Francisco based media monitoring group, found that 83% of the guests on

the televised interview shows (e.g. *Crossfire, Evans and Novak, Meet the Press, Face the Nation*) were male. From: press release, October 1992.

13. *Women, Men and Media*, Washington, D.C., April 1993.

14. "Brokaw Zooms in Popularity Poll," *TV Guide*, March 26, 1994, p. 35.

15. The finding that men receive more obituaries than women is consistent with past research. Cf. Bernard Spilka, Gerald Lacey, and Barbara Gelb, "Sex Discrimination After Death: A Replication, Extension and a Difference." *Omega* (1979–80), Vol. 10 (3), pp. 227–33; R. Kastenbaum, S. Peyton, and B. Kastenbaum, "Sex Discrimination after Death." *Omega* 7 (1977), pp. 351–59.

16. Larry Rohter, "Tough 'Front-Line Warrior'," *New York Times*, February 12, 1993, pp. 1/A22 at p. A22.

17. "The Polish Thatcher," *New York Times*, June 10, 1993, p. A27. Things haven't changed much. Three and a half decades earlier Vijaya Lakshmi Pandit complained to *The Scotsman* (Glasgow, August 29, 1955), "When my public activities are reported, it is very annoying to read how I looked, if I smiled, if a particular reporter liked my hair style."

18. Eds. Gloria T. Hull, Patricia Bell Scott, Barbara Smith (Old Westburg, N.Y.: The Feminist Press, 1982).

19. P. 49.

20. P. 36.

21. These elections included: Ann Richards/Clayton Williams, Governor, Texas, 1990; Barbara Mikulski/Alan Keyes, Senate, Maryland, 1992; Carol Moseley Braun/Richard Williamson, Senate, Illinois, 1992; Gloria O'Dell/Bob Dole, Senate, Kansas, 1992; Patty Murray/Rod Chandler, Senate, Washington, 1992; Barbara Boxer/Bruce Herschensohn, Senate, California, 1992; Dianne Feinstein/John Seymour, Senate, California, 1992; Lynn Yeakel/Arlen Specter, Senate, Pennsylvania, 1992; Kay Bailey Hutchinson/Robert Krueger, Senate, Texas, 1993; and Kim Campbell/Jean Charest, Prime Minister, Canada, 1993.

22. "The 1992 Elections: Congress; New in the United States Senate," *New York Times*, November 5, 1992, p. B6.

23. Roberto Suro, "The 1990 Elections: Governor—Texas; Fierce Election for Governor Is Narrowly Won by Richards," *Washington Post*, November 7, 1990, p. 2.

24. Peter O'Neil, *The Vancouver Sun*, May 12, 1993, p. A7.

25. Dan Morain and Ralph Frammolino, "Feinstein, Boxer Appear to Lead in Income, Texas," *Los Angeles Times*, October 2, 1992, p. A3.

26. David Jackson and Sam Attlesey, "Krueger, Hutchison Trade Accusations over Crime Stance," *The Dallas Morning News*, May 13, 1993, p. A30.

27. Wanda Motley and Russell E. Eshleman, Jr., "Yeakel Works to Woo Black Voters; Specter Stumps Across Phila. A Baptist Minister Deserted Specter after 12 Years for Yeakel," *Philadelphia Inquirer*, October 12, 1992, p. B3.

28. "If It Were Mr. Baird," *New York Times* January 25, 1993, p. A17.

29. Tony Kornheiser, "Achieving Critical Mess," *Washington Post*, March 20, 1994, p. F5.

30. Jeffrey Schmalz, "Whatever Happened to AIDS?," *The New York Times Magazine*, November 28, 1993, pp. 56–60, 81, 85–86 at p. 58 and 81.

31. David Shribman, "Before Big Hair and Beauty Contests, Texas Women Got Their Nails Dirty," *The Houston Chronicle*, May 30, 1993, p. 4.

32. Jim Urban, "Folksy Mayor Ends Term in Pittsburg," *Philadelphia Inquirer* January 3, 1994, p. B3.

33. Larry Rohter, "Woman in the News: Clinton Picks Miami Woman, Veteran State Prosecuter, to be His Attorney General," *New York Times,* February 12, 1993, p. 1.

34. Robert Reinhold, "The 1992 Campaign: Senate Race; Republican Imperils Tide of Democrats in California," *New York Times,* October 24, 1992, p. A1.

35. Glenn F. Bunting, "Boxer a Risk-Taker Who Challenges Status Quo," *Los Angeles Times,* October 6, 1992, p. A1.

36. Mark Kennedy "How Campbell Relates to Others Is Crucial to Election Victory; But Winning the Confidence of Canadians Will be a Far Bigger Task," *The Gazette* (Montreal), Monday, June 14, 1993, p. A10.

37. P. D1.

38. June 14, 1993.

39. Ibid., p. D1.

40. William Hamilton, *Washington Post,* February 25, 1994, p. D1.

41. *Wichita Eagle,* October 22, 1992, p. 4D.

42. Matt Truell, *Wichita Eagle,* October 19, 1992, p. 1C.

43. Ken Herman and Mike Hailey, November 4, 1990, p. A1.

44. *Los Angeles Times,* October 20, 1992, p. 5.

45. Thomas Hardy and Steve Johnson, *Chicago Tribune,* October 23, 1992, p. 1.

46. Peter O'Neil, *The Vancouver Sun,* "Campbell Gloves-off Campaign a Softie," June 4, 1993, p. A1.

47. Douglas P. Shuit and Bill Stall, "Seymour and Feinstein Trade Biting Charges," *Los Angeles Times,* October 20, 1992, p. A3.

48. Wanda Motley, "Yeakel Starts Tour with a Giant Send-off," *Philadelphia Inquirer,* October 17, 1992, p. B3.

49. Mark Matassa, "Anything but a Sure Thing 'Year of the Woman' Turns into Tough Race for Murray," *Seattle Times,* October 27, 1992, p. A1.

50. In electoral simulations conducted in the 1970s, college students assumed that an equally qualified female candidate would lose to her male opponent. See Virginia Sapiro, "If U.S. Senator Baker Were a Woman: An Experimental Study of Candidate Image." *Political Psychology* 3 (Spring/Summer, 1982), pp. 61–83.

51. *Winning with Women,* op.cit., p. 7.

52. Lori Stahl, "Political Etiquette; Approach to Hutchinson a Sticky Issue for Krueger," *The Dallas Morning News,* May 16, 1993, p. A1.

53. Tracy Wilkinson, "Feinstein Tours Riot Area; Seymour presses for Debate," *Los Angeles Times,* September 9, 1992, p. A3.

54. "The Debate Nobody Watched," *Baltimore Evening Sun,* October 22, 1992, p. 15A.

55. Editorial: "The Shy Senator," *Baltimore Morning Sun,* October 6, 1992, p. A10.

56. Nathan Gorenstein, ". . . feisty, dynamic Yeakel," *Philadelphia Inquirer,* November 4, 1992. "Specter beats back Yeakel bid the two ran neck and neck most of the night, but the incumbent senator pulled ahead late. The heavy turnout and Clinton's victory were not enough to help Yeakel win," p. A1.

57. "The Debate Nobody Watched," op.cit., p. 15.

58. James S. Keat, "A Missed Chance for a Real Debate," *Baltimore Morning Sun,* October 17, 1992, p. 8A.

59. Thomas Hardy, "Braun Denies Wrongdoing in Medicaid Flap," *Chicago Tribune,* September 30, 1992, p. 1.

60. Quoted by Linda Witt, Karen M. Paget, Glenna Matthews, *Running as a Woman: Gender and Power in American Politics* (New York: The Free Press, 1994), p. 237.

61. Eric Pianin and E. J. Dionne, Jr., "Party Loyalty Appears to Spare Many Democrat Incumbents," *Washington Post,* November 4, 1992, p. A30.

62. Nathan Gorenstein, *Philadelphia Inquirer,* November 4, 1992, p. Al.

63. Ken Herman, "Shake Hands with Governor Richards, Williams Concedes Tough Loss Come-From-Behind Win for Democrat," *The Houston Post,* November 7, 1990, p. Al.

64. Thomas Hardy, "Election '92—Clinton Elected President, Carol Moseley Braun Sweeps to Historic Senate Victory," *Chicago Tribune,* November 4, 1992, p. 1.

65. Mike Royko, "Braun's Troubles Should Benefit Us," North Sports Final edition, *Chicago Tribune,* October 7, 1992, p. 3.

66. Lori Rodriguez, "Democrats Give Senate Race Away," *The Houston Chronicle,* May 8, 1993, p. A25.

67. Philip Seib, "Hutchinson Will Find That Middle of the Road Is Dangerous Place," *The Dallas Morning News,* June 7, 1993, p. A19.

68. Michael A. Messner, Margaret Carlisle Duncan, and Kerry Jensen, "Separating the Men from the Girls: The Gendered Language of Televised Sports," *Gender and Society* (March 1993), Vol. 7, No. 1, pp. 121–37 at p. 130.

69. Carol Goar, "Contradiction Lies at Restless Heart of Campbell Campaign," *The Toronto Star,* May 27, 1993, p. A19.

70. Felicity Barringer with Michael Wines, "The Accused in the Palm Beach Case: Quiet, Different and Somewhat Aloof," *New York Times,* May 11, 1991, p. 6.

71. For an account, see Nan Robertson, *The Girls in the Balcony* (New York: Random House, 1992), pp. 242 ff.

72. Ibid., p. 245.

73. Julia M. Klein, "Anna Quindlen Turns the Page," *Philadelphia Inquirer,* September 19, 1994, p. E3.

74. Quoted by Robertson, op.cit., pp. 218–19.

75. "The Full Story of Cianfrani and the Reporter," October 16, 1977, p. 1F.

76. Ibid., p. 2F.

77. Ibid., p. 3F.

78. Ibid., p. 4F.

79. Interview, February 25, 1994.

80. Judy Oppenheimer, "Should Reporters Sleep with Sources?," *Village Voice,* December 5, 1977, p. 11.

81. Ibid., p. 11.

82. Laura Foreman, "My Side of the Story," *The Washington Monthly* (May 1978), pp. 49–54 at p. 50.

83. Claudia Dreifus, "Cokie Roberts, Nina Totenberg and Linda Wertheimer," *New York Times Magazine,* January 2, 1994, pp. 14–17 at p. 17.

84. Nan Robertson, op.cit., p. 239.

85. Ibid., p. 236.

86. Karen Schmidt and Colleen Collins, "Showdown at Gender Gap," *American Journalism Review* (July/August 1993), pp. 39–42 at p. 39.

Chapter 9 The Stories We Tell

1. "Appendix" to Elizabeth Cady Stanton's *The Woman's Bible* (Boston: Northeastern University Press, 1993), pp. 215–17.

2. Samuel G. Freedman, "Of History and Politics: Bitter Feminist Debate," *New York Times,* June 6, 1986, pp. B1/B4 at B1.

3. Ibid., p. B1.

4. Samuel G. Freedman, op.cit., p. B4.

5. It is in Rosenberg's offer of proof that we find the sorts of qualified statements that ordinarily characterize scholarship. In cross-examination, Kessler-Harris's arguments were undercut precisely because she had overgeneralized from the evidence she offered and because her testimony was inconsistent with what she had written as a scholar.

As she herself admitted, "To refute Rosenberg's argument, I found myself constructing a rebuttal in which subtlety and nuance were omitted, and in which evidence was marshalled to make a point while complexities and exceptions vanished from sight." Alice Kessler-Harris, "Equal Employment Opportunity Commission v. Sears, Roebuck and Company: A Personal Account," *Feminist Review* 25 (1987), p. 62.

6. Thomas Haskell and Sanford Levinson, "Academic Freedom and Expert Witnessing: Historians and the *Sears* case." *Texas Law Revies* 66 (1988), pp. 1629–59 at p. 1630.

7. Alice Kessler-Harris, op.cit., pp. 46–69 at p. 47.

8. Ibid., p. 64.

9. (Boulder: Westview Press, 1993).

10. Carol Sternhell, "Life in the Mainstream: What Happens When Feminists Turn Up on Both Sides of the Courtroom?" *Ms.* (July 1986), pp. 48–51, 86–89 at p. 88.

11. Ibid.

12. (Chapel Hill: University of North Carolina Press, 1992), p. 243.

13. June Gallop, Marianne Hirsch, and Nancy K. Miller, "Criticizing Feminist Criticism," in *Conflicts in Feminism,* Eds. Marianne Hirsch and Evelyn Fox Keller (New York: Routledge, 1990), pp. 349–69, at p. 361.

14. CF. Wendy W. Williams, "Equality's Riddle: Pregnancy and the Equal Treatment/Special Treatment Debate." *Review of Law and Social Change,* XIII (1984–85), pp. 325–80.

15. Naomi Wolf, *The Beauty Myth* (New York: W. Morrow, 1991), p. 182.

16. Christina Hoff Sommers, *Who Stole Feminism? How Women Have Betrayed Women* (New York: Simon and Schuster, 1994), p. 12.

17. "Who Stole Feminism," *Book World* (August 28, 1994), p. 10.

18. Christina Hoff Sommers, op.cit., p. 205. Sir William Blackstone, *Commentaries on the Laws of England in Four Books* (Philadelphia: Geo. T. Bisel, 1922), p. 417.

19. From *Man Cannot Speak For Her,* Vol. II, compiled by Karlyn Kohrs Campbell (New York: Greenwood Press, 1989). Carrie Chapman Catt, Presidential Address, 1902, to national convention of NAWSA.

20. "Marie Curie May Be Reburied," *New York Times,* March 9, 1994, p. B4.

21. Gloria Steinem, *Moving Beyond Words* (New York: Simon and Schuster, 1994).

22. Susan Campbell, "Part Hunk and Part Bimbo—Call Him a Himbo," *Philadelphia Inquirer,* June 16, 1994, p. G3.

23. Timothy Phelps and Helen Winternitz, *Capitol Games* (New York: Hyperion, 1992), p. 411.

24. "Hymn to the Ideal Woman," *New York Times,* June 18, 1994, p. 13.

25. Statement before the Senate Judiciary Committee, March 9, 1993. Transcript from videotape.

26. Excerpts from Senate hearings on Ginsburg nomination to Supreme Court. *New York Times,* July 21, 1993.

27. Sam Howe Verhovek, "From Demure Survivor to G.O.P. Star," *New York Times,* June 7, 1993, p. A15.

28. Londa Schiebinger, *The Mind Has No Sex? Women in the Origins of Modern Science* (Cambridge: Harvard University Press, 1989), pp. 10–11.

29. Ibid., p. 26.

30. Sharon Bertsch McGrayne, *Nobel Prize Women in Science, Their Lives, Struggles and Momentous Discoveries* (New York: Carol Publishing, 1993), p. 3.

31. Ibid., p. 37.

32. Malcolm Gladwell, "The Healy Experiment," *The Washington Post Magazine,* June 21, 1992, p. 11.

33. Susan Swan, "Women on the Verge." *Mirabella* (August 1993), p. 71.

34. Amanda Spake, "Thirty Years at the White House," *The Washington Post Magazine,* October 21, 1990, p. 19.

35. Ibid.

36. Interview with Kit Seelye on June 2, 1993.

37. Marjorie Williams and Al Kamen, "Woman of the Hour," *The Washington Post Magazine,* June 11, 1989, p. 25.

38. David Margolick, "Judge Ginburg's Life: A Trial by Adversity," *New York Times,* June 25, 1993, pp.1 and A19.

39. Ibid.

40. Ibid.

41. Ibid.

42. Jeffrey Rosen, "The Book of Ruth," *The New Republic* (August 2, 1993), p. 19.

43. Dianne Feinstein, "A Woman for All Reasons," San Diego Women's Leadership Council, April 22, 1992.

44. Interview with Elizabeth H. Rutter, June 30, 1993.

45. Sharon Bertsch McGrayne, op.cit., p. 323.

46. Ibid., p. 331.

47. Ibid., p. 330.

48. Ibid., p. 316.

49. Fawn Germer, "Stripper Outrages Lawmaker," *Rocky Mountain News,* June 3, 1993, p. 5A.

50. Fawn Germer, "'I Expected to be Subjected to Ridicule, Lawmaker Says," *Rocky Mountain News,* June 4, 1993, p. 23A.

51. Charles Truehart, "Fuming Feminists Light Senator's Fuse," *Washington Post,* September 20, 1990, p. D1.

52. Ira Berkow, "Sports of the Times: Barfield Is More than a Ballplayer," *New York Times,* July 20, 1989, p. 9.

53. Kermit Lipez, "Disrobing Gender Bias in the Courts," *Legal Times,* June 28, 1993, p. 22.

54. The Congresswoman tells the story at the National Women's Political Caucus and it is retold in the August 1993 issue of *Vogue*. But there Woolsey is reported to have told the Congressman, "I suppose you think that's a compliment." We reconfigure our recollections of experience as we integrate those experiences into our accounts of ourselves.

55. Sheila Weller, "America's Most Sexist Judges.' *Redbook* (February 1994), pp. 83–87 at p. 85.

56. "Developing Your Personal Image," National Women's Political Caucus, July 8, 1993, Los Angeles.

57. Sharon Bertsch McGrayne, op.cit., p. 352.

58. Loretta M. Itri, M.D., NWPC, Plenary Session, July 10, 1993.

59. The story is also told in Marjorie Margolies-Mezvinsky, *A Woman's Place: The Freshmen Who Changed the Face of Congress* (New York: Crown Publishers, 1994), p. 41.

60. Interview, National Women's Political Caucus, July 1993, Los Angeles.

61. Molly Moore, *A Woman At War: Storming Kuwait with the U.S. Marines* (New York: Charles Schribner's Sons, 1993), p. 73.

62. "Business Meetings: When to Butt In." *Working Woman* (December 1992), p. 20.

63. Anne Groer, "Women & Power & Politics." *Self* (March 1993), p. 148.

64. Telephone interview August 12, 1993.

65. Susan Chira, "Ginsburg's Spirit is Echoed by Other Pioneers," *New York Times,* August 2, 1993, p. A16.

66. Ibid.

67. Ibid.

68. Elaine Shannon, "The Unshakable Janet Reno," *Vogue,* August 1993, p. 260.

69. *Congressional Record* (Washington, D.C.: U.S. Government Printing Office, Tuesday, March 30, 1993), Vol. 139, No. 42 S3979.

70. Interview with Debbie Stabenow at National Women's Political Caucus Convention, July 8–11, 1993, Los Angeles.

71. Susan Faludi, *Backlash* (New York: Crown, 1991), p. 437.

72. Ibid., p. 451.

73. Johnson is treated on p. 440, American Cyanamid on pp. 440–53.

74. 499 U.S. 187.

75. *UAW v. Johnson Controls, Inc.,* 886 F.2d 871, 920 (7th Cir. 1989), reversed (1991) 499 U.S. 187.

76. For a subtle and significant discussion of the issues that remain unresolved in fetal rights cases, see Cynthia R. Daniels, "Competing Gender Paradigms: Gender Difference, Fetal Rights and the Case of Johnson Controls." *Policy Studies Review* (Winter 1991/1992), vol. 10, pp. 51–68; Cynthia R. Daniels, *At Women's Expense: State Power and the Politics of Fetal Rights* (Cambridge, Mass.: Harvard University Press, 1993), pp. 57–95.

Bibliography

ABA Commission Summary. Chicago: American Bar Association, 1988.

Abcarian, Robin. "Oven Mitts Are Off in Clash of the First Ladies." *Los Angeles Times,* July 24, 1992, 1E.

Aburdene, Patricia, and John Naisbitt. *Megatrends for Women.* New York: Villard Books, 1992.

Abzug, Bella. *Gender Gap.* Boston: Houghton Mifflin, 1984.

Action, William. *The Functions and Disorders of the Reproductive Organs.* Philadelphia: P. Blakiston Son, and Co., 1895.

Addams, Jane. *Jane Addams: A Centennial Reader.* New York: Macmillan, 1960.

Addison, Joseph. *Spectator.* Ed. George A. Aitkin. London: John C. Nimmo, 1898.

Africa, Martha Fay. "Forum on Hiring." *Lawyer Hiring and Training Report* 13, no. 5, May 1993.

American Psychological Association. Brief for *Amicus Curiae* in Support of Respondent. *Price Waterhouse v. Ann B. Hopkins.* June 18, 1988.

Anderson, Paul. *Janet Reno: Doing the Right Thing.* New York: John Wiley & Sons, 1994.

Angier, Natalie. "Radical New View of Role of Menstruation." *New York Times,* September 21, 1993, C1, C10.

Anthony, Susan B., and Ida Harper, eds. *The History of Woman Suffrage.* Rochester, NY: Susan B. Anthony, 1902.

Antilla, Susan. "Workplace Discrimination? Don't Try It Around Her." *New York Times,* February 13, 1994, F7.

Anderson, M. *The Menopause.* London: Faber and Faber, 1983.

Apple, R. W. Jr. "No Choice But to Confront the Economy." *New York Times,* August 21, 1992, A1.

Arkin, W., and L. R. Dobrofsky. "Military Socialization and Masculinity." *Journal of Social Issues* 34 (1978): 151–68.

Associated Press Stylebook and Libel Manual. New York: AP, 1980.

"At last, Justice for the Baton Twirler: Feminism Came Late to Ginsburg. Behind Her Firm Belief Is a Careful Approach." *Philadelphia Inquirer,* July 4, 1993, D1.

Atkins, T. Virginia, Martha C. Jenkins, and Mishelle H. Perkins. "Portrayal of Persons in Television Commercials Age Fifty and Older." *Psychology* 27, and 28 (1991): 30–37.

Baker, Elizabeth Faulkner. *Protective Labor Legislation: With Special Reference to Women in the State of New York*. 1925. Reprint, New York: Columbia University's Studies in History, Economics, and Public Law, AMS Press, 1969.

Ball, Karen. "Debate Erupts Into Vicious Feud Before Midwestern Votes." *AP*, May 16, 1992, 15:07 EST.

Baltes, P., and K. Schaie. "The Myth of the Twilight Years." *Psychology Today*, March 1974, 37.

Balz, Dan. "Clinton's Wife Finds She's Become Issue: Arkansas Lawyer Denies Impropriety but Vows to Rethink Her Role." *Washington Post*, March 17, 1992, 1.

―――. "Values Night Looks Like Family Feud." *Washington Post*, August 20, 1992, A1.

Balz, Dan, and David Broder. "Fighting Doubts and Each Other." *Washington Post*, March 22, 1992, 1.

Balzar, John. "Hillary's Role Resurrects an Old Arkansas Dilemma." *Los Angeles Times*, October 4, 1992, 26.

Banner, Lois W. *Elizabeth Cady Stanton: A Radical for Woman's Rights*. Boston: Little, Brown, 1979.

―――. *In Full Flower: Aging Women, Power, and Sexuality*. New York: Vintage Books, 1993.

Barbieri, Susan M. "Office Politics in the Nicer '90s." *Working Woman*, August 1993, 36.

Barman, Sharon. "The Business of Beauty." *Working Woman*, August 1993, 50.

Barnett, Rosalind C., and Grace K. Baruch. "Correlates of Fathers' Participation in Family Work." In *Fatherhood Today: Men's Changing Role in the Family*, edited by Phyllis Bronstein and Carolyn P. Cowen. New York: John Wiley & Sons, 1988.

Barringer, Felicity with Michael Wines. "The Accused in the Palm Beach Case: Quiet, Different and Somewhat Aloof." *New York Times*, May 11, 1991, 6.

Bart, Pauline B., and Marilyn Grossman. "Menopause." In *The Woman Patient: Medical and Psychological Interfaces*. Vol. 1, edited by Malkah T. Notman and Carol C. Nadelson, 337–54. New York: Plenum Press, 1978.

Bartlett, Donald, and James Steele. "The Full Story of Cianfrani and the Reporter." *Philadelphia Inquirer*, October 16, 1977, F1.

Bassuk, Ellen. "The Rest Cure: Repetition or Resolution of Victorian Women's Conflict." In *The Female Body*, edited by Susan Rubin Suleiman. Cambridge: Harvard University Press, 1986.

Bateson, Gregory, Don D. Jackson, Jay Haley, and John Weakland. "Toward a Theory of Schizophrenia." *Behavioral Science* 1 (1956): 251–64.

Baxter, Sandra, and Marjorie Lansing. *Women and Politics*. Ann Arbor: University of Michigan Press, 1981.

Beard, Mary R. *Women as a Force in History*. New York: The Macmillian Co., 1946.

Beck, Joan. "Other Women Shine as Hillary Becomes a Stepford Mother." *Chicago Tribune*, July 16, 1992, C27.

Bell, John. "In Search of a Discourse on Aging: The Elderly on Television." *The Gerontologist* 32 (1992): 305–11.

Bem, Sandra Lipsitz. *The Lenses of Gender*. New Haven: Yale University Press, 1993.

―――. "Probing the Promise of Androgyny." In *The Psychology of Women*, edited by Mary Roth Walsh, 206–25. New Haven: Yale University Press, 1987.

Berg, Christine, and Philippa Berry. "Spiritual Whoredom: An Essay on Female Prophets in the Seventeenth Century." In *Feminist Literary Theory: A Reader*, edited by Mary Eagleton. Oxford: Basil Blackwell, 1986.

Bergmann, Barbara R. *The Economic Emergence of Women*. New York: Basic Books, 1986.

Berkow, Ira. "Sports of the Times: Barfield is More Than a Ballplayer." *New York Times*, July 20, 1989, 9.

Bernard, Jesse. "Talk, Conversation, Listening, Silence." *The Sex Game*. New York: Atheneum, 1972.

Bettelheim, Bruno. "The Problem of Generations." *Daedalus* 91 (winter 1962): 68–96.

———. *The Uses of Enchantment*. New York: Knopf, 1976.

Binstock, R. H. "The Aged as Scapegoat." *The Gerontologist* 23 (1983): 136–43.

Birch, Thomas, ed. *The Works of Mrs. Catherine Cockburn, Theological, Moral, Dramatic, and Poetical with an Account of the Life of the Author*. London: J. and P. Knapstoa, 1751.

Blackstone, William. *Commentaries on the Laws of England*. 1765. Reprint, Philadelphia: George T. Bisel Co., 1922.

Blumstein, Philip, and Pepper Schwartz. *American Couples: Money, Work, and Sex*. New York: William Morrow, 1983.

Bobo, Jacqueline. "The Color Purple: Black Women as Cultural Readers." In *Female Spectators: Looking at Film and Television*, edited by E. Deidre Pribram, 90–109. London: Verso, 1988.

Boccaccio, Giovanni. *The Corbaccio*. Translated by Anthony K. Cassell. Urbana: University of Illinois Press, 1975.

Bogan, Meg. *The Women Troubadours*. New York: W. W. Norton, 1976.

Borgatta, Edgar F., and J. Stimson. "Sex Differences in Interaction Characteristics." *Journal of Social Psychology* 60 (1963): 89–100.

Boxer, Barbara with Nicole Boxer. *Stranger in the Senate: Politics and The New Revolution of Women in America*. Washington, DC: National Press Books, 1994.

Bradley, P. H. "The Folk-Linguistics of Women's Speech: An Empirical Examination." *Communication Monographs* 48 (1981): 73–90.

Bradstreet, Anne. *The Tenth Muse. In The Works of Anne Bradstreet*. Edited by Jeannine Hensley. Cambridge: Harvard University Press, 1967.

Bradwell v. Illinois, 83 U.S. (16 Wall) 130 (1872).

Brewer, M. B., V. Dull, and L. Lui. "Perceptions of the Elderly: Stereotypes as Prototypes." *Journal of Personality and Social Psychology* 41 (1981): 656–70.

Bridges A., and H. Hartman. "Pedagogy by the Oppressed." *Review of Radical Political Economics* 6 (1975): 75–79.

"Briefs." *Media Report to Women* 21 (fall 1993): 8.

Brock, David. "The Real Anita Hill." *The American Spectator*, March 1992, 18–30.

Broder, John. "First Lady Takes the Gloves Off." *Los Angeles Times*, August 19, 1992, A1.

Brodie, Janet Farrell. *Contraception and Abortion in 19th Century America*. Ithaca and London: Cornell University Press, 1994.

"Brokow Zooms in Popularity Poll." *TV Guide*, March 26, 1994, 35.

Bromberg-Ross, JoAnn. "Storying and Changing: An Examination of the Consciousness-Raising Process." *Folklore Feminists Communication* 6 (1975): 9–11.

Broner, E. M. "Hymn to the Ideal Woman." *New York Times*, June 18, 1994, 13.

Broughton, John M. "Women's Rationality and Men's Virtues: A Critique of Gender Dualism in Gilligan's Theory of Moral Development." *Social Research* (autumn 1983): 597–642.

Broverman, Inge K., Donald M. Broverman, et al. "Sex-Role Stereotypes and Clinical Judgments of Mental Health." *Journal of Consulting and Clinical Psychology* 3 (1970): 1–7.

Brownmiller, Susan. *Femininity*. New York: Fawcett Columbine, 1984.

Brownstein, Ronald. "Age Issue Would Likely Shape Bush-Clinton Debate." *Los Angeles Times,* April 24, 1992, 1A.

Bultena, Gordon L., and Edward A. Powers. "Denial of Aging: Age Identification and Reference Group Orientations." *Journal of Gerontology* 3 (1978): 748–54.

Bunting, Glenn F. "Boxer a Risk-Taker Who Challenges Status Quo." *Los Angeles Times,* October 6, 1992, A1.

Burgoon, Michael, J. P. Dillard, and N. E. Doran. "Friendly or Unfriendly Persuasion." *Human Communication Research* 10 (1983): 283–94.

Burros, Marion. "Even Women at Top Still Have Floors to Do." *New York Times,* May 31, 1993, 11.

Burrows, George Man. *Commentaries on the Causes, Forms, Symptoms, and Treatment, Moral and Medical, of Insanity.* 1828. Reprint, New York: Arno Press, 1976.

Burwell v. Eastern Airlines, 633 F2d 361 (4th Cir. 1980).

"Business Meetings: When to Butt In." *Working Woman,* December 1992, 20.

Butler, Benjamin F. *Butler's Book.* Boston: A. M. Thayer and Co., 1892.

Butler, Robert N. "Care of the Aged in the United States of America." In *Textbook of Geriatric Medicine and Gerontology.* 4th ed., edited by J. C. Brocklehurst, R. C. Tallis, and H. M. Fillit, 993–99. Edinburgh: Churchill Livingstone, 1992.

Byars, Jackie. "Gazes/Voices/Power: Expanding Psychoanalysis for Feminist Film and Television Theory." In *Female Spectators: Looking at Film and Television,* edited by E. Deidre Pribram, 110–31. London: Verso, 1988.

Callahan, Joan C. *Menopause: A Midlife Passage.* Bloomington: Indiana University Press, 1993.

Cameron, Deborah, Fiona McAlindin, and Kathy O'Leary. "Lakoff in Context: The Social and Linguistic Functions of Tag Questions." In *Women's Studies: Essential Readings,* edited by Stevi Jackson, 421–26. New York: New York University Press, 1993.

"Campaign's New Rules for Wives of Hopefuls," *Newsday,* March 24, 1992, 86.

Campbell, Angus et al. *The American Voter.* New York: Wiley, 1960.

Campbell, Karlyn K. "Elizabeth Cady Stanton." In *American Orators Before 1900: Critical Studies and Sources,* edited by H. R. Ryan, 340–49. New York: Greenwood Press, 1987.

———. "Hearing Women's Voices." *Communication Education* 40 (1991): 33–48.

———. *Man Cannot Speak for Her.* Vols. 1–2. New York: Greenwood Press, 1989.

———. "The Rhetoric of Women's Liberation: An Oxymoron." *Quarterly Journal of Speech* 59 (1973): 74–86.

———. "What Really Distinguishes and/or Ought to Distinguish Feminist Scholarship in Communication Studies?" *Women's Studies in Communication* 11 (1988): 4–5.

Campbell, Karlyn K., and E. C. Jerry, "Woman and Speaker: A Conflict in Roles." In *Seeing Female: Social Roles and Personal Lives,* edited by S. Brehm, 123–33. New York: Greenwood Press, 1988.

Campbell, Susan. "Part Hunk and Part Bimbo—Call Him a Himbo." *Philadelphia Inquirer,* June 16, 1994, G3.

Cannon, Angie. "Clinton Has Named Women to 40% of Top Government Jobs." *Philadelphia Inquirer,* September 28, 1993, A7.

Cannon, Lou. "2 California Democrates Diverge in Senate Races, Feinstein and Boxer React to Foes' Campaigns." *Washington Post,* October 19, 1992, A1.

Cannon v. University of Chicago, 441 U.S. 677 (1957).

Cantor, Dorothy W., and Toni Bernay with Jean Stoess. *Women in Power.* Boston: Houghton Mifflin, 1992.

Caplan, Paula J. "Don't Blame Mother." *Ms.,* September/October 1993, 96.

Carlson, Margaret. "A Hundred Days of Hillary." *Vanity Fair,* June 1993, 169.

Carter, Rosalyn. *First Lady from Plains.* Boston: Houghton Mifflin, 1984.

Chavez, Anna Maria. "Turn-Off for the 90s: Educated and Career-Minded Young Women." *Los Angeles Times,* September 13, 1992, 3M.

Childs, James. "Rebirth." In *The Hollywood Screenwriters,* edited by Richard Corliss. New York: Discus Books, 1972.

Chira, Susan. "Ginsburg's Spirit Is Echoed by Other Pioneers." *New York Times,* August 2, 1993, A16.

———. "Of a Certain Age, and in a Family Way." *New York Times,* January 2, 1994, 5.

Chodorow, Nancy. *The Reproduction of Mothering.* Berkeley: University of California Press, 1978.

Chrysostom, Saint [John]. "Homily IX." In *A Select Library of The Nicene and Post-Nicene Fathers,* edited by Philip Schaff, 436. XIII. New York: The Christian Lit. Co., 1889.

Civil Rights Act of 1964, Title VII, 42 U.S.C. at 2000e et seq.

Clarke, Edward H., M.D. *Sex in Education; or, A Fair Chance for the Girls.* 1873. Reprint, New York: Arno Press and The New York Times, 1972.

Cleveland Board of Education v. LaFleur, 414 U.S. 632 (1974).

Clift, Eleanor. "Not the Year of the Women." *Newsweek,* October 25, 1993, 31.

Clymer, Adam. "Hillary Clinton, on Capitol Hill, Wins Raves, If Not a Health Plan." *New York Times,* September 29, 1993, A1.

CNN/Time Survey. *Time.* March 9, 1992, 54.

CNN/*USA Today.* National Survey, September 29–30, 1993.

Coates, Jennifer. "Gossip Revisited." In *Women's Studies,* edited by Stevi Jackson, 426–29. New York: New York University Press, 1993.

Cohen v. Chesterfield County School Board, 474 U.S. 395 (1973).

Condit, Celeste. "What Makes Our Scholarship Feminist? A Radical Liberal View." *Women's Studies in Communication* 11 (1988): 6–8.

Condit, Celeste, and John Louis Lucaites. *Crafting Equality: America's Anglo-African Word.* Chicago: University of Chicago Press, 1993.

Condit v. United Air Lines, Inc., 558 F2d 1176 (4th Cir. 1977).

"Confidence in Institutions." National Opinion Research Center Survey. Survey conducted by *The American Enterprise,* November/December 1993.

Conrad, Charles. "Agon and Rhetorical Form: The Essence of 'Old Feminist' Rhetoric." *Central States Speech Journal* 32 (1981): 45–53.

———. "The Transformation of the 'Old' Feminist Movement." *Quarterly Journal of Speech* 67 (1981): 284–97.

Constantini, Edward, and Kenneth H. Craik. "The Social Background, Personality and Political Careers of Female Party Leaders." *Journal of Social Issues* 28 (1972): 235.

Cook, Blanche Wiesen. *Eleanor Roosevelt*. New York: Penguin, 1992.

Cook, Fay Lomax. "Ageism: Rhetoric and Reality." *The Gerontologist* 32 (1992): 292–95.

Costrich Norma, Joan Feinstein, Louise Kidder, Jeanne Marecek, and Linda Pascale. "When Stereotypes Hurt: Three Studies of Penalties for Sex-Role Reversals." *Journal of Experimental and Social Psychology* 11 (1975): 520.

Crocker, Jennifer, and Kathleen M. McGraw. "What's Good for the Goose Is Not Good for the Gander: Solo Status as an Obstacle to Occupational Achievement for Males and Females." *American Behavioral Scientist* 27 (January/February 1984): 357–70.

Crockett, W. H., and M. L. Humbert. "Perceptions of Aging and the Elderly." In *Annual Review of Gerontology and Geriatrics*. Vol. 7, edited by C. Eisdorfer. New York: Springer, 1987.

Daniels, Cynthia R. *At Women's Expense: State Power and the Politics of Fetal Rights*. Cambridge: Harvard University Press, 1993.

———. "Competing Gender Paradigms: Gender Difference, Fetal Rights and the Case of Johnson Controls." *Policy Studies Review* 10 (winter 1991/92): 51–68.

Darling, Lynn, "Age, Beauty and Truth." *New York Times,* January 23, 1994, 1, 5.

Darnton, John. "With a Mace, Madam Rules a Most Unruly House." *New York Times,* September 27, 1993, A4.

David, R. H., and J. A. David. *TV's Image of the Elderly: A Practical Guide for Change*. Lexington, MA: Lexington Books, 1986.

Dawes, Amy. "SAG: Women Shortshrifted." *Variety,* August 8, 1990, 3.

"The Debate Nobody Watched." *Baltimore Evening Sun,* October 22, 1992, A15.

deBeauvoir, Simone. *The Coming of Age*. New York: G. P. Putnam's Sons, 1972.

———. *Force of Circumstance*. Translated by R. Howard. New York: G. P. Putnam's Sons, 1965.

———. *The Second Sex*. Translated and edited by H. M. Parshley. 1952. Reprint, New York: Vintage Books, 1989.

deBeauvoir, Simone, and Betty Friedan. "Sex, Society, and the Female Dilemma: A Dialogue Between Simone deBeauvoir and Betty Friedan." *Saturday Review,* June 14, 1975, 18.

Degler, Carl N. *At Odds: Women and the Family in America from the Revolution to the Present*. New York: Oxford University Press, 1980.

———. *In Search of Human Nature: The Decline and Revival of Darwinism in American Social Thought*. New York: Oxford University Press, 1991.

Demos, John Putnam. *Entertaining Satan*. New York: Oxford University Press, 1982.

Deutsch, F. M., C. M. Zalenski, and M. E. Clark. "Is There a Double Standard of Aging," *Journal of Applied Social Psychology* 16 (1986): 771–85.

Diaz v. Pan American World Airways, 442 F2d 385 (1971).

Donovan, Josephine. "The Silence is Broken." In *The Feminist Critique of Language,* edited by Deborah Cameron, 41–56. London: Routledge, 1990.

Doress, Paula Brown, and Diana Laskin Siegel. Foreword to *You Don't Need a Hysterectomy,* by Ivan K. Strausz, M.D. Reading, MA: Addison-Wesley, 1993.

Dorthard v. Rawlinson, 433 U.S. 321 (1977).

Dow, Bonnie J. "The 'Womanhood' Rationale in the Woman Suffrage Rhetoric of Frances E. Willard." *Southern Communication Journal* 56 (1991): 298–307.

Dowd, Maureen, "The Faces Behind the Face That Masks Clinton's Smile." *New York Times,* October 25, 1992, 1–1.

———. "Hillary Rodham Clinton Strikes a New Pose and Multiplies Her Images." *New York Times,* December 12, 1993, 3E.

———. "The 1992 Campaign Trail: From Nixon, Predictions on the Presidential Race." *New York Times,* February 6, 1992, 18A.

Dreifus, Claudia. "Cokie Roberts, Nina Totenberg and Linda Wertheimer." *New York Times Magazine,* January 2, 1994, 14–17.

Drevenstedt, J. "Perceptions of Onsets of Young Adulthood, Middle Age, and Old Age." *Journal of Gerontology* 31 (1976): 53–57.

DuBois, Ellen et al. "Feminist Discourse." *Buffalo Law Review,* 34 (1985): 68.

Edmunds, Lavinia. "Barbara Mikulski." *Ms.,* January 1987, 63.

Edsall, Thomas B. "G.O.P. Plans a 'Family Values' Offensive Pitch to White Middle Class, Casts Bush as Defender of Social Norms." *Washington Post,* August 19, 1992, A1.

Edwards, Tryon, ed. *The New Dictionary of Thoughts.* New York: Standard Book Co., 1966.

EEOC v. Delta Air Lines, Inc., 441 F. Supp. 626 (S.D. Tex. 1977).

EEOC v. Sears, Roebuck & Co., 628 F. Supp. 1264 (1986).

Eisenstadt v. Baird, 405 U.S. 438 (1972).

Eisner, Jane R. "Hillary's Work on Health Care Has Silenced All Those Bashers." *Philadelphia Inquirer,* September 24, 1993, A23.

"Elephant in the Tent: Who's a Republican?" *Los Angeles Times,* August 19, 1992, B6.

Elshtain, Jean Bethke. *Public Man, Private Woman: Women in Social and Political Thought.* Princeton, NJ: Princeton University Press, 1981.

Ely, Robin J. "Organizational Demographics and Women's Gender Identity at Work." Working Paper, John F. Kennedy School of Government, Harvard University, 1992.

Epstein, Cynthia Fuchs. *Women in Law.* Urbana and Chicago: University of Illinois, 1993.

Erasmus, Desiderius. *The Praise of Folly.* Translated by Clarence H. Miller. New Haven: Yale University Press, 1979.

Evans, Sara. *Personal Politics: The Roots of Women's Liberation in the Civil Rights Movement & the New Left.* New York: Vintage Books, 1979.

"Excerpts from an Interview with Clinton After the Air Strikes." *New York Times,* January 14, 1993, A10.

"Excerpts from Senate Hearings on Ginsburg Nomination to Supreme Court." *New York Times,* July 21, 1993, A12.

Exline, R., D. Gray, and D. Schutte. "Visual Behavior in a Dyad as Affected by Interview Control and Sex of Respondent." *Journal of Personality and Social Psychology* 1 (1965): 201–9.

"Factsheet on Women's Political Progress." Washington, DC: National Women's Political Caucus, June, 1993.

Faludi, Susan. *Backlash: The Undeclared War Against American Women.* New York: Crown, 1991.

Feinstein, Dianne. "A Woman for All Reasons." San Diego Women's Leadership Council, San Diego, April 22, 1992.

Fenelon, Abbe. *Women's Realities, Women's Choices.* Hunter College Women's Studies Collective. New York: Oxford University Press, 1983.

Feshbacker, D. "Sex Differences in Empathy and Social Behavior in Children." In *The Development of Prosocial Behavior,* edited by N. Eisenberg, 315–38. New York: Academic Press, 1982.

Fineman, Martha Albertson. *The Illusion of Equality: The Rhetoric and Reality of Divorce Reform.* Chicago: University of Chicago Press, 1991.

Finley, Lucinda M. "Transcending Equality Theory: A Way Out of the Maternity and the Workplace Debate." *Columbia Law Review* 86 (October 1986): 1135.

Fisher, Helen E. "Mighty Menopause." *New York Times,* October 21, 1992, A23.

Fiske, T. Susan, and Steven T. Neuberg. "A Continuum of Impression Formation from Category-Based to Individuating Processes: Influences of Information and Motivation on Attention and Interpretation." In *Advances in Experimental Social Psychology* 23, San Diego: Academic Press, 1990.

Foreman, Laura. "My Side of the Story," *The Washington Monthly,* May 1978, 49–54.

Foss, K. A., & S. K. Foss. *Women Speak: The Eloquence of Women's Lives.* Prospect Heights, IL: Waveland, 1991.

Fox-Genovese, Elizabeth. *A Critique of Individualism.* Chapel Hill: University of North Carolina Press, 1992.

Frasca, Kathy. "Issues Forum: 'The Religious Right,' A View From the Field," National Women's Political Caucus, Los Angeles, July 9, 1993.

Fraser, Antonia. *The Weaker Vessel.* New York: Knopf, 1984.

Freedman, Samuel. "Of History and Politics: Bitter Feminist Debate." *New York Times,* June 6, 1986, B1, B4.

Friedan, Betty. *The Fountain of Age.* New York: Simon and Schuster, 1993.

———. *The Second Stage.* New York: Dell, 1981.

Friedman, Jane M. *America's First Woman Lawyer: The Biography of Myra Bradwell.* Buffalo, NY: Prometheus Books, 1993.

Friedman, Sally. "Keeping a Close Watch on First Ladies." *New York Times,* November 1, 1992, 13NJ, 3.

Frontiero v. Richardson, 411 U.S. 677, 684 (1973).

Gallop, June, Marianne Hirsch, and Nancy K. Miller. "Criticizing Feminist Criticism." In *Conflicts in Feminism,* edited by Marianne Hirsch and Evelyn Fox Keller, 349–69. New York: Routledge, 1990.

Gardiner, Judith Kegan. "On Female Identity and Writing by Women." In *Writing and Sexual Difference,* edited by Elizabeth Abel, 177–91. Chicago: University of Chicago Press, 1982.

Garrard, Mary D. "Artemesia and Susanna." In *Feminism and Art History,* edited by Norma Broude and Mary D. Garrard. New York: Harper & Row, 1982.

Garrow, David J. "Justice Souter Emerges." *New York Times Magazine,* September 25, 1994, 39.

Gelb, Joyce, and Marian Lief Palley. *Women and Public Policies.* Rev. ed. Princeton, NJ: Princeton University Press, 1987.

Gelder, Lindsy Van. "Countdown to Motherhood, When Should You Have a Baby?" *Ms.,* December 1986, 37.

Gelpi, B. C. "The Politics of Androgyny." *Woman's Studies* 2 (1974): 151–60.

"Gender and the 1993 Gubernatorial Campaigns." Sponsored by the National Women's Political Caucus, National Press Club, November 10, 1993.

General Electric v. Gilbert, 429 U.S. 125 (1976).

Gerbner, George. "Women and Minorities on Television: A Study in Casting and

Fate." Report to the Screen Actors Guild and the American Federation of Radio and Television Artists, June 1993.

Germer, Fawn. "'I Expected to be Subjected to Ridicule,' Lawmaker Says," *Rocky Mountain News,* June 4, 1993, 23A.

———. "Stripper Outrages Lawmaker." *Rocky Mountain News,* June 3, 1993, 5A.

Gilley, H. M., and C. S. Summers. "Sex Differences in Use of Hostile Verbs." *Journal of Psychology* 76 (1970): 33–37.

Gilligan, Carol. "Feminist Discourse." *Buffalo Law Review* 36 (1985): 39.

———. *In a Different Voice: Psychological Theory and Women's Development.* Cambridge: Harvard University Press, 1982.

Gilman, Charlotte Perkins. *The Living of Charlotte Perkins Gilman. An Autobiography.* New York: D. Appleton-Century Co., 1935.

———. *The Man-Made World: Or, Our Androcentric Culture.* 1911. New York: Johnson Reprint, 1971.

———. *Women and Economics: A Study of the Economic Relation Between Men and Women as a Factor in Social Evolution.* 1898. Reprint, New York: Source Book Press, 1970.

Gilmore, Susan. "The Big Winners: Women, Murray's Win: Clincher Was Her Gender." *Seattle Times,* September 16, 1992, A1.

Ginsburg, The Honorable Ruth Bader. "American University Commencement Address," May 10, 1981. In *The American University Law Review* (Summer 1981): 891–902.

Gladwell, Malcolm. "The Healy Experiment." *Washington Post Magazine,* June 21, 1992, W9.

Gleick, Elizabeth. "General Janny Baby," *People,* March 29, 1993, 40–41.

Goar, Carol. "Contradiction Lies at Restless Heart of Campbell Campaign." *Toronto Star,* May 27, 1993, A19.

Goldmark, Josephine. "The World's Experience Upon Which Legislation Limiting the Hours of Labor for Women Is Based." In Goldmark, *Fatigue and Efficiency: A Study in Industry.* New York: Charities Publication Committee, 1912.

Gompers, Samuel. "Woman's Work, Rights and Progress." *American Federationist* 20 (August 1913): 625.

Goodman, Ellen. "Cheers for the First Volunteer." *Philadelphia Inquirer,* February 27, 1993, A9.

Grace, Stephanie. "Hillary Clinton Stresses Children's Issues." *Los Angeles Times,* July 16, 1992, A6.

Gray, Jerry. "And Now, Whoppers vs. Waffles." *New York Times,* October 7, 1993, B8.

Greenberg, B. S., and D. D'Alessio. "Quantity and Quality of Sex in the Soaps." *Journal of Broadcasting and Electronic Media* 29 (summer 1985): 309–21.

Greenberg, Stan, James Carville and Frank Greer. "The General Election Project—Interim Report." April 27, 1994. A Memo.

Greene, Chris. "Women: The Emotional or the Expressive Sex?" *Psychology Today,* June 1986, 12.

Greer, Germaine. *The Change: Women, Aging and the Menopause.* New York: Fawcett Columbine, 1991.

Grimke, Sarah Moore. *Letters on the Equality of the Sexes and the Condition of Woman, Addressed to Mary Parker, President of the Boston Female Anti-Slavery Society.* Boston: Isaac Knapp, 1838.

Griswold v. Connecticut, 381 U.S. 479 (1965).

Groer, Anne. "Women & Power & Politics." *Self,* March 1993, 148.

Gross, Jane. "Big Grocery Chain Reaches Landmark Sex-Bias Accord." *New York Times,* December 17, 1993, 1, B10.

Grove City College v. Bell, 465 U.S. 555 (1984).

Grove, Lloyd. "The Limited Life of a Political Wife." *Washington Post,* March 19, 1992, C1.

———. "The Short List for Spouses." *Washington Post,* July 8, 1992, C1.

Guazzo, Steeven. *The Civile Conversation of M. Steeven Guazzo.* Translated by George Pettie. 1581. Reprint, London: Constable, 1925.

Gunnison, Robert B. "A Senate Sweep for Feinstein, Boxer 'Washington, Ready or Not, Here We Come,'" *San Francisco Chronicle,* November 4, 1992, A1.

Hall, Judith A. *Nonverbal Sex Differences.* Baltimore: Johns Hopkins University Press, 1984.

Hamilton, William. "California's Warm Brown." *Washington Post,* February 25, 1994, D1.

Harding, Susan. "Women and Words in a Spanish Village." In *Toward an Anthropology of Women,* edited by Rayne Reiter. New York: Monthly Review Press, 1975.

Hardy, Thomas. "Braun Denies Wrongdoing in Medicaid Flap." *Chicago Tribune,* September 30, 1992, 1.

———. "Election '92—Clinton Elected President, Carol Moseley Braun Sweeps to Historic Senate Victory." *Chicago Tribune,* November 4, 1992, 1.

Hardy, Thomas, and Steve Johnson. "Debate Has Familiar Ring, Williamson Chides Braun, But Neither Can Draw Blood." *Chicago Tribune,* October 23, 1992, 1.

Harrington, Walt. "A Woman Between Two Worlds." *Washington Post Magazine,* July 9, 1989, 35.

Harris, Barbara J. *Beyond Her Sphere: Women and the Professions in American History.* Westport, CT: Greenwood Press, 1942.

Harris, Louis and Associates. *Survey of Legislators and Regulators, 1988. Table 9–1.* In Robert Blendon and Tracey Stelzer Hyams eds. *Reforming the System* New York: Faulkner and Gray, 1992.

Harris v. Forklift Systems, Inc., 114 S. Ct. 367 (1993).

Harrison, Brian. *Separate Spheres: The Opposition to Women's Suffrage in Britain.* New York: Holmes & Meier Publishers, 1978.

Harrison, Colin. "Here's Baby. Dad Stays Home. Dad Gets Antsy." *New York Times,* August 31, 1993, A17.

Haskell, Thomas, and Sanford Levinson. "Academic Freedom and Expert Witnessing: Historians and the *Sears* Case." *Texas Law Review* 66 (1988): 1629–59.

Hayes v. Shelby Memorial Hospital, 726 F2d 1543 (11th Cir. 1984).

Hedges, Elaine, and Shelly Fisher Fishkin, eds. *Listening to Silences: New Essays in Feminist Criticism.* New York: Oxford University Press, 1994.

Heerboth, Joel R., and Nervella V. Ramanaiah. "Evaluation of the BSRI Masculine and Feminine Items Using Desirability and Stereotype Ratings." *Journal of Personality Assessement* 49 (1985): 264–70.

Heilman, Madeline E. "The Impact of Situational Factors on Personnel Decisions Concerning Women: Varying the Sex Composition of the Applicant Pool." *Organizational Behavior and Human Performance* 26 (December 1980): 386–95.

Heilman, Madeline E., and Melanie H. Stopeck. "Attractiveness and Corporate

Success: Different Causal Attributes for Males and Females." *Journal of Applied Psychology* 70 (May 1985): 379–88.

Henderson, Katherine Usher, and Barbara F. McManus. *Half Humankind.* Urbana and Chicago: University of Illinois Press, 1985.

Herman, Ken. "Shake Hands with Governor Richards, Williams Concedes Tough Loss Come-From-Behind Win for Democrat." *Houston Post,* November 7, 1990, A1.

Herman, Ken, and Mike Hailey. "Richards Spars with Williams Over Tax Issue . . ." *Houston Post,* November 4, 1990, A1.

Herschberger, Ruth. *Adam's Rib.* 1948. Reprint, New York: Harper & Row, 1970.

Hill, George Birkbeck. *Johnsonian Miscellanies.* Vol 2. 1784. Reprint, New York: Barnes and Noble, 1966.

"The History of Susanna." *The Apocrypha.* New York: Tudor Publishing Co., 1937.

Hochschild, Arlie Russell. *The Managed Heart: Commercialization of Human Feeling.* Berkeley: University of California Press, 1983.

———. *The Second Shift: Working Parents and Revolution.* New York: Viking Books, 1989.

Holloway, Marguerite. "A Lab of Her Own." *Scientific American,* November 1993, 94–103.

Hopkins v. Price Waterhouse, 825 F2d 458 (1987).

Hopkins v. Price Waterhouse, 737 F. Supp. 1202 (D.D.C.) affirmed 920 F2d 967 (D.C. Cir. 1990).

Howell, Deborah. "True Confessions—My Life as a White Male." *Media Studies Journal* (winter/spring 1993): 197–203.

Hoyt v. Florida, 368 U.S., 61 (1961).

Hull, Gloria T., Patricia Bell Scott, and Barbara Smith, eds. *All the Women Are White, All the Blacks Are Men, But Some of Us Are Brave.* Old Westbury, NY: Feminist Press, 1982.

Hymowitz, Carol, and Michaela Weissman. *A History of Women in America.* New York: Bantam, 1978.

Ifill, Gwen. Election Debriefing. Annenberg School, University of Pennsylvania, December 1992.

———. "Role in Health Expands Hillary Clinton's Power." *New York Times,* September 22, 1993, A24.

Infante, Dominick A. "Motivation to Speak on a Controversial Topic." *Central States Speech Journal* 34 (1983): 96–103.

Itri, Loretta M. Plenary Session I. National Women's Political Caucus, July 10, 1993.

Jackson, David, and Sam Attlesey. "Krueger, Hutchison Trade Accusations Over Crime Stance." *Dallas Morning News,* May 13, 1993, A30.

Jamieson, Kathleen Hall. *Eloquence in an Electronic Age.* New York: Oxford University Press, 1988.

Japp, Phyllis. "Esther or Isaiah?: The Abolitionist-Feminist Rhetoric of Angelina Grimke." *Quarterly Journal of Speech* 71 (1985): 335–48.

J.E.B. v. Alabama ex. rel. T.B., 511 U.S. (1994).

Jeffreys, Margot. "The Elderly in Society." In *Textbook of Geriatric Medicine and Gerontology.* 4th ed., edited by J. C. Brocklehurst, R. C. Tallis, and H. M. Fillot, 971–79. Edinburgh: Churchill Livingstone, 1992.

Jerome, Jim. "Saved by Love." *Redbook,* February 1994, 88–91, 138–40.

Jespersen, Otto. *Language: Its Nature, Development and Origin*. London: Allen and Unwin, 1922.

Jones, Landon Y. "The Governor: On His Music, His Marriage and His Future." In "Road Warriors." *People*, July 20, 1992, 68ff.

Judis, John B. "Voice from the Right." *Washington Post*, Book World, September 5, 1992, x2.

Juvenal. *Satires*. Vol. 6. Translated by Niall Rudd. Oxford: Clarendon Press, 1991.

Kalcik, Susan. "'. . . Like Ann's Gynecologist or the Time I Was Almost Raped.' Personal Narratives in Women's Rap Groups." *Journal of American Folklore* 88 (1975): 3–11.

Kaminer, Wendy. "Feminism's Identity Crisis." *The Atlantic* 272 (October 1993): 51–68.

Kanter, Rosabeth Moss. *Men and Women of the Corporation*. New York: Basic Books, 1977.

Kass, John, and Elaine Popovich. "Democratic Convention '92: As Women Step Up, Hillary Steps Back." *Chicago Tribune*, July 14, 1992, C1.

Kastenbaum R., S. Peyton, and B. Kastenbaum. "Sex Discrimination After Death." *Omega* 7 (1977): 351–59.

Kay, Herma Hill, ed. *Sex-Based Discrimination: Text, Cases and Materials*. St. Paul, MN: West Publishing Co., 1988.

Keen, Judy. "Leading the Crusade: First Lady Takes Role as Saleswoman." *USA Today*, September 28, 1993. 1A.

Kennedy, Florynce. *Color Me Flo*. Englewood Cliffs, NJ: Prentice Hall, 1976.

Kennedy, Mark. "How Campbell Relates to Others Is Crucial to Election Victory; But Winning the Confidence of Canadians Will Be a Far Bigger Task." *The Gazette* (Montreal), June 14, 1993, 10.

Kerber, Linda K. *Women of the Republic: Intellect and Ideology in Revolutionary America*. Chapel Hill: University of North Carolina Press, 1980.

Kerber, Linda K., Catherine G. Green, Eleanor Maccoby, Zella Luria, Carol B. Stack, and Carol Gilligan. "On *In a Different Voice*: An Interdisciplinary Forum." *Signs* 11 (winter 1986): 304–33.

Kessler-Harris, Alice. "Equal Employment Opportunity Commission v. Sears, Roebuck and Company: A Personal Account." *Feminist Review* 25 (1987): 46–69.

———. *Out of Work: A History of Wage-Earning Women in the United States*. New York: Oxford University Press, 1982.

———. *A Woman's Wage: Historical Meanings and Social Consequences*. Lexington: University of Kentucky Press, 1990.

———. "Written Textimony of Alice Kessler-Harris, EEOC v. Sears" (No. 79-C-4373).

Kierkegaard, Soren. *The Concept of Anxiety*. Princeton, NJ: Princeton University Press, 1980.

King, Colbert I. "Once This Was the Party of Frederick Douglass." *Washington Post*, August 19, 1992, A19.

King, Larry. "With Hillary on the Home Front." *USA Today*, October 4, 1993, 1A.

Kinnear, Mary. *Daughters of Time: Women in the Western Tradition*. Ann Arbor: University of Michigan Press, 1982.

Klein, Dianne. "Republicans Say She's a Radical." *Los Angeles Times*, August 25, 1992, 1E.

Klein, Edward. "Masako's Sacrifice." *Vanity Fair*, June 1993, 76.

Klein, Julia M. "Anna Quindlen Turns the Page." *Philadelphia Inquirer,* September 19, 1994, E3.

Klerman, G., and M. Weissman. "Depressions Among Women." In *Psychology of Women: Selected Readings,* edited by Juanita Williams. New York: W. W. Norton, 1985.

Kornheiser, Tony. "Achieving Critical Mess." *Washington Post,* March 20, 1994, F5.

Kramarae, Cheris. *Women and Men Speaking: Frameworks for Analysis.* Rowley, MA: Newbury House, 1981.

Kuhn, Maggie. Congressional Committee on Aging, September 8, 1977.

Kurtz, Howard. "TV Coverage Ignores G.O.P. Script: Viewers Miss Fiery Speech." *Washington Post,* August 18, 1992, A15.

Lakoff, Robin T. *Language and Woman's Place.* New York: Harper & Row, 1975.

Lamb, Michael E. "The Changing Role of Fathers." In *The Father's Role: Applied Perspectives,* edited by Michael E. Lamb, 9ff. New York: John Wiley & Sons, 1986.

Lambert, Bruce. "Richard Salant, 78, Who Headed CBC News in Expansion, Is Dead." *New York Times,* February 17, 1993, B8.

Landsbaum, Mark. "First Lady Backs Sen. Seymour." *Los Angeles Times,* March 25, 1992, 7B.

Landsberg, Mitchell. "Reno Is Strictly No Frills." *Philadelphia Inquirer,* February 14, 1993, A12.

Lederer, Edith M. "Birth of a Controversy: Is New Mother Too Old?" *Philadelphia Inquirer,* December 29, 1993, 1.

"Leers, Smears, and Governor Clinton." Editorial, *New York Times,* January 28, 1992, A20.

Lehrman, Karen. "The Feminist Mystique." *New Republic,* March 16, 1992, 31.

Lerner, Gerda. *The Creation of Feminist Consciousness: From the Middle Ages to Eighteen-seventy.* New York: Oxford University Press, 1993.

———. *The Grimke Sisters from South Carolina: Pioneers for Woman's Rights and Abolition.* New York: Schocken Books, 1971.

———, ed. *The Female Experience: An American Documentary.* Indianapolis: Bobbs-Merrill Educational Pub., 1977.

Levin, Michael. "Exalted by Their Inferiority: Arguments Against the Female Franchise." In *Research in Social Movements, Conflicts and Change,* 13, (1991): 199—220.

Levine, Susanne Braun. "News-speak and 'Genderlect'—(It's Only News If You Can Sell It)." *Media Studies Journal* 7 (1993): 118.

Lewin, Tamar. "Feminists Wonder If It Was Progress to Become 'Victims'." *New York Times,* May 10, 1992, E6.

Lewis, Anthony. "Abroad at Home; Merchants of Hate." *New York Times,* August 21, 1992, A25.

———. "If It Were Mr. Baird." *New York Times,* January 25, 1993, A17.

Linton, Ralph. *The Study of Man.* New York: Appleton-Century-Crofts, 1936.

Lipez, Kermit. "Disrobing Gender Bias in the Courts." *Legal Times,* June 28, 1993, 22.

Lockheed, M. E., and K. P. Hall. "Conceptualizing Sex as a Status Characteristic: Applications to Leadership Training Strategies." *Journal of Social Issues* 32 (1976): 111–24.

London Observer. "Indira Gandhi's Guiding Star." November 4, 1984, 10.

Lott, Bernice. "The Devaluation of Women's Competence." In *Seldom Seen, Rarely Heard: Women's Place in Psychology,* edited by Janis S. Bohan, 171–91. Boulder, CO: Westview Press, 1992.

Love, Susan. "Changing the Face of American Politics." Speech, National Women's Political Caucus, Los Angeles, July 10, 1993.

Lurie, Nicole, Jonathan Slater, Paul McGovern, Jacqueline Ekstrum, Lois Quam, and Karen Margolis. "Preventive Care for Women." *New England Journal of Medicine* (August 12, 1993): 478–82.

Lyndon, Christopher. "Role of Women Sparks Debate by Congresswoman and Doctor." *New York Times,* July 26, 1970, 35.

Mabee, Carleton. *Sojourner Truth: Slave, Prophet, Legend.* New York: New York University Press, 1993.

Macartney, C. A. *Maria Theresa and the House of Austria.* London: English University Press, 1969.

MacDonald, M. L. "Assertion Training for Women." In *Social Skills Training,* edited by J. P. Curran and P. M. Monti. New York: Guilford, 1981.

MacKinnon, Catherine. "Feminist Discourse." *Buffalo Law Review* 34 (1985): 74.

Maclennan v. American Airlines, Inc., 440 F. Supp. 466 (E.D. Va. 1977).

Madigan, Charles. "Reagan to Nation: Trust Bush, Republicans Begin with Bow to Right." *Chicago Tribune,* August 18, 1992, C1.

Maraniss, David. "Clinton Camp Sizes Up the Rhetoric." *Washington Post,* August 19, 1992, A26.

Marcus, Ruth. "Republicans Aim Barbs at Hillary Clinton." *Washington Post,* August 19, 1992, A21.

Margolick, David. "Judge Ginsburg's Life a Trial by Adversity." *New York Times,* June 25, 1993, 1, A19.

Margolies-Mezvinsky, Marjorie. *A Woman's Place: The Freshmen Who Changed the Face of Congress.* New York: Crown, 1994.

"Marie Curie May Be Reburied." *New York Times,* March 9, 1994, B4.

Marshall, Susan E. "In Defense of Separate Spheres: Class and Status Politics in the Antisuffrage Movement." *Social Forces* 65 (December 2, 1986): 334.

Martin, Antoinette. "Governor Who?" *Detroit Free Press Magazine,* July 4, 1993, 6–17.

Martin, Emily. "The Egg and the Sperm: How Science Has Constructed a Romance Based on Stereotypical Male-Female Roles." *Signs: Journal of Women in Culture and Society* 16 (1991): 485–501.

———. *The Woman in the Body: A Cultural Analysis of Reproduction.* Boston: Beacon Press, 1987.

Marvin, Carolyn. "The Body of the Text: Literacy's Corporeal Constant." *Quarterly Journal of Speech* 80 (May 1994): 129–49.

Matassa, Mark. "Anything But a Sure Thing 'Year of the Woman' Turns Into Tough Race for Murray." *Seattle Times,* October 27, 1992, A1.

Mathews, Donald, and Jane Sherron De Hart. *Sex, Gender and the Politics of ERA.* New York: Oxford University Press, 1990.

McGrath, Ellen. *Women and Depression.* Report for the American Psychological Association, Washington, D.C., 1990.

McGrayne, Sharon Bertsch. *Nobel Prize Women in Science, Their Lives, Struggles and Momentous Discoveries.* New York: Carol Publishing, 1993.

McGrory, Mary. "Putting Women to Work." *Washington Post,* August 21, 1992, A2.

McKinlay, John B., Sonja M. McKinlay, and Donald Brambilla. "The Relative Contributions of Endocrine Changes and Social Circumstances to Depression in Mid-Aged Women." *Journal of Health and Social Behavior* 28 (1987): 345–63.

"The McLaughlin Group." PBS, October 1, 1993.

McPhee, Carol, and Ann Fitzgerald. *Feminist Quotations.* New York: Thomas Y. Crowell, 1979.

Meeker, B. F., and P. A. Weitzel-O'Neill. "Sex Roles and Interpersonal Behavior in Task-Oriented Groups." *American Sociological Review* 42 (1977): 91–105.

Meir, Golda. *My Life.* New York: G. P. Putnam's Sons, 1975.

Menkel-Meadow, Carrie. "Portia in a Different Voice: Speculations on a Women's Lawyering Process." *Berkeley Women's Law Journal* (fall 1985): 39–63.

Meritor Savings Bank v. Vinson, 477 U.S. 57 (1986).

Merton, Robert K. "The Self-Fulfilling Prophecy." *The Antioch Review* 8 (June 1948): 193–210.

Messner, Michael A., Margaret Carlisle Duncan, and Kerry Jensen. "Separating the Men from the Girls: The Gendered Language of Televised Sports." *Gender and Society* 7 (March 1993): 121–37.

Meyerowitz, Joanne. *Women Adrift: Independent Wage Earners in Chicago, 1880–1930.* Chicago: University of Chicago Press, 1988.

Middleton, Faith. "She's Prepared." *Ms.,* November 1986, 29.

Mikulski, Barbara. "Women and Leadership." Speech at the University of Pennslyvania, February 28, 1994.

Miles, Margaret R. *Carnal Knowing: Female Nakedness and Religious Meaning in the Christian West.* Boston: Beacon Press, 1989.

Mill, John Stuart. *The Subjection of Women.* London: Longman, Green, Reader, and Dyer, 1869.

Miller, Alan. "Marilyn Quayle Has Unconventional Style." *Los Angeles Times,* August 19, 1992, A4.

Mills, Kay. *A Place in the News: From the Women's Pages to the Front Page.* New York: Dodd, Mead, 1988.

Minkler, Meredith. "Gold in Gray: Reflections on Business' Discovery of the Elderly Market." *The Gerontologist* 29 (1989): 17–23.

Mitchell-Kernan, Claudia. "Signifying." In *Mother Wit from the Laughing Barrell,* edited by Alan Dundes. Englewood Cliffs, NJ: Prentice-Hall, 1973.

Moore, Carey A. "Susanna: A Case of Sexual Harassment in Ancient Babylon." *Bible Review* (June 1992): 21–29, 52.

Moore, Molly. *A Woman At War: Storming Kuwait with the U.S. Marines.* New York: Charles Scribner's Sons, 1993.

Morain, Dan, and Ralph Frammolino. "Feinstein, Boxer Appear to Lead in Income, Taxes." *Los Angeles Times,* October 2, 1992, A3.

Morgan, Edward S. *The Puritan Family: Religion and Domestic Relations in Seventeenth Century New England.* New York: Harper & Row, 1966.

Morgan, Robin. *The Word of a Woman: Feminist Dispatches 1968–1992.* New York: W. W. Norton, 1993.

Morris, Celia. "Changing the Rules and the Roles: Five Women in Public Office." In *The American Woman 1992–93,* edited by Paula Ries and Anne Stone, 95–126. New York: W. W. Norton, 1992.

Morrison, Patt. "Time for a Feminist First Lady?" *Los Angeles Times,* July 14, 1992, A1.

Mosbacher, Georgette. *Feminine Force: Release the Power Within to Create the Life You Deserve.* New York: Simon and Schuster, 1993.

"A Mother's Work Never Seems to End." *Wall Street Journal,* July 29, 1994, B1.

Motley, Wanda. "Yeakel Starts Tour with a Giant Send-off." *Philadelphia Inquirer,* October 17, 1992, B3.

Motley, Wanda, and Russell E. Eshleman, Jr. "Yeakel Works to Woo Black Voters; Specter Stumps Across Phila. A Baptist Minister Deserted Specter After 12 Years for Yeakel." *Philadelphia Inquirer,* October 12, 1992, B3.

Mueller, Carol. "Nurturance and Mastery: Competing Qualifications for Women's Access to High Public Office." In *Research in Politics and Society.* Vol. 2, edited by Gwenn Moore and Glenna Spitze, 211–32. Greenwich, CT: JAI Press, 1986.

Muller v. Oregon, 208 U.S. 412, 422, (1908).

Munda, Constantia. "The Worming of a Mad Dog." 1617. Reprinted in *Half Humankind: Contexts and Texts of the Controversy about Women in England, 1540–1640,* edited by Katherine Usher Henderson and Barbara F. McManus. Urbana: University of Illinois Press, 1985.

Myrdal, Alva, and Viola Klein. *Women's Two Roles: Home and Work.* London: Routledge & Kegan Paul, 1968.

Nasch, Norma. "The Emerging Legal History of Women in the United States: Property, Divorce, and the Constitution." *Signs* 12 (autumn 1986): 112–15.

Nation, Cary A. *The Use and Need of the Life of Cary A. Nation.* Topeka: F. M. Steves & Sons, 1905.

The Nation. cxx: 3111. February 18, 1925, 173.

Neugarten, B. L. "Age Groups in American Society and the Rise of the Young-Old." *Annals of the American Academy of Political and Social Science* 415 (1974): 187–98.

The New Diversity: Women and Minorities on Corporate Boards. Chicago: Henrick and Struggles, 1993.

Newman, Jody. "Perception and Reality: A Study Comparing the Success of Men and Women Candidates." Washington, DC: National Women's Political Caucus, 1994.

Nichols, Nancy A. "Whatever Happened to Rosie the Riveter?" *Harvard Business Review* (July/August 1993): 60.

"The 1992 Elections: Congress; New in the United States Senate." *New York Times,* November 5, 1992, B6.

Noble, Barbara Presley. "Little Discord on Harassment Ruling." *New York Times,* November 14, 1993, 25.

Oberleder, Muriel. "Study Shows Mindset About Aging Influences Longevity." *Washington Post,* September 15, 1982, B5.

"O'Dell Fires a Barrage at Dole." *Wichita Eagle,* October 22, 1992, D4.

O'Faolain, Julia, and Lauro Martines, eds. *Not in Gods' Image.* New York: Harper Torchbooks, 1973.

Oil, Chemical and Atomic Workers v. American Cyanamid Co., 741 F2d 444 (D.C. Cir. 1984).

Okin, Susan Moller. *Women in Western Political Thought.* Princeton, NJ: Princeton University Press, 1979.

Olive, David, ed. *Political Babble.* New York: John Wiley & Sons, 1992.

Olsen, Frances. "The Sex of Law." Unpublished. As cited in Christine A. Littleton, "Reconstructing Sexual Equality." *California Law Review* (July 1987) Vol. 75, No. 4, 1332.

O'Malley, Kathey, and Dorothy Collin. "O'Malley and Collin." *Chicago Tribune,* July 15, 1992, C18.

O'Neill, June, and Solomon Polachek. "Why the Gender Gap in Wages Narrowed in the 1980s." *Journal of Labor Economics* 11, Pt. 1 (January 1993): 205–28.

O'Neil, Peter. "Campbell Gloves-Off Campaign a Softie." *Vancouver Sun,* June 4, 1993, A1.

———. "Campbell Sees Double Standard in Media Spin on Her Campaign." *Vancouver Sun,* May 12, 1993, A7.

Oppenheimer, Judy. "Should Reporters Sleep with Sources?" *Village Voice,* December 5, 1977, 11.

Orr v. Orr, 440 U.S. 268 (1979).

Packard, Elizabeth Parsons Ware. *Modern Persecution or Insane Asylums Unveiled as Demonstrated by the Report of the Investigating Committee of the Legislature of Illinois.* Vol. 1. Hartford, CT: Case, Lakewood and Brainard, 1875.

Paglia, Camille. *Sex, Art, and American Culture.* New York: Vintage Books, 1992.

Palmore, E. B. *Ageism: Negative and Positive.* New York: Springer, 1990.

Paludi, Michele A., and Lisa A. Strayer. "What's in an Author's Name? Differential Evaluations of Performance as a Function of Author's Name." *Sex Roles* 12 (1985): 353–61.

Pardes, Ilana. *Countertraditions in the Bible: A Feminist Approach.* Cambridge: Harvard University Press, 1992.

Parker, Pat. *Movement in Black.* Ithaca, NY: Firebrand Books, 1989.

Parker, Rozsika, and Griselda Pollock. *Old Mistresses: Women, Art and Ideology.* New York: Pantheon Books, 1981.

Parsons, Talcott, and Robert F. Bales. *Family, Socialization and Interaction Process.* Glencoe, IL: Free Press, 1955.

Path, Vicki. "Developing Your Personal Image." Speech, National Women's Political Caucus, Los Angeles, July 8, 1993.

Perry, James M., and Jeffrey H. Birnbaum. "We the President." *Wall Street Journal,* January 28, 1993, 1.

Peterson, Houston, ed. *A Treasury of the World's Great Speeches.* New York: Simon and Schuster, 1954.

Phelps, Timothy M., and Helen Winternitz. *Capitol Games.* New York: Harper Perennial, 1992.

Pianin, Eric, and E. J. Dionne, Jr. "Party Loyalty Appears to Spare Many Democratic Incumbents." *Washington Post,* November 4, 1992, A30.

Pogrebin, Letty Cottin. "Hillary Clinton and the Year of the Woman," *El Nuevo Herald,* June 16, 1992, 7A.

Pomeroy, Sarah B. *Goddesses, Whores, Wives, and Slaves.* New York: Schocken Books, 1975.

Povich, Elaine S. "Millions Left Out of GOP Pitch to Moms at Home." *Chicago Tribune,* August 23, 1992, 1D.

Powell, Gary N., and D. Anthony Butterfield. "If 'Good Managers' Are Masculine, What Are 'Bad Managers'?" *Sex Role* 10 (1984): 477–84.

Prather, J., and L. S. Fiddell. "Sex Differences in the Content and Style of Medical Advertisements." *Social Science & Medicine* 9 (1975): 23.

Pregnancy Discrimination Act of 1978, 42 U.S.C. at 2000e (k).

Pribram, E. Deidre, ed. *Female Spectators: Looking at Film and Television*. London: Verso, 1988.

Price Waterhouse v. Hopkins, 490 U.S. 228 (1989).

Profet, Margie. "Menstruation as a Defense Against Pathogens Transported by Sperm." *Quarterly Review of Biology* 68 (September 1993): 335–81.

Purnick, Joyce. "Let Hillary Be Hillary." *New York Times*, July 15, 1992, A20.

Putnam-Jacobi, Mary. *Common Sense Applied to Woman Suffrage*. New York: G. P. Putnam's Sons, 1894.

Quayle, Marilyn. "Workers, Wives and Mothers." *New York Times*, September 11, 1992, A35.

Quindlen, Anna. "G.O.P. Fairy Tale Has a Sleeping Barbara and Wicked Hillary." *Chicago Tribune*, August 18, 1992, C13.

———. "Public and Private: The Two Faces of Eve." *New York Times*, July 15, 1992, A21.

Quinn, Sally. "Tabloid Politics, the Clintons, and the Way We Now Scrutinize Our Potential Presidents." *Washington Post*, January 26, 1992, C1.

Quintilanus, Marcus Fabius. *Institutes of Oratory*. London: Bahn, 1856.

Quong Wing v. Kirkendall, 223 U.S. 59 (1912).

Raymond, Robert. *The Patriotic Speaker*. New York: A. S. Barnes and Burr, 1864.

Reagan, Nancy. *My Turn: The Memoirs of Nancy Reagan*. New York: Dell, 1990.

Redman, Selina, Gloria R. Webb, Deborah J. Hennrikus, Jill J. Gordon, and Robert W. Sanson-Fisher. "The Effects of Gender on Diagnosis of Psychological Disturbance." *Journal of Behavioral Medicine* 14 (1991): 527–40.

Reed v. Reed, 404 U.S. 71 (1971).

Reinhold, Robert. "The 1992 Campaign: Senate Race; Republican Imperils Tide of Democrats in California." *New York Times*, October 24, 1992, 1.

Reno, Janet. Statement before the Senate Judiciary Committee, March 9, 1993.

Rensberger, Boyce. "Contraception the Natural Way: Herbs Have Played a Role From Ancient Greece to Modern-Day Appalachia." *Washington Post*, July 25, 1994, A3.

Rhode, Deborah L. *Justice and Gender*. Cambridge: Harvard University Press, 1989.

Ries, Paula, and Ann J. Stone, eds. *The American Woman, 1992–93: A Status Report*. New York: W. W. Norton, 1992.

Rinehart, Sue Tolleson. *Gender Consciousness and Politics*. New York: Routledge, 1992.

Roberts, Sam. "Just Casually Clenched, Holtzman Looks Ahead." *New York Times*, September 30, 1993, B7.

Robertson, Nan. *The Girls in the Balcony*. New York: Random House, 1992.

Rodin, Judith, and Ellen Langer. "Aging Labels: The Decline of Control and the Fall of Self-Esteem." *Journal of Social Issues* 16 (1980): 12–29.

Rodriguez, Lori. "Democrats Give Senate Race Away." *Houston Chronicle*, May 8, 1993, A25, 2.

Roe v. Wade, 410 U.S. 113 (1973).

Rohter, Larry. "Tough 'Front-Line Warrior'," *New York Times*, February 12, 1993, A1, A22.

———. "Woman in the News: Clinton Picks Miami Woman, Veteran State Prosecutor, to be His Attorney General." *New York Times*, February 12, 1993, 1.

Rose, James. "Women Are No Longer the Second Strings in Classical-Music Circles." *Philadelphia Inquirer.* May 10, 1994, F5.

Rosen, Jeffrey. "The Book of Ruth." *The New Republic,* August 2, 1993, 19.

Rosenberg, Rosalind. "Offer of Proof Concerning the Testimony of Dr. Rosalind Rosenberg, EEOC v. Sears" (No. 79-C-4373).

Rosenthal, Andrew. "Women Are Triumphant in Their Gains, But Cautious." *New York Times,* July 14, 1992, A9.

Rosser, Sue V. "Is There Androcentric Bias in Psychiatric Diagnosis?" *Journal of Medicine and Philosophy* 17 (April 1992): 215–32.

Rossi, Alice S., ed. *The Feminist Papers: From Adams to deBeauvoir.* New York: Columbia University Press, 1973.

Royko, Mike. "Braun's Troubles Should Benefit Us." *Chicago Tribune,* October 7, 1992, 3.

Rubin, Carol Hogfoss. "Occupation as a Risk Identifier for Breast Cancer." *American Journal of Public Health* 83 (1993): 1311.

Russ, Joanna. *How to Suppress Women's Writing.* Austin: University of Texas Press, 1983.

Ruthven, K. K. "Feminist Literary Studies." In *Feminist Literary Theory: A Reader,* edited by Mary Eagleton. London: Basil Blackwell, 1986.

Sadker, Myra P., and David M. Sadker. "Sexism in the Classroom of the 80's." *Psychology Today,* March 1985, 54–57.

Safire, William. Essay: "The Hillary Problem." *New York Times,* March 26, 1992, 23A.

———. "The First Lady Stages a Coup." *New York Times,* March 2, 1987, 17A.

———. "Macho Feminism, R.I.P." *New York Times,* January 27, 1992, A21.

———. "113 Days is Enough." *New York Times,* March 12, 1987, 31A.

———. "The Polish Thatcher." *New York Times,* June 10, 1993, A27.

"Salute America's First Ladies." *Good Housekeeping.* April 1992, 2.

Sampson, E. E. "Psychology and the American Ideal." *Journal of Personality and Social Psychology* 35 (1977): 767–82.

Samuels, Patrice Duggan. "Enterprise." *Lear's,* January 1994, 17–18.

Sandelowski, Margarete J. "Failures of Volition: Female Agency and Infertility in Historical Perspective." In *Ties that Bind,* edited by Jean F. O'Barr, Deborah Pope, and Mary Wyer, 35–59. Chicago: University of Chicago Press, 1990.

Sanger, David E. "Silent Empress, Irate Nation (and Contrite Press)." *New York Times,* December 24, 1993, A4.

Sapinsley, Barbara. *The Private War of Mrs. Packard.* New York: Paragon, 1991.

Sapiro, Virginia. "If U.S. Senator Baker Were a Woman: An Experimental Study of Candidate Image." *Political Psychology* 3 (spring/summer 1982): 61–83.

Schiebinger, Londa. *The Mind Has No Sex? Women in the Origins of Modern Science.* Cambridge: Harvard University Press, 1989.

Schlafly, Phyllis. *The Power of the Positive Woman.* New York: Harcourt Brace Jovanovich, 1977.

Schmalz, Jeffrey. "Whatever Happened to AIDS?" *New York Times Magazine,* November 28, 1993, 56.

Schmidt, Karen, and Colleen Collins. "Showdown at Gender Gap." *American Journalism Review* (July/August 1993): 39–42.

Schneider, Karen. "Women Seeking Office Say Rumors Hint of Double Standard." *Philadelphia Inquirer,* September 20, 1992, A8.

Schwartz, Felice N. "Management Women and the New Facts of Life." *Harvard Business Review* (January/February 1989): 68.

Scott, Anne Firor. *Natural Allies: Women's Associations in American History.* Urbana and Chicago: University of Illinois Press, 1993.

Scott, Joan W. "Deconstructing Equality-Versus-Difference." In *Conflicts in Feminism,* edited by Marianne Hirsch and Evelyn Fox Keller. New York: Routledge, 1990.

———. *Gender and the Politics of History.* New York: Columbia University Press, 1988.

Secor, C. "Androgyny: An Early Reappraisal." *Women's Studies* 2 (1974): 161–69.

Seib, Philip. "Hutchinson Will Find That Middle of the Road Is a Dangerous Place." *Dallas Morning News,* June 7, 1993, A19.

Seidenburg, R. "Drug Advertising and Perceptions of Mental Illness." *Mental Hygiene* 55 (1971): 21.

Shales, Tom. "Campaign '92: The Muck Starts Here." *Washington Post,* January 28, 1992, E1.

Shannon, Elaine. "The Unshakable Janet Reno." *Vogue,* August 1993, 260.

Shappert, S. M. "Office Visits to Psychiatrists: United States 1989–90." In *Advanced Data* 237. Hyattsville, MD: National Center for Health Statistics, 1993.

Sheehy, Gail. *The Silent Passage.* New York: Pocket Books, 1993.

———. "What Hillary Wants." *Vanity Fair,* May 1992, 140.

Showalter, Elaine. "Managing Female Minds." In *Women's Studies: Essential Readings,* edited by Stevi Jackson, 378–80. New York: New York University Press, 1993.

Shribman, David. "Before Big Hair and Beauty Contests, Texas Women Got Their Nails Dirty." *Houston Chronicle,* May 30, 1993, 4.

Shuit, Douglas P., and Bill Stall. "Seymour and Feinstein Trade Biting Charges." *Los Angeles Times,* October 20, 1992, A3.

"The Shy Senator." *Baltimore Morning Sun,* October 6, 1992, A10.

Smith, E. K. "The Effect of Double-Bind Communications Upon the State-Anxiety of Normals." Abstract in *Dissertation Abstracts International* 34 (1973): 427B, University Microfilms no. 73–16, 583.

Smith, T. W. "The Polls: Gender and Attitudes Towards Violence." *Public Opinion Quarterly* 48 (1984): 384–96.

Smith-Rosenberg, Carroll. *Disorderly Conduct: Visions of Gender in Victorian America.* New York: Oxford University Press, 1985.

———. "Puberty to Menopause: The Cycle of Femininity in Nineteenth-Century America." *Feminist Studies* 1 (1973): 58–73.

Smith-Rosenberg, Carroll, and C. Rosenberg. "The Female Animal: Medical and Biological Views of Woman and Her Role in Nineteenth-Century America." In *Women and Health in America,* edited by J. W. Leavitt, 12–27. Madison: University of Wisconsin, 1984.

Solomon. Barbara Miller. *In the Company of Educated Women.* New Haven: Yale University Press, 1985.

Sommers, Christina Hoff. *Who Stole Feminism? How Women Have Betrayed Women.* New York: Simon and Schuster, 1994.

———. "Who Stole Feminism?" *Washington Post,* Book World, August 28, 1994, 10.

Sontag, Susan. "The Double Standard of Aging." In *No Longer Young: The Older*

Woman in America, edited by P. Bart. Detroit: Wayne State Institute of Gerontology, 1975.

Sowerman, Esther. "Esther Hath Hanged Haman." 1617. Reprinted in *Half Humankind: Contexts and Texts of the Controversy about Women in England, 1540–1640,* edited by Katherine Usher Henderson and Barbara F. McManus. Urbana: University of Illinois Press, 1985.

Spake, Amanda. "Thirty Years at the White House." *Washington Post Magazine,* October 21, 1990, 19.

Spee, Friedrich. *Cautio Criminalis seu De Processibus Contra Sagas Liber. Ad Magistratus.* Austria: Sumptibus Ioannis Gronaei, 1632.

Speers, W. "A Blessing for the Movie Biz: Cardinal Rejects Censorship." *Philadelphia Inquirer,* October 2, 1992, c2.

Spelman, Elizabeth. *Inessential Woman: Problems of Exclusion in Feminist Thought.* Boston: Beacon Press, 1988.

Spence, J. T., and R. L. Helmreich. *Masculinity and Femininity: Their Psychological Dimensions, Correlates, and Antecedents.* Austin: University of Texas Press, 1978.

Spender, Dale. *Man Made Language.* London: Routledge and Kegan Paul, 1980.

Spilka, Bernard, Gerald Lacey, and Barbara Gelb. "Sex Discrimination After Death: A Replication, Extension and a Difference." *Omega* 10 (1979–80): 227–33.

Sprogis v. United Airlines, Inc., 444 F2d 1194 (7th Cir. 1971), cert. denied 404 U.S. 991.

Stahl, Lori. "Political Etiquette; Approach to Hutchison a Sticky Issue For Krueger." *Dallas Morning News,* May 16, 1993, A1.

Stanley, Alessandra. "Family Values and Women: Is G.O.P. a House Divided?" *New York Times,* August 21, 1992, A1.

———. "A Softer Image for Hillary Clinton." *New York Times,* July 13, 1992, B1.

Stanton, Elizabeth Cady. *The Woman's Bible.* 1895. Reprint, Boston: Northeastern University Press, 1993.

Stanton, Elizabeth Cady, Susan B. Anthony, and Matilda Joslyn Gage. *History of Woman Suffrage.* 3 Vols. 1881–87. Reprint, New York: Source Book, 1970.

State v. Hall, 187 So. 2d 861, 863 (Miss.), appeal dismissed 385 U.S. 98 (1966).

Stein, P. J., and S. Hoffman. "Sports and Male Role Strains." *Journal of Social Issues* 34 (1978): 136–50.

Steinem, Gloria. *Moving Beyond Words.* New York: Simon and Schuster, 1994.

———. *Revolution From Within.* Boston: Little, Brown, 1993.

Sterling, Joyce. "Women on the Bench." In *Women in Law.* 2d ed., quoted by Cynthia Fuchs Epstein, 454. Urbana: University of Illinois Press, 1993.

Sternhell, Carol. "Life in the Mainstream." *Ms.* July 1986, 48–51, 86–89.

Stewart, Lea P., Pamela J. Cooper, and Sheryl A. Friedley. *Communication Between the Sexes.* Scottsdale, AZ: Gorsuch Scarisbrick, 1986.

Stibbs, Anne, ed. *A Woman's Place: Quotations About Women.* New York: Avon Books, 1992.

Stimpson, C. R. "The Androgyne and the Homosexual." *Woman's Studies* 2 (1974): 237–48.

Storms, M. D. "Sex Role Identity and Its Relationship to Sex Role Attributes and Sex Role Stereotypes." *Journal of Personality and Social Psychology* 37 (1979): 1779–89.

Strausz, Ivan K., M.D. *You Don't Need a Hysterectomy.* Reading, MA: Addison-Wesley, 1993.

Suro, Roberto. "The 1990 Elections: Governor—Texas; Fierce Election for Governor Is Narrowly Won by Richards." *Washington Post,* November 7, 1990, 2.

Swacker, Marjorie. "The Sex of the Speaker as a Sociolinguistic Variable." *Language and Sex,* edited by Barrie Thorne and Nancy Henley, 76–83. Rowley, MA: Newbury House, 1975.

Swan, Susan. "Woman on the Verge." *Mirabella,* August 1993, 71.

Swetnam, Joseph. "The Arraignment of Lewd, Idle, Forward, and Unconstant Women. . . ." 1615. Reprinted in *Half Humankind: Contexts and Texts of the Controversy about Women in England, 1540–1640,* edited by Katherine Usher Henderson and Barbara F. McManus. Urbana: University of Illinois Press, 1985.

Szasz, Thomas. *The Manufacture of Madness.* New York: Harper & Row, 1970.

Tannen, Deborah. *You Just Don't Understand.* New York: William Morrow, 1990.

Tarr-Whelan, Linda. "Realigning Priorities." *Social Policy* 23 (summer 1993): 8–13.

Tattlewell, Mary, and Joan Hit-Him-Home. "The Women's Sharp Revenge." 1640. Reprinted in *Half Humankind: Contexts and Texts of the Controversy about Women in England, 1540–1640,* edited by Katherine Usher Henderson and Barbara F. McManus. Urbana: University of Illinois Press, 1985.

Taylor, Kristin Clark. *The First to Speak: A Woman of Color Inside the White House.* New York: Doubleday, 1993.

Taylor, Ruth Ashton. Interview by Shirley Biagi. January 11, 1992. In the Oral History Collection of Columbia University, Women in Journalism Oral History Project of the Washington Press Club Foundation.

Terry, Don. "Swift Change of Image for Illinois Candidate." *New York Times,* March 21, 1994, A12.

Thurow, Lester. "62 Cents to the Dollar: The Earnings Gap Doesn't Go Away." *Working Mother,* October 1984, 42.

Tilghman, Shirley M. "Science vs. Women—A Radical Solution." *New York Times,* January 26, 1993, A23.

Toner, Robin. "Republicans Send Bush Into the Campaign Under a Banner Stressing 'Family Values': Bestow Nomination." *New York Times,* August 20, 1992, A1.

Truehart, Charles. "Fuming Feminists Light Senator's Fuse." *Washington Post,* September 20, 1990, D1.

———. "Woman to Succeed Mulroney." *Washington Post,* June 14, 1993, 1, A15.

Truell, Matt. "Dole, O'Dell Spar Over Senator's Status as 'Insider'." *Wichita Eagle,* October 19, 1992, C1.

Tumulty, Karen. "Stark Says 'Pillow Talk' Taught Health Care to Colleague." *Los Angeles Times,* March 17, 1994, A20.

UAW v. Johnson Controls, Inc., 499 U.S. 187 (1991).

UAW v. Johnson Controls, Inc., 886 F2d 871 (7th Cir. 1989).

United Press International Stylebook. New York: UPI, 1977.

Urban, Jim. "Folksy Mayor Ends Term in Pittsburgh." *Philadelphia Inquirer,* January 3, 1994, B3.

U.S. Department of Labor. *Report on the Glass Ceiling Initiative.* Washington, DC: GPO, 1991.

U.S. v. Nicholas, 97 F2d 510 (2d Cir. 1938).

Ussher, Jane M. *Women's Madness: Misogyny or Mental Illness*. Amherst: University of Massachusetts Press, 1991.

Vaid, Urvashi, Naomi Wolf, Gloria Steinem, and bell hooks. "Let's Get Real About Feminism: The Backlash, The Myths, The Movement." *Ms.* September/October, 1993, 34–43.

Valenstein, E. S. "The Practice of Psychosurgery: A Survey of the Literature (1971–1976)." In *Psychosurgery*, United States Commission for the Protection of Human Subjects of Biomedical and Behavioral Research. Washington, DC: GPO, 1977.

Vaux, A. "Variations in Social Support Associated with Gender, Ethnicity, and Age." *Journal of Social Issues* 41 (1983): 89–110.

Verhovek, Sam Howe. "From Demure Survivor to G.O.P. Star." *New York Times,* June 7, 1993, A15.

Voboril, Mary. "Elect Him, You Get Hillary." *Philadelphia Inquirer,* May 4, 1992, C1.

"Vote for My Wife: The Candidates' Women Carry More Political Weight Each Time," *El Mundo,* September 2, 1992, 14.

Voter Research & Surveys (VRS). Election day exit polls, 1992.

Walker, Martin. "Britain's Uncommon Leader." *Washington Post,* Book World, October 31, 1993, 2.

Walker, Nancy. *A Very Serious Thing: Women's Humor and American Culture.* Minneapolis: University of Minnesota Press, 1988.

Warner, Judith. *Hillary Clinton: The Inside Story.* New York: Penguin Books, 1993.

Weaver, David, Roy W. Howard, and G. Cleveland Wilhoit. "The American Journalist in the 1990s." Arlington, VA: The Freedom Forum World Center, November 17, 1992.

Webster, Daniel. *The Writings and Speeches of Daniel Webster.* Boston: Little, Brown, 1903.

Webster, Wendy. "Not a Man to Match Her: The Marketing of a Prime Minister." London: *The Women's Press,* 1990.

Weingler v. Druggists Mutual Insurance Co., 446 U.S. 142 (1980).

Weller, Sheila. "America's Most Sexist Judges." *Redbook,* February 1994, 83–87.

The Wellesley College Center for Research on Women, *How Schools Shortchange Girls.* Washington, DC: AAUW Educational Foundation, 1992.

Wessley, Stephen E. "The Guglielmites: Salvation Through Women." In *Medieval Women,* edited by Derek Baker. Oxford: Basil Blackwell, 1978.

"Why The Gay Ban Hurts All Military Women." *Glamour.* August 1993, 98.

Wier, Johann. *Whitches, Devils and Doctors in the Renaissance,* edited by George Mora. Binghamton, NY: Medieval and Renaissance Texts and Studies, 1991.

Wildermuth, John, and Harriet Chiang. "Feinstein and Boxer Staffs Claim Victories." *San Francisco Chronicle,* November 4, 1992, A3.

Wilkinson, Tracy. "Feinstein Tours Riot Area; Seymour Presses For Debate." *Los Angeles Times,* September 9, 1992, A3.

Williams, Marjorie. "First Ladies." *Washington Post Magazine,* November 1, 1992, W11.

Williams, Marjorie, and Al Kamen. "Woman of the Hour." *Washington Post Magazine,* June 11, 1989, 25.

Williams, Wendy. "Equality's Riddle: Pregnancy and the Equal Treatment/Special

Treatment Debate." *New York University Review of Law & Social Change* 13 (1984–85): 325.

Wilson, Robert A. *Feminine Forever*. New York: M. Evans, 1966.

Wilson v. Southwest Airlines, 517 F. Supp. 292 (N.D. Tex. 1981).

Wilson, Thomas. *The Art of Rhetoric*. University Park: Pennsylvania State University Press, 1994.

"Winning With Women." A survey commissioned by EMILY'S LIST, the National Women's Political Caucus, and the Women's Campaign Fund. Washington, D.C., 1991.

Winston, Kenneth, and Mary Jo Bane. *Gender and Public Policy: Cases and Comments*. Boulder, CO: Westview Press, 1993.

Winthrop, John. *The Short Story of the Rise, Reign, and Ruine of the Antinomians. . . .* (London, 1644). In *The Female Experience: An American Documentary*, edited by Gerda Lerner. Indianapolis: Bobbs-Merrill Educational Pub. 1977.

Witt, Linda, Karen M. Paget, and Glenna Matthews. *Running as a Woman: Gender and Power in American Politics*. New York: Free Press, 1994.

Wolf, Naomi. *The Beauty Myth: How Images of Beauty Are Used Against Women*. New York: Doubleday, 1992.

———. *Fire with Fire: The New Female Power and How It Will Change the 21st Century*. New York: Random House, 1993.

Woo, Junda. "Widespread Sexual Bias Found in Courts." *Wall Street Journal*, August 20, 1992, B1.

Wood, Robert G., Mary E. Corcoran, and Paul N. Courant. "Pay Differences Among the Highly Paid: The Male-Female Earnings Gap in Lawyers' Salaries." *Journal of Labor Economics* 11 (July 1993): 417–40.

Woolf, Virginia. *A Room of One's Own*. New York: Harcourt, Brace and World, 1929.

Wyman, Elizabeth, and Mary E. McLaughlin. "Traditional Wives and Mothers." *Counseling Psychologist* 8 (1979): 25.

Young, Robin. "The Selling of Motherhood. *Newsweek*, October 1, 1990, 12.

Zepplin, H., R. A. Sills, and M. W. Halth. "Is Age Becoming Irrelevant? An Exploratory Study of Perceived Age Norms." *International Journal of Aging and Human Development* 24 (1987): 241–56.

Index

275